Starstruck

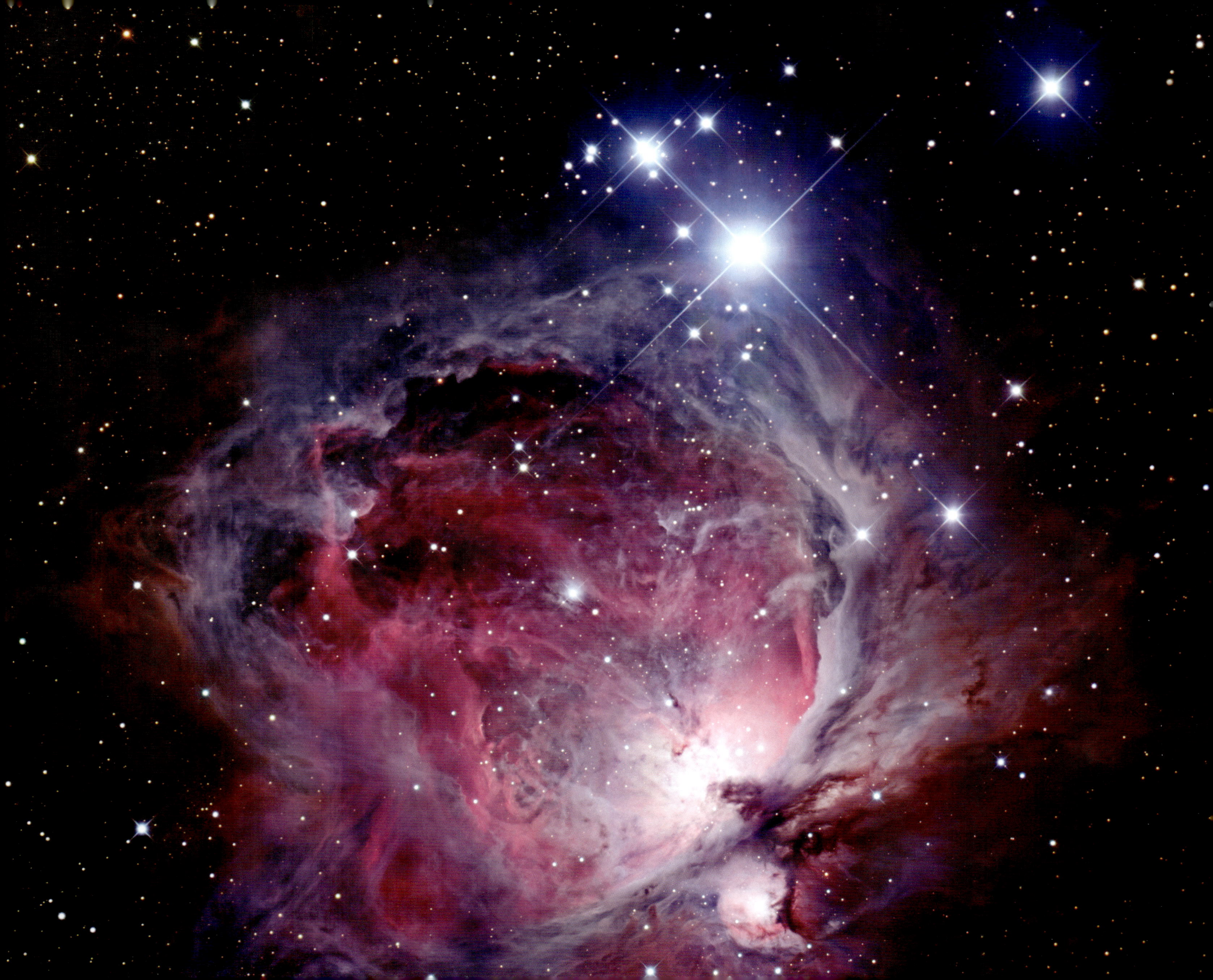

Starstruck

The Fine Art of Astrophotography

Edited & organized by Anthony Shostak

Essays by Jerry T. Bonnell · Dennis di Cicco · Weston Naef · Eric Wollman

Bates College MUSEUM *of* ART

 First published on the occasion of the exhibition "Starstruck," June 8 through December 15, 2012.

ISBN 978-0-9832176-0-2 (alk. paper)

Library of Congress CIP Data, QB121.S73 2012, 779'.9523—dc23, 2012017859

Design and production by Boyd Hill, Natick, Massachusetts. hilldesignart.com

Typeface family: ITC Galliard Pro, Pro Italic, and Bold. ITC Galliard is an adaptation of Matthew Carter's 1978 phototype design for Mergenthaler Linotype. It was modeled on the work of Robert Granjon, the sixteenth-century French letter cutter.

FSC
www.fsc.org
MIX
Paper from responsible sources
FSC® C018210

Paper: Chorus Art Silk Text, 100 pound paper, FSC certified, 55 percent recycled fibers, 30 percent post-consumer, elemental chlorine free, acid free; and Touché® Cover, .024" black, which contains 40 percent total recycled fiber and 20 percent post-consumer waste.

This catalogue was printed by the Flagship Press using 100 percent wind power.

Frontispiece: detail of catalogue no. 29, and cover: detail of catalogue no. 47.

Contents

Textual Illustrations

Acknowledgements

Planning Starstruck began auspiciously in 2008 at a brainstorming session, graciously hosted by artist Greg Mort and his wife, Nadine, at their home in Port Clyde, Maine; a home built, coincidentally, by Russell W. Porter, artist, explorer, and influential amateur astronomer. The July night was crisp and brilliantly clear with stars, visible through the windows, that spread like diamonds on deep blue velvet. The evening was shared with local astrophotographers Jacob Gerritsen and Peter Kalajian, and their spouses Terry and Kristie, respectively. The warm glow of enthusiasm that ignited the meeting set the tone for the entire evolution of Starstruck.

Among the many people I was fortunate to work with, I am especially grateful for the contributions of our jurors and essayists. Jerry T. Bonnell, Astrophysicist/Staff Scientist—Compton Gamma Ray Observatory Science Support Center, and Dennis di Cicco, senior editor, *Sky & Telescope* magazine, brought a wealth of knowledge about astronomy and imaging, which informed our discussions about the photographs. Weston Naef, curator emeritus of photography at the J. Paul Getty Museum, transcended his role of juror and became a curatorial consultant, contributing his deep understanding of photography and exhibition planning while making vital connections to historical and contemporary photography. Eric Wollman, Professor of Physics, Bates College, enthusiastically volunteered his time and considerable expertise to shaping interpretive programming, mapping astronomical subjects, and writing his informative essay. Without their efforts, Starstruck would scarcely resemble the remarkable exhibition it has become.

A special word of thanks goes to artists Jim Burnell, Ken Crawford, Paul Howell, Steve Mazlin, and Kirk Rogers, and to Will Ash, Bates College Imaging and Computing Center, and Seth Shostak (no relation), SETI Institute in Mountain View, California, who offered advice and expertise which helped me surmount a variety of hurdles. I also thank Andy Graham of Designtex in Portland, Maine, for his assistance with printing several images for the exhibition. Gene Clough, lecturer in Geology and Physics at Bates College, has, for years offered planetarium tours, and manned telescopes on the quad, showing the splendors of Saturn and Jupiter to passersby, and he freely gave me advice and assistance with educational programming related to Starstruck.

Stephanie Kelly-Romano, assistant professor of Theater and Rhetoric; Bill Matthews, Professor of Music; Hiroya Miura, Associate Professor of Music; Katalin Vecsey, Lecturer in Theater and Rhetoric, all signed on early to create a program of powerful, provocative performances complementing the exhibition. Thanks, too, to Nicole Hastings, Assistant in Instruction, Physics, for her enthusiastic support and link to the Bates Astronomy Club.

I am deeply grateful to our collaborators in educational programs, the Central Maine Astronomical Society and the Southern Maine Astronomers, for their commitment, time, and enthusiastically shared knowledge about the night sky.

It has been a distinct pleasure to work with Boyd Hill, our graphic designer for the project. His knowledge of art history and interest in astronomy informed the design of the exhibition and resulted in a handsome and enjoyable catalogue. I look forward to new projects on which we can again work together.

Starstruck would not have been possible without substantial financial support, much of which came from the sustained efforts of Hieu Nguyen, Claire Hammen, and their colleagues in the

Bates College Department of Advancement, as well as Lindsay Hancock. The Friends of the Bates College Museum of Art have also been instrumental in supporting the Starstruck project, and special words of thanks go to Dominique and Jane, for their generosity.

Bates College Museum of Art interns and gallery assistants over the past few years have helped with various aspects of Starstruck, and special recognition goes to Julia Foxworth '13, Cara Garcia-Bou '13, Jee Hye Kim '12, Aisling Ryan '14, and Charlotte Weidlein '09. I also acknowledge Tom Hayward, Humanities Reference Librarian, Bates College Ladd Library, who helpfully contributed his assistance with the bibliography.

My past and present colleagues at the Museum, Catherine Jones, Dan Mills, Anne Odom, and Alexandra Rollins, have all contributed more than can be succinctly described, and I am grateful to them all. My remaining colleague, Bill Low, has taught me more about exhibitions and museums than anyone, and I count myself lucky to have worked with him nigh on twenty years. Finally, I am deeply thankful to my wife, Crystal Nicholas, for her understanding, patience, and appreciation for a fine night spent looking up at the stars.

Anthony Shostak, Curator of Education
March 20, 2012
Lewiston, Maine

ANTHONY SHOSTAK
Curator of Education
Bates College Museum of Art

Introduction

"Certainly it is a great thing to add to the countless multitude of stars visible hitherto by natural means and expose to our eyes innumerable others never seen before ..."

GALILEO GALILEI[1]

STARS ADRIFT on the velvety sable of night skies, the life-giving sun, and the enigmatic, changing face of the moon have inspired humankind since prehistoric times, informing our understanding of the universe and igniting countless creative imaginations. For millennia, a glimpse of stars in a dark sky has provoked an immediate emotional response and sense of wonder, resulting in cultural impacts that are seen in a multitude of examples from ancient creation myths to current cinema. The firmament's inviting splendor has also captured the eye of photographers from the earliest history of their medium. Today, photographic technologies such as image capture, processing, and printing capabilities, give artists immense freedom within the confines dictated by their celestial subjects. Their creations are nothing less than overwhelming, depicting humbling, glorious delights that are often invisible to both the naked eye and even the telescope, but are revealed only through photographic means. Unfortunately, these sights are becoming ever more elusive to earth-bound astronomers, skygazers, and artists. Even as we enter an unprecedented era of telescope and camera evolution, light pollution from urban and suburban sprawl dims the stars planet wide, estranging us from our most ancient of cultural well-springs and threatening to wither the future of a blooming branch of fine arts.

The firmament's inviting splendor has also captured the eye of photographers from the earliest history of their medium.

Starstruck's seed was planted with was a chance glance through my bedroom window one late fall night. The Great Nebula in Orion was clearly visible, and my encounter with it rekindled my long interest in depictions of night. My preliminary research quickly revealed that astronomical subjects are the source of inspiration for a large body of artists working in all corners of the world (fig 1.2). I also learned that, while many science museums have presented this material, it has not often been the subject of art museum exhibitions—none at the time I began this work. Starstruck is a step toward recognizing the astrophotography as a genre of the fine arts and its contributions to the history of photography.

Astrophotography is a curious genre, with few clear boundaries between the art mainstream, scientific endeavor, and what might be called a high-tech outsider art community. Since it is so varied, it is difficult to develop a useful definition. And, once one has developed a sense of the field, there are other issues that prevent it from fitting neatly into a mental box. Even something as straightforward as authorship of images is not always readily distinguishable, with some artists working as teams, others appropriating images outright, and still others using their own images augmented with data from astrophysical research missions. Complicating things further is the fact that many astrophotographers operate at or beyond the art mainstream's periphery. Indeed, there are large number astrophotographers for whom standard conventions of printing and editioning are eschewed, making exhibiting their work challenging.

But the task is worth the effort. Astrophotography is a rich and varied genre, providing fascinating images that invite investigation and learning as much as enjoyment. For many viewers, part of the delight of the art form is tied to understanding the phenomena depicted. Those familiar with stargazing know how stunning the colorful images that astrophotographers produce are in comparison to the ghostly pale grays and greens afforded by telescope eyepieces. Indeed, the intricate and subtle detail made possible by careful layering of narrowband and full-spectrum exposures and skillful choices regarding contrast, color, and saturation while processing digital exposure data often reveals structures invisible to the eye.

As real as all of the structures in these images are, it would be an error to think that the photographs represent how these deep space subjects actually look (fig 1.1). These are, with rare exception, artistic depictions. But how is there room for artistry when, in the case of deep sky

Fig. 1.1. Sean Walker and Dennis di Cicco, *Hydrogen Alpha Sky with House*, 2011. Image courtesy the artists.

This composite image was made with extensive exposures recording the hydrogen-alpha emission band of light. By "going ultra-deep," with their exposure, the artists reveal exceptionally faint structures over an immense angular view of the sky. These are so prominent in the composition that they overwhelm even those familiar with the night sky. The Pleiades, the Orion Nebula, Castor and Pollux, and other familiar objects are depicted, but seem to be in wholly new celestial surroundings. The artists carefully mapped their exposure to a second photograph of the house and landscape, depicting their super-wide view to scale with its terrestrial surroundings. The convergence in the vertical walls of the house, accentuated by lens distortion, creates a heightened sense of verticality, further disorienting the viewer.

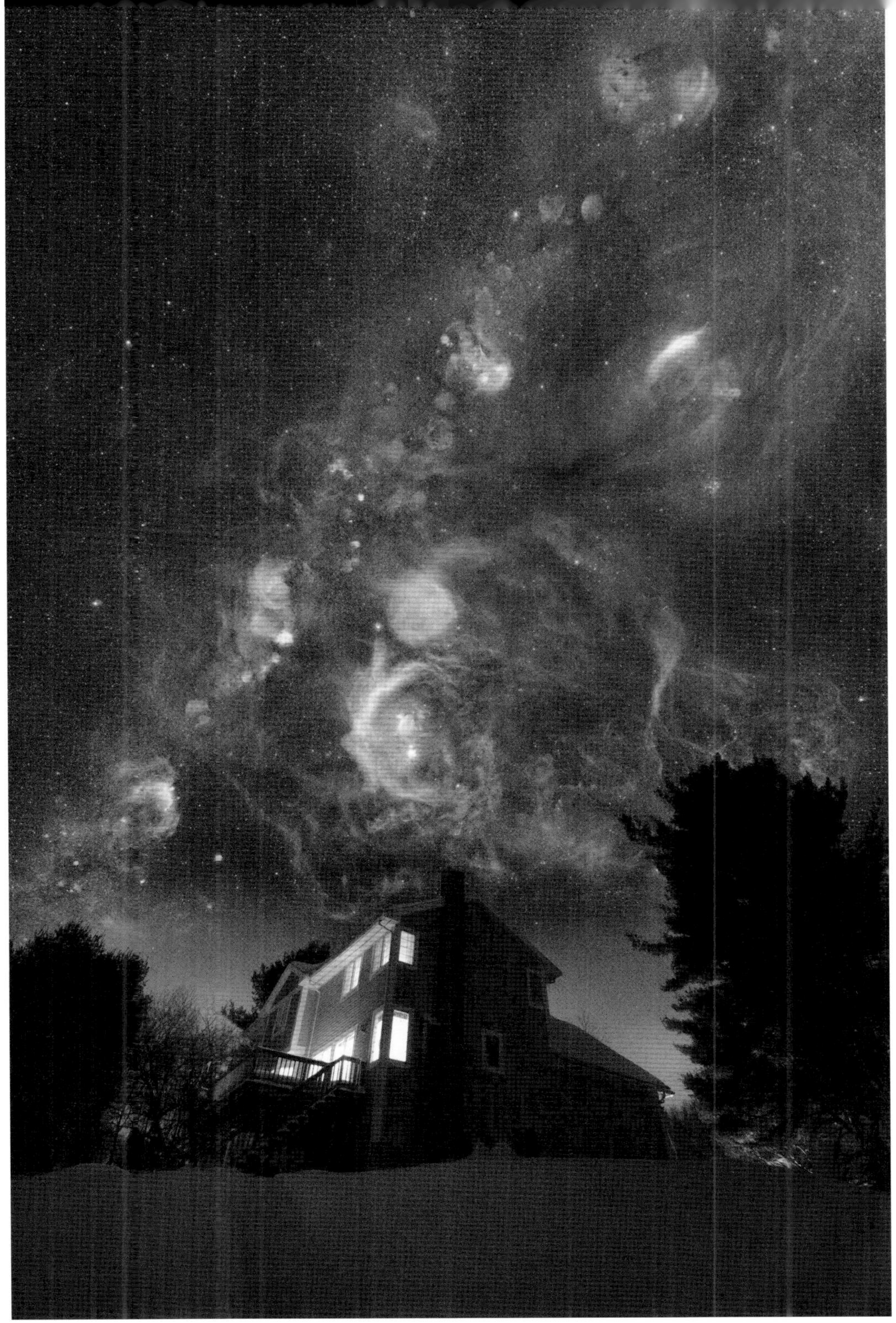

Astrophotography's subjects and the veracity it brings to them invite us to contemplate many of life's most interesting, urgent, and elusive questions.

objects, the photographer has no control over vantage or lighting, and the subject is essentially unchanging? It is a common question, and, at first examination it seems to close the issue for any debate. But a modicum of consideration shows that the question is entirely misleading. The making of these images is nothing if not a vast and complicated series of idiosyncratic, essentially aesthetic decisions. For instance, to show greater detail, an artist photographing an emission nebula might choose to represent light given off by ionized oxygen as blue, even though it actually radiates in the green portion of the spectrum. Similarly, the artist might represent light from sulphur ions and hydrogen alpha ions with reds and greens, respectively, despite the fact that they both radiate in the red swath of the spectrum. And, although the light recorded by the artist *is* of a particular color, technically speaking, it is in reality too dim or diffuse for the human eye to perceive in color, so that even if one were physically close to these vast and distant subjects, they would be seen as gray, if at all. Astrophotographers also must, at times, represent light that is beyond the limits of human color vision—what color is ultraviolet? As humans we not only do not know, *we cannot possibly even imagine it* since it lies beyond our senses. And so, even on a purely technical level, astrophotographers make artistic choices no less difficult or meaningful than any other artist in any medium. To this, one must add the consideration of the possibilities presented by concepts which may underpin an astrophotographer's works.

As an art form, astrophotography is remarkably powerful, being both visually compelling and intellectually engaging. Astrophotography's subjects and the veracity it brings to them invite us to contemplate many of life's most interesting, urgent, and elusive questions. Where did the universe come from? How did it organize into what we see today? What will happen to it all? How did life begin? Is there intelligent life elsewhere? If we cannot realistically travel the unimaginable distances to the stars, how do we continue to live here on Earth? Assuming humankind survives to the end of our sun's lifespan, what then? …

Like a telescopic view displaying rich, mesmerizing detail in a narrow field of view of the far wider scintillating sky, Starstruck neither shows the whole of astrophotography, nor does it pretend to. Rather, it focuses on a sampling of remarkable images spanning a range of subjects, from terrestrial views of aurorae to deep sky objects, which hint at the vastness of both the genre and the subject. In organizing the exhibition and its attendant catalogue, I opted to include a

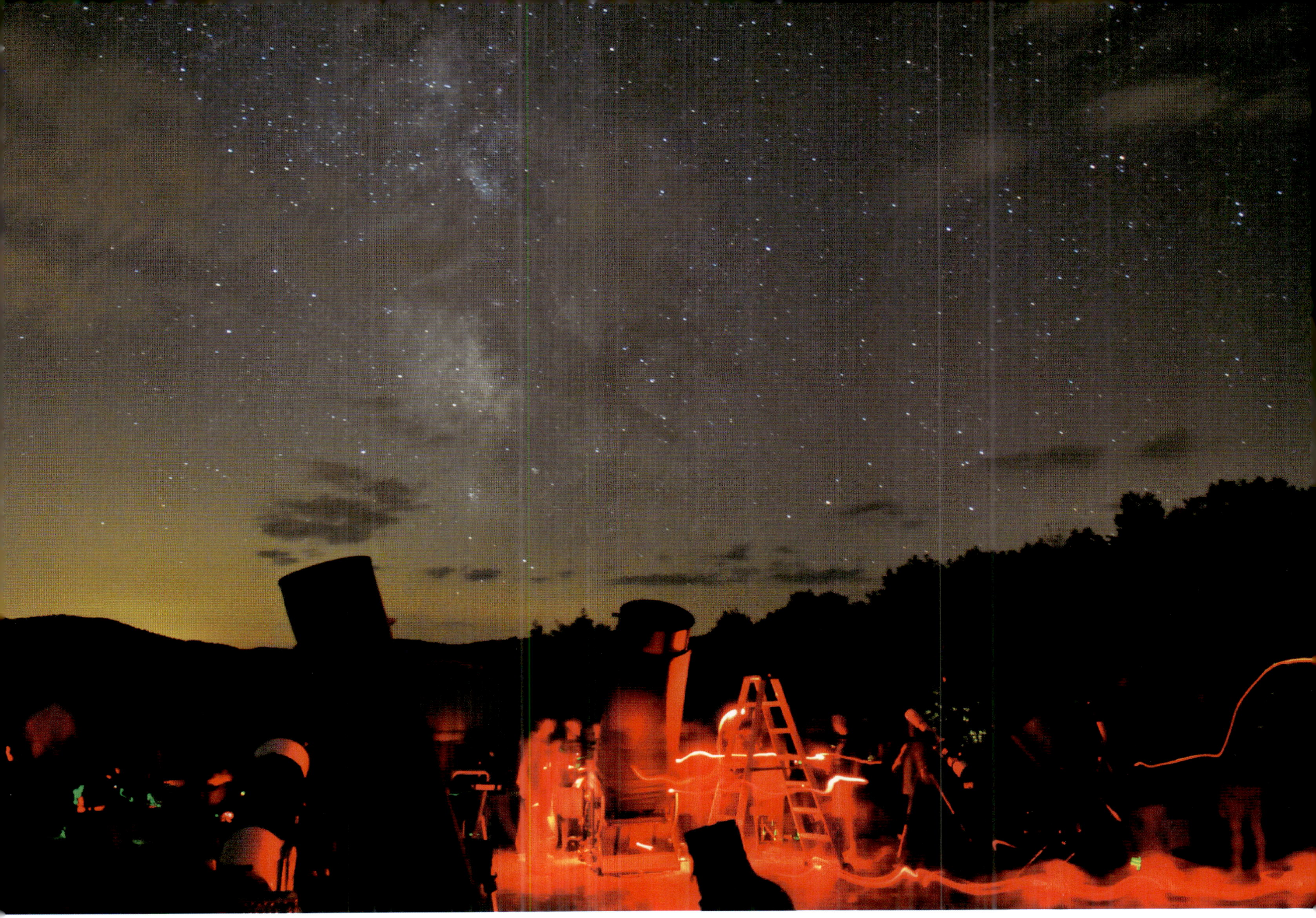

Fig. 1.2. Dennis di Cicco, *Stellafane Star Party, Springfield, Vermont*, 2010. Image courtesy the artist.

number of artists, mostly from within the art mainstream, whose works served as a core around which the rest of the exhibition was built through a juried format open to artists from around the world. Many deserving artists, far outnumbering those who are represented in Starstruck, could not be included in the exhibition, and their images are worth further study. I hope Starstruck sparks sufficient interest in the genre that these artists' works gain further recognition elsewhere, and that it urges the viewer to take steps to protect and preserve the precious resource of the night sky so that we may continue, for future millennia, to be inspired, mystified, and challenged by the stars.

Notes

1. Galilei 1989, p. 35. For full citations, please refer to the Selected Bibiography, p. 65 ff.

DENNIS DI CICCO
Senior Editor, *Sky & Telescope*

From Daguerreotypes to Digital Detectors: Photography by Amateur Astronomers

Many of the stunning photographs in the Starstruck exhibition were, unsurprisingly, made with state-of-the-art equipment and imaging-processing techniques that were unimaginable just a few short years ago. But the almost universal human desire to capture images of the night sky goes back further. Much further. Proof of that can be found in a map of the cosmos in the Paleolithic cave paintings at Lascaux, France (estimated to be 17,000 years old), and the countless petroglyphs scratched into the desert patina of rocks found around the globe. Thus, it is pretty safe to say that the starry sky has always captivated the human imagination. And it has certainly inspired artists throughout the ages and, more recently, photographers.

Prior to the invention of the telescope in the early seventeenth century, images of the heavens recorded only what could be seen with the unaided eye—star patterns, eclipses of the Sun and Moon, and the occasional bright comet or exploding supernova. When the telescope came along it dramatically transformed our concept of heaven's vault. The Milky Way was no longer just a hazy band of light, but rather a vast river of tiny stars. The wandering planets became little worlds unto themselves. Venus not only waxed and waned in size, but also experienced phases like the Moon. Jupiter had a retinue of four bright moons that appeared as "stars" shuffling about from night to night. The Sun had spots that constantly changed, and the Moon was covered with mountains, valleys, and amazing craters. All these sights fascinated observers and they were quickly recorded with sketches and drawings. From our modern perspective some of these early renderings border on art, but almost without exception they originated as scientific records of what was seen through the crude telescopes of the day.

Throughout the seventeenth century telescopes grew in size and quality, and by the beginning of the eighteenth century they were opening a window on a hitherto unknown aspect of the night sky—the vast realm of star clusters and nebulae. At first it was a mere trickle of discoveries by a handful of observers who generally chanced upon these so-called deep-sky objects. But by the late 1700s, the number had grown into the thousands, thanks especially to systematic searches conducted by the German-born English observer William Herschel and assisted by his sister Caroline.

Another revolution—one that would forever change our view of the universe—was just beginning.

Our concept the universe was again changing, and so too was the way it was being presented to the public. Whereas almost any skilled draftsman could portray a scene involving stars, most deep-sky objects appeared fuzzy and ill defined in a telescope eyepiece. Accordingly, sketches and drawing of them were strongly dependent on the telescope as well as the observer's eyesight and talents as an artist. Then too, there were the interpretations of the engravers who prepared the drawings for publication. It is not unusual to find early illustrations of the same deep-sky object appearing wildly different depending on the circumstances behind their creation.

Improvement came in measured steps that accompanied the introduction of ever-larger telescopes, and by the mid-nineteenth century, astronomy books often included illustrations of nebulae and star clusters that were accurate enough to be easily recognized today as specific objects. By then, however, another revolution—one that would forever change our view of the universe—was just beginning.

The Birth of Astrophotography

The history of photography is rich and fascinating, involving a wide variety of people, places, and scientific disciplines from chemistry to physics and optics. The cornerstone of modern photography was laid in January 1839 when two independent experimenters publicly displayed pictures that they had made by the action of light on sensitized materials. The first was Louis Jacques Mandé Daguerre in Paris, a talented painter who had no formal scientific training. When word of Daguerre's achievements crossed the English Channel a few days later, William Henry Fox Talbot, a member of the landed gentry with a bent for scientific pursuits, quickly made public the results of his early experiments in an effort to claim priority.

Initially, both men guarded the details of their processes with hopes of using them for financial gain. But the following August, in exchange for a modest pension from the French government entwined with a healthy dose of national pride, Daguerre made his process freely available to the world. The secrets were disclosed on August 19, 1839, to an overflow crowd at a meeting of the French Academy of Sciences in Paris. Despite the relatively slow methods of communication at the time, the process went viral. Within days, Daguerreotypes were appearing throughout Europe. The first detailed description of Daguerre's process sailed into New York harbor aboard the steamer *British Queen* on September 20 that same year, and before month's end, experimenters were displaying Daguerreotypes in New York, Philadelphia, and Boston.

Astronomy, interestingly enough, had gotten into the photography act even before the details of Daguerre's process were made public. At the request of astronomer Francois Arago, one of Daguerre's early confidants and secretary of the French Academy of Sciences, Daguerre tried to capture an image the Moon. While the experiment was deemed unsuccessful, Arago clearly foresaw the astronomical potential of the process, and he mentioned the possibility of using Daguerreotypes to map the Moon when he lobbied his political colleagues in the French Chamber of Deputies for Daguerre's pension. Even more insightful were Arago's suggestions that Daguerreotypes might eventually be used to measure the brightnesses of stars and as a way of recording astronomical spectra, which was an equally embryonic scientific discipline at the time.

Only months after details of Daguerre's process reached American shores, John W. Draper, a noted New York scientist and physician who was trained as a chemist, made a Daguerreotype of the Moon. Although flawed, his twenty-minute exposure in March 1840 recorded recognizable light and dark areas on the lunar surface. Regarded as the first successful astronomical Daguerreotype, it was presented to the Lyceum of Natural History of New York, where it was later destroyed during a devastating fire in 1866.[1]

Despite Draper's "success," and being championed by several high-profile astronomers, the early photographic processes were not sensitive enough to be of much use to the research needs of professional astronomers. Furthermore, progress on this front was very slow at first, and only modest astronomical milestones were reached in the decade that followed. By 1851, just the Sun, Moon, Jupiter, several solar eclipses, and the brilliant star Vega had been captured on Daguerreotypes. Interesting results for sure, but not anything that would change the course of astronomical research.

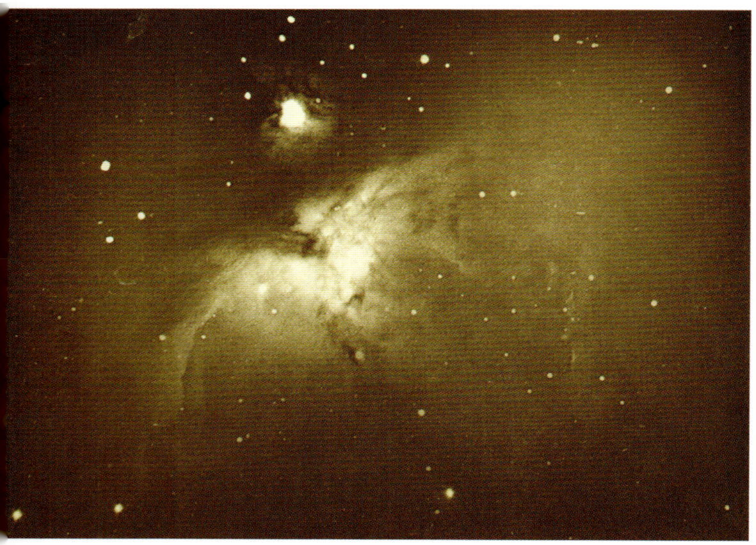

Fig. 2.1. Andrew Ainslie Common, *Orion Nebula*. English engineer and amateur astronomer Andrew Ainslie Common made this one-hour exposure of the Orion Nebula on February 28, 1883, with a thirty-six-inch reflecting telescope that he built himself. His thirty-seven-minute exposure of the nebula made a month earlier with the same equipment proved pivotal in the history of astronomical photography since it was the first to record fainter stars than could be seen by eye with the world's largest telescopes. Image courtesy *Sky & Telescope*.

Photographic processes steadily improved, and with the introduction of increasingly light-sensitive materials in the 1850s came more astronomical successes, but mainly at the hands of "gentleman scientists" in Europe and America. Notable pioneers included Warren De la Rue in England, who advanced lunar photography using wet-collodion plates with his thirteen-inch homemade reflecting telescope. In New York City, "private scientist" Lewis Morris Rutherford designed and built telescopes optimized for photography, and with them captured noteworthy images of star clusters and other celestial objects. John Draper, Henry's son and a skilled amateur astronomer, obtained the first photographic record of a star's spectrum in 1872 and the first image of the Orion Nebula in 1880, two years before his untimely death at the age of forty-five years.

It was, however, another photograph of the Orion Nebula by an amateur astronomer that marked a pivotal turning point in the history of astrophotography—one that helped transform photography from an astronomical curiosity to one of the science's greatest tools. In January 1883, English engineer Andrew Ainslie Common made a thirty-seven-minute exposure (fig. 2.1) of the nebula that recorded stars fainter than any that had ever been seen by observers using the world's largest telescopes. Common showed that photographs could now look deeper into the night sky than could the human eye. There was no turning back. Before the turn of twentieth century, the photographic plate had risen to become the detector of choice for professional astronomers around the world.

Changing Roles

The line separating amateur and professional astronomers during the late nineteenth century was often a blurry one. Nevertheless, many of astrophotography's success stories up to the early 1880s involved people who pursued astronomy as a pastime and not as a profession. That changed once professional astronomers embraced photography and their observatories began turning out images for research. Professional astronomers now had a vested interest in pushing the science of astrophotography forward.

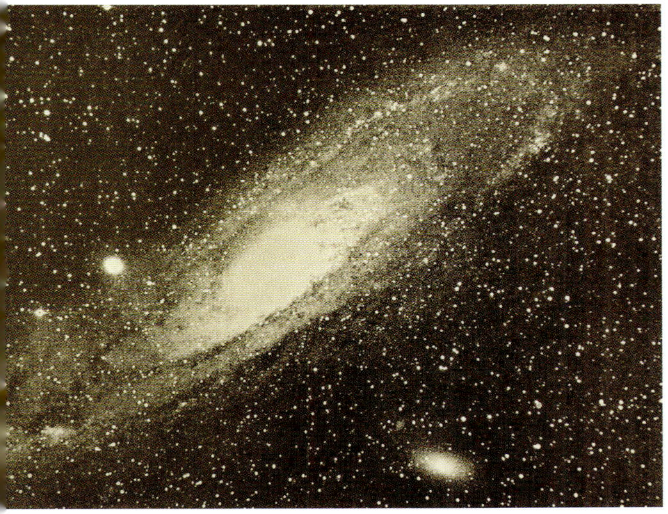

Fig. 2.2. Isaac Roberts, *Andromeda Nebula [Galaxy]*. Amateur astronomers were still making significant photographic contributions in the decade after professional astronomers began embracing photography as a serious scientific tool. This ninety-minute exposure of the Andromeda nebula (as it was then known) on October 17, 1895, was the work of English amateur Isaac Roberts whose private observatory held an excellent twenty-inch reflecting telescope. The image revealed hitherto unknown spiral structure that helped astronomers realize the "nebula" was actually a grand spiral galaxy separate from the Milky Way, and that the universe was a far bigger place than previously thought. Image courtesy *Sky & Telescope*.

There were exceptions, however, such as Isaac Roberts in England, a Welsh-born engineer and successful businessman who maintained a first-class amateur observatory at his home in Liverpool. In the late 1880s he made some of the finest deep-sky photographs to date with a twenty-inch telescope (fig. 2.2) specifically built for astrophotography. His 1888 photograph of the Andromeda Nebula, as this grand galaxy was then known, revealed unexpected spiral structure and provided another clue that helped lead to our eventual understanding that galaxies were far removed from our own Milky Way and that the universe was vastly larger than anyone had previous imagined.

Astrophotography got a huge boost following the establishment of Eastman Kodak in the 1890s. On its way to becoming a twentieth century photographic juggernaut, Kodak established one of the world's finest research laboratories. Astronomers worked hand in hand with Kodak's scientists to create emulsions and processing techniques that were specifically suited for astrophotography. Although amateurs were no longer the ones doing astronomy's pioneering photographic work, they benefitted from these professional collaborations, and, by the 1920s, astrophotography had become a well-defined aspect of the growing amateur astronomy hobby. It has remained such ever since.

Amateur astronomy, like most hobbies, experienced an uptick following World War II as many Americans found themselves with more leisure time on their hands. It was also helped along by war-surplus optical equipment and a growing national interest in the Space Age. And, like their forebears a century earlier, amateur astrophotographers once again found themselves dabbling in technology that was largely being ignored by professionals—color photography. Despite their public appeal, color photographs of celestial objects generally do not contain information beyond what astronomers can glean from black-and-white images made through various filters and with black-and-white emulsions that are sensitive to specific wavelengths of light. Furthermore, early color emulsions were also extremely slow compared to their black-and-white counterparts made for astronomical work. Early color films required relatively long exposures to record even the brightest deep-sky objects. This did not deter amateurs, however, and I know firsthand that a small, fuzzy color picture of the Orion Nebula (figs. 2.3 and 2.4) or Andromeda Galaxy displayed at a gathering of amateur astronomers in the 1960s caused quite a sensation.

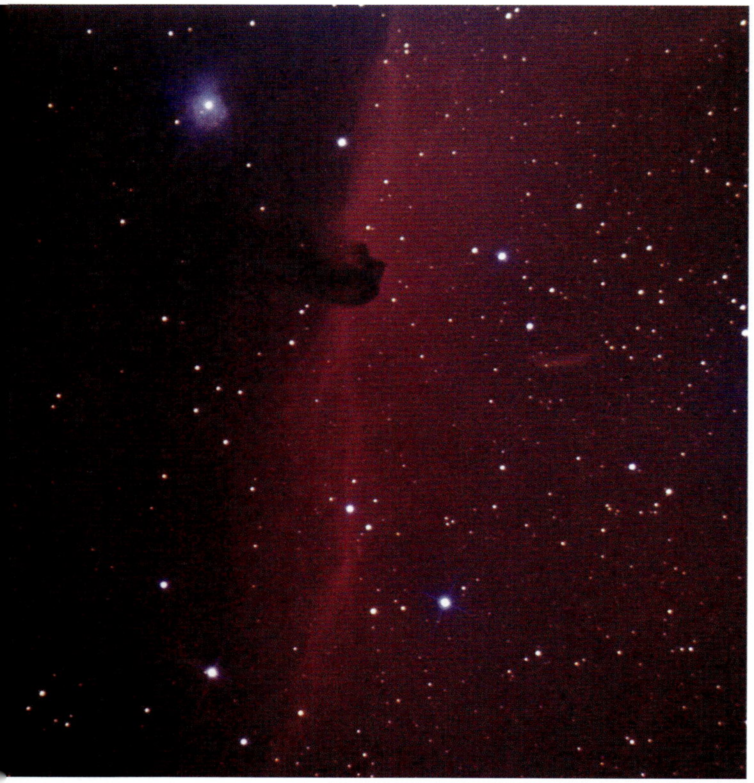

Fig. 2.3. Evered Kreimer, *Horsehead Nebula*. Image courtesy the author.

Astrophotography by amateur astronomers underwent a major growth spurt in the 1970s. Up to this point most astrophotographers had to be telescope makers or at least be capable of extensively modifying equipment, since very few commercial telescopes were out-of-the-box ready for long-exposure astrophotography.[2] Although the activity began at a time when there were limited commercial telescopes, especially ones that typical amateurs could afford, telescope-making remains alive and well today, amid an ever-growing market of inexpensive, quality telescopes. In the 1960s, Tom Johnson, an engineer and amateur astronomer who owned an electronics-manufacturing firm in Southern California, started a side business building Schmidt-Cassegrain telescopes that were designed for astrophotography. Within a few years the new company, named Celestron, overtook the electronic business.

Johnson's first Celestron telescopes were relatively expensive and thus were purchased mainly by schools and universities. But a redesign in the early 1970s retained the astrophotography capabilities while lowering prices to be within the reach of amateurs. Almost overnight the eight-inch Celestron, which cost less than $1,000, became wildly popular. The telescope was also highly portable, making it easy for astrophotographers living in light-polluted areas to head out to the countryside where they could effectively do deep-sky photography. Soon, a cottage industry sprang up of small companies making astrophotography accessories for these scopes, adding to the line already available from Celestron. Before long, other companies began making similar Schmidt-Cassegrain telescopes with astrophotography capabilities. Coupled with ever-better color films, these telescopes launched a revolution in amateur deep-sky astrophotography. It continues to this day, but not without significant changes.

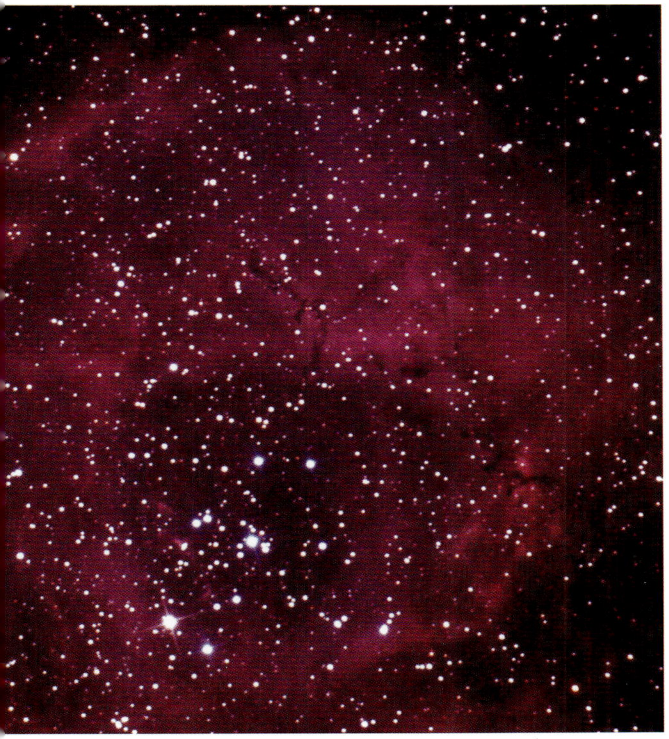

Fig. 2.4. Evered Kreimer, *Rosette Nebula*. Image courtesy the author.

The Digital Revolution

Photographic emulsions remained astronomy's workhorse for about a century until electronic CCD detectors (charge-coupled device, a reference to the inner working of the detector) entered the scene in the 1970s. Compared with astronomical plates, early CCDs, were small and had low resolution. But they were also an order of magnitude more efficient, and they had a linear response, meaning that they generate exactly twice as much signal when the object they image is twice as bright (something that emulsions do not do). Both features are extremely important for astronomy. There is also the huge advantage of the digital output from CCDs being ready-made for computer processing and, especially, analysis. While it took several generations for photography to achieve widespread acceptance in astronomy, it took barely a decade for CCDs to replace film. By the late 1980s, most professional astronomical imaging was being done with digital detectors.

Amateurs, too, got into the digital act rather quickly—at least the ones who could afford it. The first CCD cameras marketed to amateur astronomers began appearing in the early 1990s, and by the turn of the century the technology was surprisingly widespread. The price-to-performance ratio of CCD cameras improved rapidly, but the total dollar outlay to set up a digital-astrophotography system remained high. By the mid-1990s, an astronomical CCD camera cost upward of $5,000 for a decent-quality model, and cutting-edge cameras with large detectors could run two and three times that amount—a steep price for equipment that could only be used for astronomical photography. But as spectacular amateur CCD pictures began appearing in magazines, on websites, and at astronomy club meetings, it was obvious that the Golden Age of emulsion-based astrophotography had passed.

CCDs changed more than just the way that amateurs did astrophotography. Every astronomical image, whether made with film or a digital detector, contains scientific data. Because it is so easy to extract from a digital image, it is difficult to ignore, especially when a computer for analyzing the data is an integral part of the system used to take the image in the first place. Almost from the get-go, amateurs were following in the footsteps of their professional counterparts, analyzing digital images to measure the brightnesses of stars and the positions of asteroids. This created an unprecedented resurgence in collaborations between amateur and professional astronomers. Once again, the line separating professional and amateurs was becoming blurred. The first decade of the twenty-first century has seen several cutting-edge professional-amateur research projects, and virtually all of them are thanks to digital imaging.

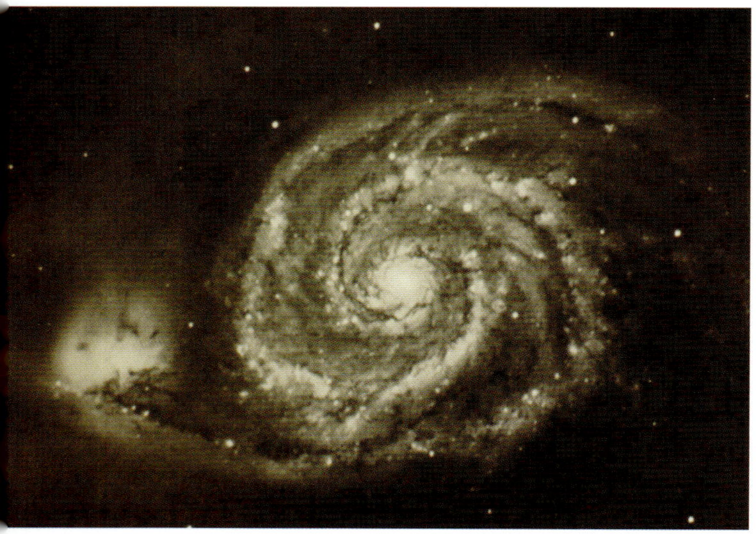

Fig 2.5. George Ritchey, *M51 (Whirlpool Galaxy)*. Image courtesy *Sky & Telescope*.

This view of the Whirlpool Galaxy was the quintessential state-of-the-art astronomical photograph when astronomer George Ritchey made it in 1910. Working with the sixty-inch reflector at Mount Wilson Observatory in California (then the world's largest operational telescope), Ritchey exposed the same plate on the nights of April 6, 7, and 8, accumulating ten and three-quarters hours of exposure before developing it. Contrast it with the author's fifty-minute exposure made with a CCD camera and sixteen-inch reflector in his suburban-Boston backyard in June 2011 when a bright supernova exploded in one of the galaxy's outer arms.

Another change brought about by the digital revolution is readily apparent as one gazes at the images in the Starstruck exhibition and realizes that many of the most spectacular ones were created by amateurs working with backyard telescopes that are tiny by professional standards. Ever since Galileo first turned a telescope skyward, astronomers have sought bigger telescopes in their quest to see fainter objects and resolve finer detail. Because of the way photographic emulsions record light, it was a simple fact that bigger telescopes, which gather more light than smaller ones, took better pictures. As such, when film ruled, the most stunning deep-sky images came from large, professional telescopes. Amateurs could not compete (figs. 2.5 and 2.6).

Digital imaging works differently. Because of the way digital exposures are processed and combined, it is possible to add multiple exposures together, effectively combining the light collected by many exposures into a single image. A small telescope does not collect as much light as a big one, but, if you combine enough exposures, it can take pictures almost as good a big one. And that is exactly what most amateurs today are doing. Whereas in the days of film an amateur photograph made with an exposure lasting, say, two hours was considered exceptional, today it is commonplace to find amateurs making digital deep-sky images with exposures totaling ten or even one hundred or more hours accumulated over the course of several nights. There are still advantages to having larger telescopes, but digital imaging has leveled the playing field somewhat; and now skilled amateurs routinely make deep-sky images that are all but indistinguishable from those made at professional mountaintop observatories (fig. 2.7).

The digital revolution has also brought about a dramatic change in the way that amateur astrophotographers work, and this too has greatly impacted the volume and quality of deep-sky photography being done today. In the days of film, ninety-five percent of the effort was spent making an exposure—getting the telescope properly aimed at the target, focusing the camera, and manually guiding the telescope to ensure that it accurately tracked the stars as they moved across the night sky during the photograph. When the shutter closed, all that remained was to develop the film (it was the rare amateur astronomer who spent time experimenting with tedious darkroom processes in an effort to improve his or her photographs). Simply put, making exposures was a hands-on activity and film-based artists had to find time to be out under a dark night sky, something that can be difficult to do when a "day job" enters the equation!

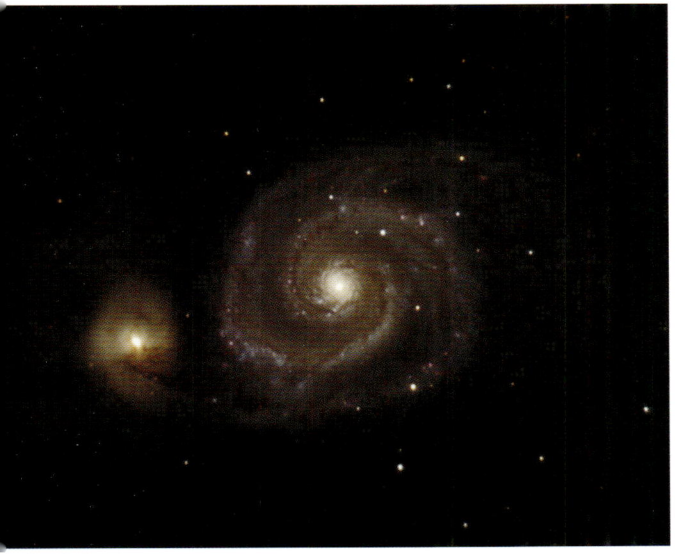

Fig 2.6. Dennis di Cicco, *Whirlpool Galaxy (M51)*. Image courtesy the author.

There has been a complete flip-flop today. Most modern telescopes are computer-controlled and can automatically point to any target in the sky. The laborious task of guiding has been relegated to small digital detectors that have replaced the human eye at the eyepiece of a guide telescope. Even the task of accurately focusing a camera on a telescope has been automated. The entire process of making a deep-sky exposure can now be done under computer control without the need for real-time human intervention. There is a very good chance that many of the exposures used to make the images in the Starstruck exhibition were done while the astrophotographers were sleeping. Furthermore, since the whole exposure process can be controlled over the internet, the telescope could be in the backyard or it could just as easily be half a world away in another hemisphere. Many of today's elite astrophotographers have set up their observatories at "telescope farms" in remote locations thousands of miles from their homes to take advantage of the best observing conditions. Making exposures still takes time, but as far as many astrophotographers are concerned, it is happening in the background. Their real work—the ninety-five percent effort today—comes when the artist sits down in front of a computer to process the images after the shutter closes.

Why Do We Do It?

As a long-time astrophotographer (I began as a teenager in the early 1960s), I am surprised by how often people ask why I take pictures of the night sky. My simple and admittedly glib answer is because astrophotography for me is mostly a hobby, and why does anyone do any hobby? Because we enjoy it, of course.

The question probably gets asked so often because of the unusual nature of astrophotography, though I think if you pry deeply enough into anyone's hobby you find at least a few oddities. It is just that, even on the surface, astrophotography has some pretty obvious ones. After all, in the days of film, it was a hobby that most of us did alone, sitting outside in the dark, staring at some nondescript star in the eyepiece of a guide telescope, while fighting mosquitoes during the summer and bitter cold during the winter. It certainly sounds like an odd form of enjoyment when we could have been inside getting a nice night's sleep in a warm bed.

One of the easily understood driving forces behind astrophotography is the prospect of creating spectacular images. Years ago, I was always thinking ahead to the final image when I sat guiding a telescope or swirling film in a developing tank, and it is the same today when I am waiting for the hourglass to stop spinning as my computer cranks away processing images. It does not matter if someone else has taken a better photograph that I could look at in a book or find on a Web site. I still want to take my own picture. No one questions why someone with a camera takes a photograph of the Eiffel Tower when there are better pictures of it on postcards.

Astrophotographers' real work—95 percent of the effort today—comes when the artist sits down in front of a computer to process the images after the shutter closes.

It is also easy to understand the motivation when astrophotography is done in the name of research. But even if scientific advancement is not the initial goal, astrophotography can lead to discoveries. Amateurs have serendipitously found exploding supernovae in distance galaxies and, closer to home, asteroids and comets on their images. And I speak from experience!

There is, however, one outward aspect of astrophotography that can be difficult to explain, and it is one that also confronts amateur astronomers who enjoy viewing the heavens at the eyepiece of a telescope. While the Sun, planets, and Moon (with its ever-changing play of shadows on crater walls and mountains) constantly provide something new to see through a telescope, deep-sky objects don't change in appearance. Many times amateur astronomers have been asked why they would look at something they have looked at a dozen times before if is always appears the same. Substitute the word "photograph" for "look" and the question is reasonably directed at astrophotographers. Although I cannot really offer a precise answer, when I have been asked this question I often turn it around and ask rhetorically if the person has ever returned to an art museum to see the same paintings. Is it that we are hoping to see something new on each return visit—some detail that we missed before? Or is it something in

Fig 2.7. Ken Crawford, *NGC 3576*.
Image courtesy of the artist.

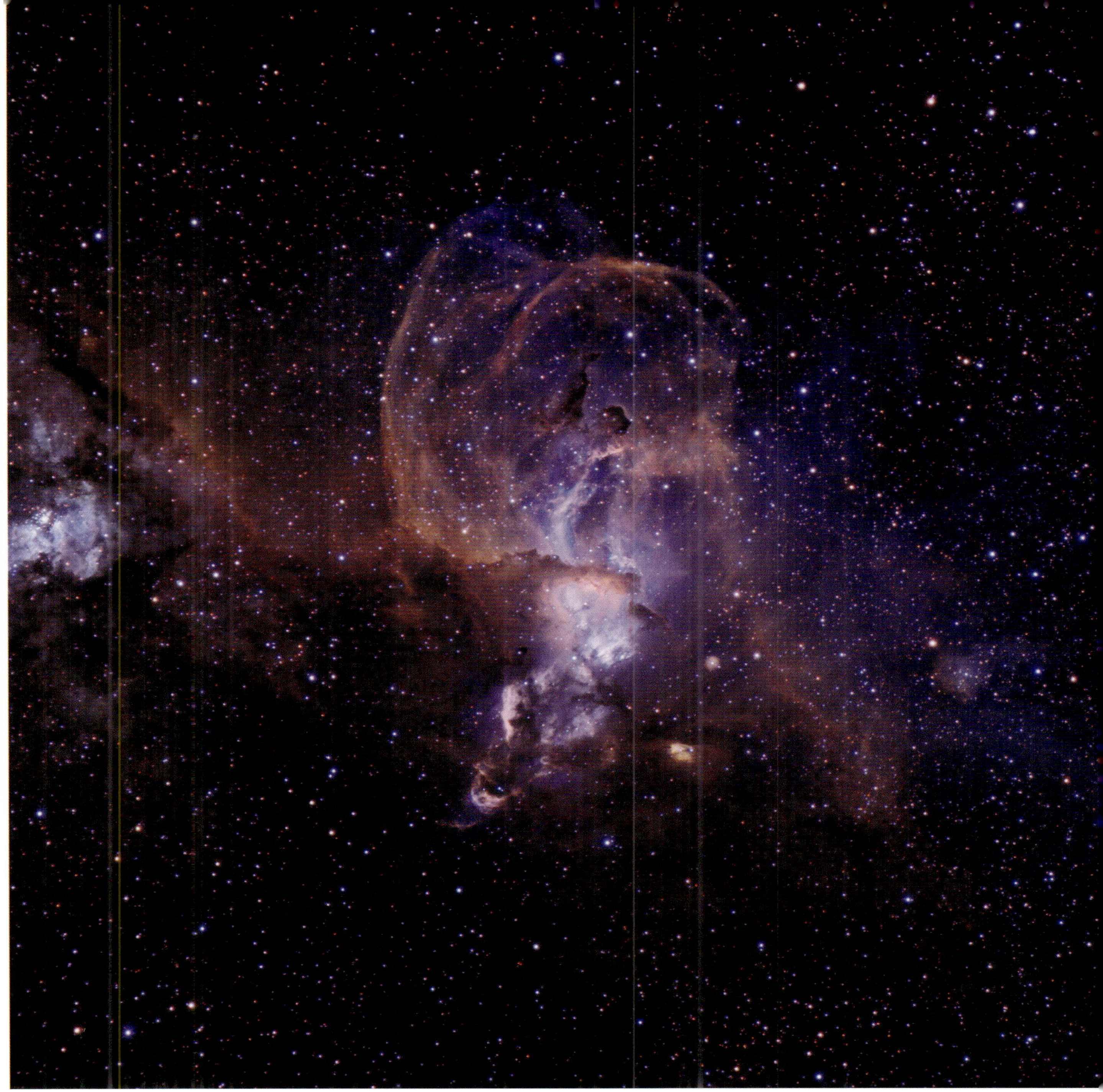

the mind's eye that lets us see something different each time? Or, finally, is it simply that we enjoy the view so much that we just like doing it over again?

Using those arguments actually makes deep-sky astrophotography seem dynamic, since continually improving astrophotography equipment and image-processing techniques make it a given that there will always be new compelling photographs to take of celestial objects. For all their international fanfare, Andrew Ainslie Common's photograph of the Orion Nebula and Isaac Roberts's shot of the Andromeda Galaxy barely gave a hint of the detail captured in the images gracing the walls of the Starstruck exhibition. And a few decades from now we might see today's images as just as much of a point of departure as we see those from a century ago. I certainly hope so.

Notes

1. In the 1960s a small Daguerreotype of the Moon turned up in a Greenwich Village used-book store not far from where Draper taught at New York University a century earlier. A decade after the discovery, an amateur astronomer made a circumstantial case for the unearthed Daguerreotype being Draper's original Moon image. Although reports of this discovery are repeated from time to time in various sources, there is no possibility that this Daguerreotype is the one made by Draper in March 1840, for two main reasons. First, descriptions of Draper's photograph clearly refer to an image with far less detail than the newly-surfaced photograph. Second, Draper photographed through a telescope that reversed the image. While the Greenwich Village Daguerreotype shows a reversed image, it should be a double-reverse one, due to Daguerreotype's inherent image reversal.

2. While the idea of making your own telescope might seem rather extreme, the telescope-making hobby has been an integral part of amateur astronomy since the 1920s, and was even deemed popular enough to have its own monthly column in *Scientific American* for many years.

WESTON NAEF
Curator Emeritus of Photographs
The J. Paul Getty Museum

Drawing with Light

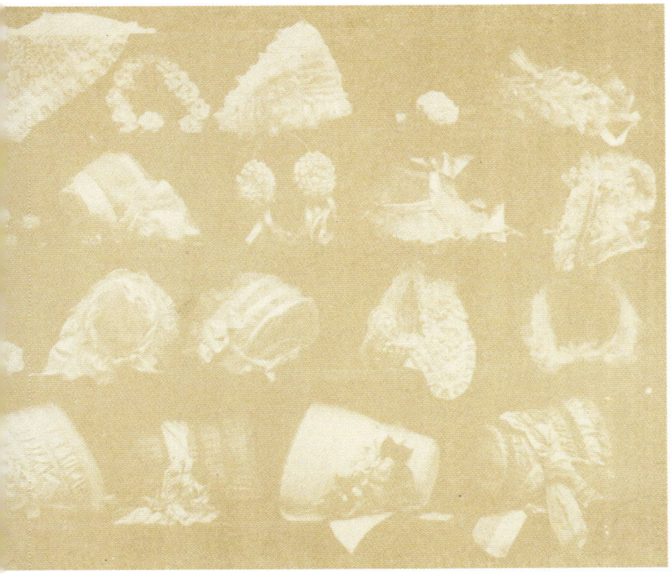

Fig. 3.1. W. H. F. Talbot (British, 1800–77), Talbotype —or—Sun Picture, in *The Art Union Monthly Journal*, June 1, 1846, salt-fixed print from a paper negative. Private Collection.

THE VERY FIRST OCCASION for a photograph to be used successfully to illustrate a magazine of art was for *The Art Union Monthly Journal* in London on June 1, 1846 to accompany an article entitled "Sun Pictures." The photograph showing a shop window with an arrangement of women's hats and bonnets (fig. 3.1) has nothing to do with this exhibition, but how it was made does. The picture was identified on its back as a "Talbotype —Sun Picture," with reference to the inventor of this branch of photography, W. H. F. Talbot. The text accompanying the illustration describes the novelty of pictures that were made by the action of the sun. The phrase "Pencil of Nature" was used to describe the process of their creation. From its very beginning photography has been associated with the universe and all subsequent visual records of the celestial spheres are dependent on the core principle of photography, which is drawing with light.

The word "photograph" soon came to be the preferred name for sun pictures as a combination of the Greek roots: photo = light and graph = to draw. If we jump ahead in time by half a century to the heyday of the Lick Observatory, practitioners of photography around 1900 were taught that photography is "... bound up with the theory of light and hence with solar energy; with spectrum analysis; with chemistry; with optics; and [through] X-rays, with electricity."[1] This description brings us close to the twentieth century and the advent of modern astro-physics, upon which much of the accompanying exhibition is dependent.

Visual records of the celestial spheres are dependent on the core principle of photography, which is drawing with light.

It was two decades before this time, in 1880, that Carleton Watkins created one of the first pictures of the sun that was inspired as much by art as by science. The occasion was the total solar eclipse of January 11, 1880 (fig. 3.2) with a path of favorable viewing across the central California coast in the vicinity of present-day Big Sur and the Santa Lucia Range of coastal mountains. Watkins traveled there with his friend George Davidson, with whom he had worked in 1879 in the high peaks of California's Nevada and Alpine counties, and who would soon inspire the new observatory named after its benefactor, the San Francisco millionaire, James M. Lick. We can imagine the photographer standing at a location that is today favored by monks and aesthetes with a group of scientists, including besides Davidson, Professor Edgar Firsby of the U.S. Naval Observatory. They waited for 3:39 in the afternoon, when the moon began its transit directly between the Earth and the Sun. There was time for just one exposure at the instant of maximum eclipse (3:50 p.m.).[2] Watkins' photograph transcends the documentary because he has so skillfully captured the surrounding context of sky, horizon, clouds, trees, and—the essence of this particular photograph and crucial to all photography—light. Since light is the activating force in photography, it is to be expected that the Sun and the shadows it casts have been favorite subjects for photographers from the very beginning of the new art.

Almost half a century after Watkins created his seminal view of the solar eclipse of January 11, 1880, where clouds were an important, but secondary element, Alfred Stieglitz cast his gaze to the sky and reversed the priorities (cat. 92, p. 84). For Stieglitz the Sun was entirely peripheral to his main concern, which was the form and emotive power of his picture. In his Equivalents Stieglitz wanted to prove that his photographs were "... [good] not due to subject matter—not to special trees, or faces, or interiors, to special privileges," but rather, good as the result of the photographer's eye.[3] For Stieglitz the iconic symbolism of the sun (if pictured) would literally be an overpowering visual element that defeated his quest to make meaningful pictures out of the commonplace ever-changing earthbound atmosphere.

Fig. 3.2. Carleton Watkins, (American, 1829–1916), *Solar Eclipse from Mt. Santa Lucia, January 11, 1880*, albumen print.
The J. Paul Getty Museum (88.XM.92.83).

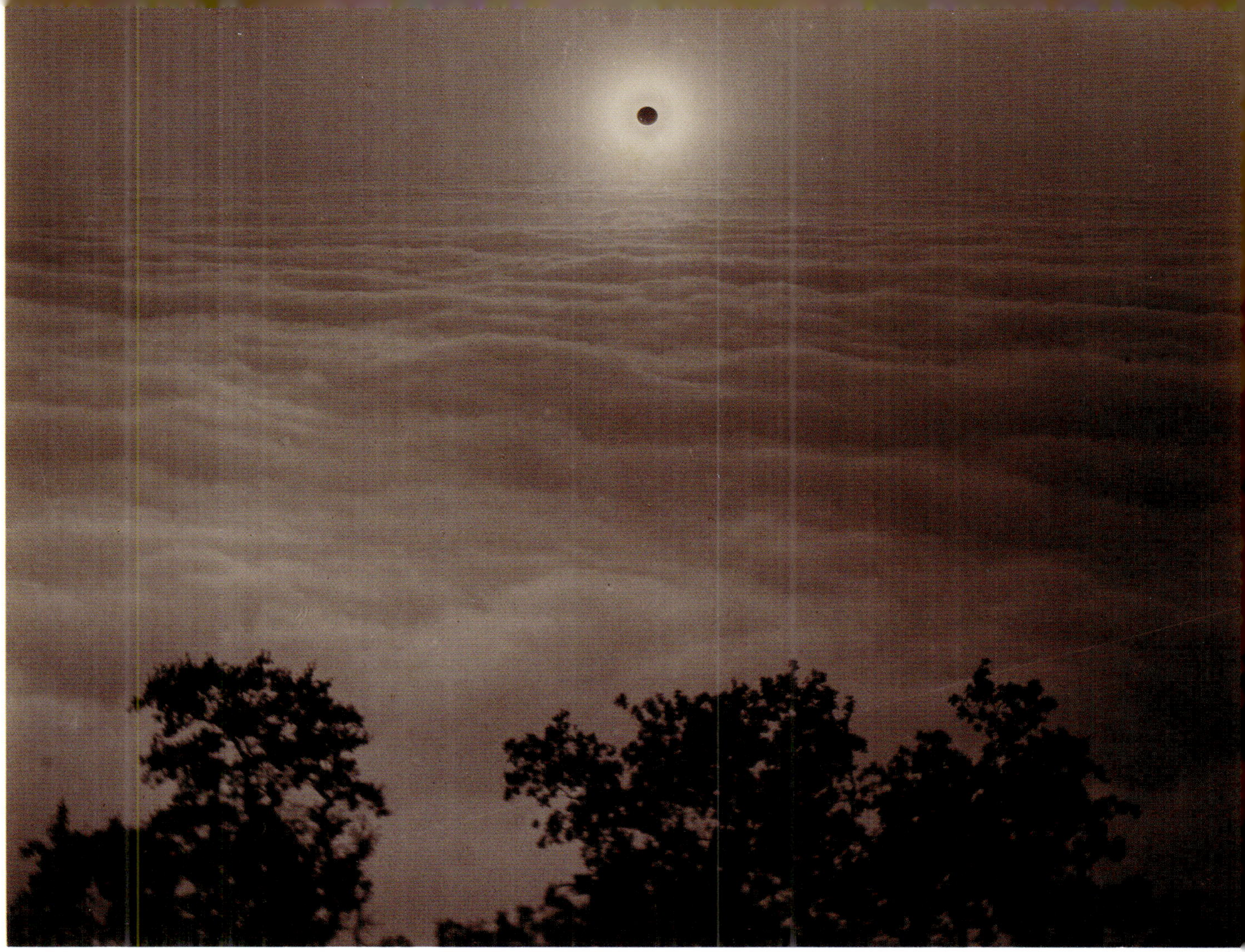

Many artists and photographers of the twenty-first century followed in the footsteps of their forebears. Let's jump ahead to the 1990s, when the artist-photographer Linda Connor found her way the to the first mountain top observatory in California that came into being on Mount Hamilton in Santa Clara County near San Jose under the guidance of Watkins's friend, George Davidson, with money from their mutual friend, James M. Lick. Around the time the Lick Observatory opened, California was in the path of another total eclipse that occurred on January 1, 1889, but the event was not visible from the new observatory.

More than a century after the 1889 eclipse, Connor began a self-assigned project of studying the glass negatives in the Lick Observatory archive that date from the early years of astronomy in California. One of the negatives she located was made during the eclipse of January 1, 1889 (fig. 3.3) and was a multiple exposure showing the moon at several different positions as it passed across the sun during more than two minutes of the eclipse, which was observed from a location in Lake County, two hundred miles north of Mount Hamilton. Connor's work with the Lick negatives is a homage to their maker, E. E. Barnard, who improvised a forty-nine inch focal-length camera on the site.[4] The 1889 eclipse negative merges time and light in two ways. The original negative was made over a span measured in minutes, but Conner's rediscovery and revival of the picture took more than a century.

Connor's objective was to create new prints using photographic materials similar to those available when the negative was made. The material was known as "printing out paper" because the image was developed, not by the action of liquid chemicals on the light-sensitive silver ("development"), but rather by the chemical ("actinic") power of sunlight itself. Connor's method was to fit the glass negative into a wood printing frame with the special printing-out-paper sandwiched between the glass negative and a clear glass cover. The loaded printing frame was positioned in direct sunlight for the time necessary for sunlight to form the image, an elapsed time that ranged from a few minutes to a quarter-hour depending on the negative.

Another aspect of Connor's work with old glass astronomical negatives was her attraction to broken ones (fig. 3.4). In the Lick Observatory archives she encountered a film box labeled "Broken Plates." One negative-at-a-time she reassembled the fragments in the printing frame so that they could be held in place by the wood spring-loaded back. Connor's process reflects a deep understanding of the materials of photography and the role of light in creating the photograph: "As for light, it is one of the things I am most conscious of … It is often the animating element, as well as providing symbolic and metaphoric possibilities," she said.[5] Connor incorporated prints from the found-negatives into her own body of work by pairing the celestial pictures with her own photographs that she sees as earthbound equivalents of the celestial imagery.

The question "What is light?" has preoccupied thinkers for centuries. Sir David Brewster was one of the first to write an entire book on the physics of human eyesight. His *A Treatise on*

Fig. 3.3. Linda Connor, *Solar Eclipse, January 1, 1889*, 1998. Printed from Lick Observatory Archive negative. Original photograph by E. E. Barnard. Copyright UC Regents/Lick Observatory. Image courtesy of the artist and Lick Observatory.

Optics of 1847 describes the properties of light and vision as performed by the human eye in the process of seeing.[6] Brewster and others asked questions for which there were no clear answers like: "Why does [light] render objects visible, exhibit different colors, produce chemical changes in bodies and so forth?"[7]

Photographers over the decades have been inspired in a variety of ways by the idea that a subtle elastic medium of light could be the raw material for their art. While living in Southern

California in 2002 on a study-leave from his native Germany, Hans-Christian Schink began a body of work totally dependent on the daily voyage of the earth around the sun. He asked himself the question, What would be the result if a sheet of black-and-white negative film were exposed with the camera shutter open for one hour (hereafter referred to as "1h") with the lens facing directly into the Sun across a landscape scene (cat. 67, p. 103)?[8] He correctly predicted based on his prior experience that the Sun would be recorded on the negative as a black line, but he also wondered whether the landscape would be recorded in recognizable detail, which it turned out not to be in his first trials. The idea he had in mind would only work if he could succeed in recording the landscape context of the picture along with the sun as it was at that location. It was not until Schink experimented with very old fashioned films that were sensitive only to green and blue light ("orthochromatic" in the language of photography) that the landscape was recorded with the range of gray tones that he hoped to achieve. Schink's 1h project has in common with Connor's rediscovery of the Lick Observatory's glass negative of the total solar eclipse of January 1, 1889 two facts: that the light and time are central visual elements, and that their resulting pictures are like drawings made by the pencil of nature.

> *"The results of my travels are images that could never actually be perceived by the naked human eye."*
>
> HANS-CHRISTIAN SCHINK

Connor and Schink also have in common the necessity of travel to far flung locations to realize fully the envisioned concept of recording how light is received from the heavens from different spots on the planet. Connor travels widely to find just the right light in pictures made with her own cameras, and she juxtaposes these camera-images with prints from the series of total eclipse negatives exposed at locations all over the globe by the Lick Observatory team. Schink's 1h project was organized according to subjects found in the northern hemisphere (cat. nos. 67–78, "Sun" section), and those found in the southern hemisphere (cat. nos. 79–90, "Sun" section), which required extensive travel in quest of the perfect opportunity for the Sun to draw its own signature in the context of a diverse earthly terrain. Schink succinctly phrased his experience when he remarked, "The results of my travels are images … that could never actually be perceived by the naked human eye."[9] His 1h series incorporates what the photographer could see and what was invisible to his eyes at the time the exposures were made.

Accident and chance could be the most creative forces in the universe, so Leonardo da Vinci is reported to have said. Forces outside the control of the artist's contributed to Connor's work

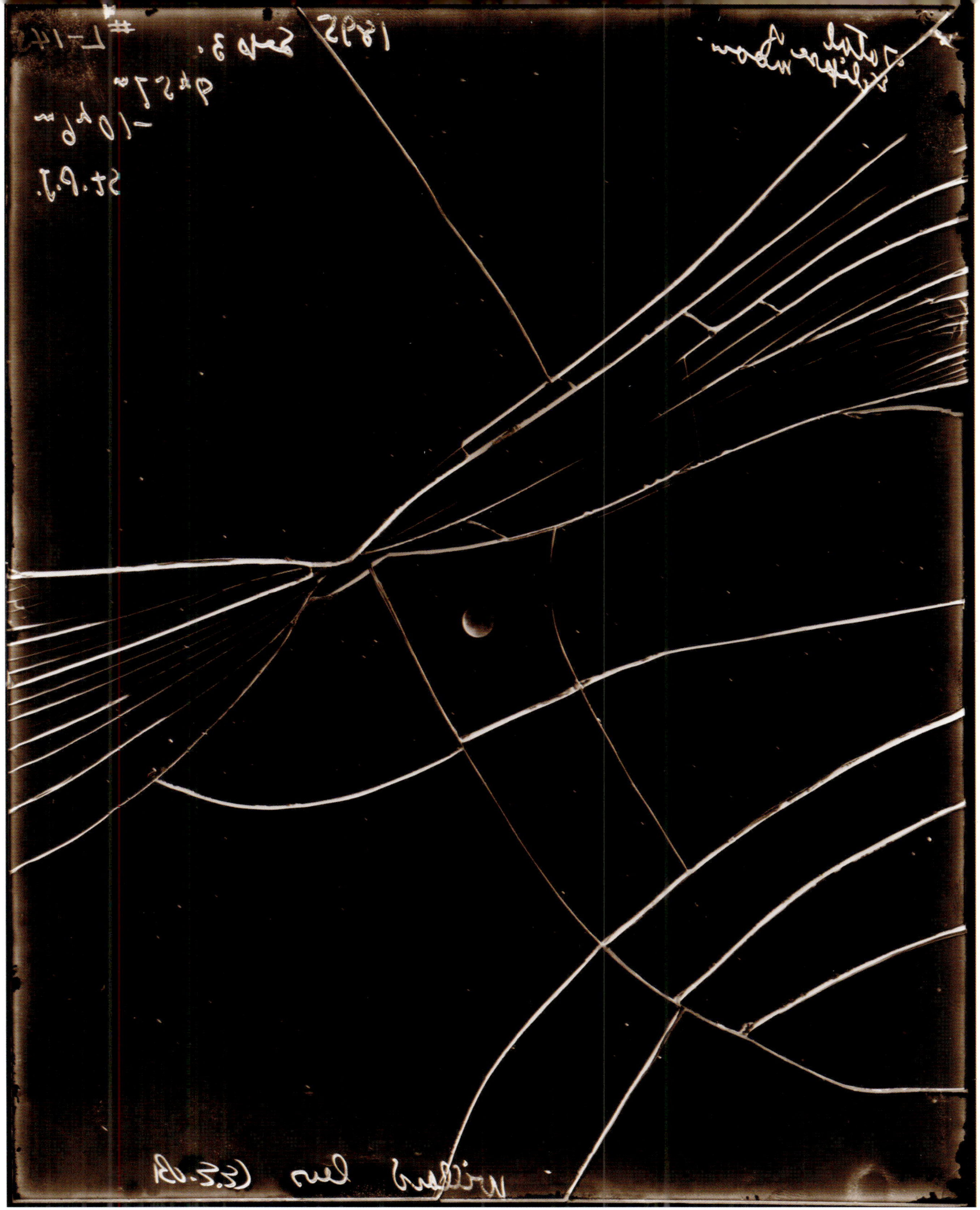

Fig. 3.4. Linda Connor, *Lunar Eclipse, September 3, 1895, Mount Hamilton*, 1998. Printed from Lick Observatory Archive negative. Original photograph by E. E. Barnard. Image courtesy of the artist and Lick Observatory.

with glass plate negatives, some of which were broken and consequently required reassembly by hand for printing with the breaks becoming dynamic elements of the new images. For Schink, chance played an important role. Despite his careful advanced planning to identify the right places to position his tripod-mounted camera in Patagonia, the Mojave Desert, or the Arctic Circle, they were never the ones he imagined before reaching the locations. After arriving in the general vicinity of his destination, Schink says "I would simply find the right [position] along the way."[10]

Inspired by images of the Sun like those created by Connor and Schink, another California artist-photographer, Jacqueline Woods, allows accident and chance to become the controlling elements in her Black Sun series (cat. nos. 101–106, "Sun" section), which were made without the use of a camera or direct reliance on the Sun itself. Both Schink and Woods admit to having been influenced by Minor White's 1955 photograph, Black Sun. The solarization process used in realizing her prints is a close cousin to the Sabatier effect that Schink harnessed to create his negatives. Very early in the history of photography it was discovered that extreme overexposure of the negative, such as Schink's one hour exposures, resulted in the light-source (usually the Sun), was the darkest instead of being the brightest part of the image, resulting in the overall image being reversed from negative to positive.

Photography is frequently a jousting match between truth and fiction.

The same reversal effect can be achieved in cameraless photography with a sheet of printing paper being exposed to light while soaked with developing chemicals. Photography is frequently a jousting match between truth and fiction. For Woods the Sun is nothing more than a rounded piece of paper that blocks light from reaching the surface of the printing paper during exposure before liquid chemicals are invoked. Choice of the printing paper is central to the results Woods achieves—her materials were manufactured ten, twenty, thirty or more years before being exposed to light, thus introducing the element of time passing as an element of the content. Ironically, by comparison to Connor and Schink, who travel endlessly in quest of their subjects, Woods never has to leave the darkroom to make her Black Sun pictures. The often painterly effects are the result of a brief explosion of human and chemical energy in the darkroom, the prints are created from vintage materials that it took a lifetime to secure. Woods's Black Sun pictures are works exclusively of the imagination where the Sun and the Moon (fig. 3.5) are never seen like this, and they require a deep knowledge of photographic materials and their chemistry in order to bring the process of accident and chance under control.

The theory behind the pictures of Connor, Schink and Woods could be explained thus: "The ray of white light contains within it powers of which earlier philosophers had but a faint idea. … [Light] has the power of decomposing and determining the recomposition of chemical components … This principle—actinism—is regarded by some as the fundamental principle upon which photography is based. That principle is a component part of light, but not light itself," wrote an early theorist.[11] From the time when the first Sun pictures were published by *The Art Union Monthly Journal* on June 1, 1846, history and practice have been inseparable when it comes to artists who represent and visually interpret the celestial spheres. For Connor, Schink, and Woods, it has been the miraculous process of drawing with light.

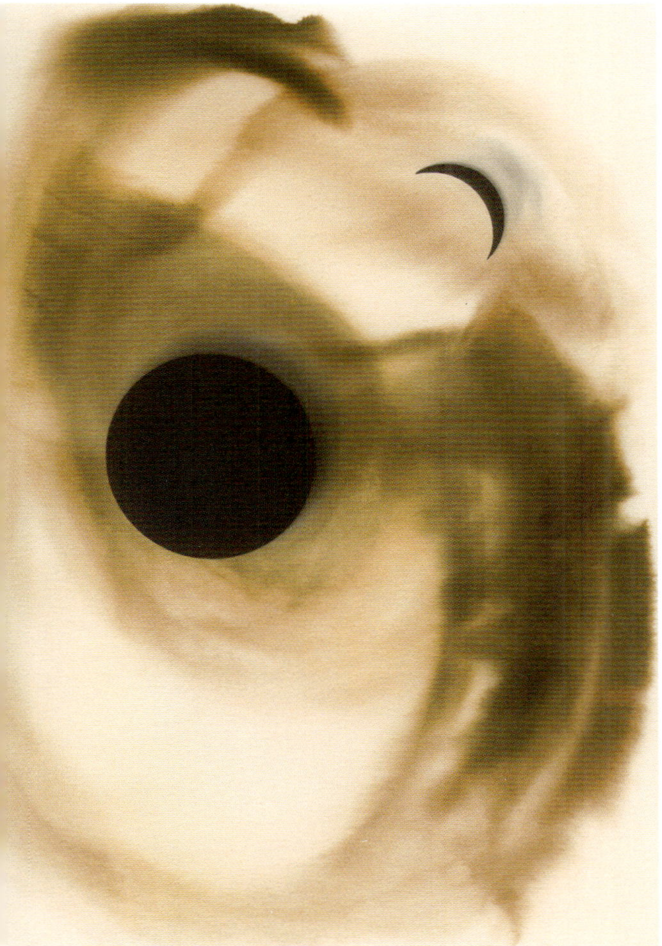

Fig 3.5 Jacqueline Woods, *Black Sun #12*, 2011, gelatin silver print. Image courtesy of the artist.

Notes

1. Story 1898, p. 5.
2. Naef, Weston. *The J. Paul Getty Museum Handbook of the Photographs Collection.* Los Angeles: The J. Paul Getty Museum, 1995, p. 96.
3. Travis, David B. "A Mind Above the Clouds: Alfred Stieglitz's Equivalents Photographs," in *At the Edge of the Light: Thoughts on Photography and Photographers, on Talent and Genius.* Jaffrey, New Hampshire: David R. Godine, 2003.
4. With thanks for this information to Tony Misch, Lick Observatory.
5. Connor, Linda. *Odyssey.* San Francisco: Chronicle Books, 2008, [p. 154].
6. Brewster, Sir David. *A Treatise on Optics.* Philadelphia: Lea & Blanchard, 1847, p. 11.
7. Sutton, Thomas A., and George Dawson. *A Dictionary of Photography.* London: Sampson Low, Son & Marston, 1867, p. 196.
8. Schink, Hans-Christian. *1h.* Ostfildern, Germany: Hatje Kantz Verlag, 2011.
9. Ibid., p. 87.
10. Ibid.
11. Story, op. cit., pp. 30–1.

JERRY T. BONNELL
Co-Editor/Author, Astronomy Picture of the Day
and Astrophysicist/Staff Scientist,
Compton Gamma Ray Observatory Science Support Center

A Confluence of Science & Art: Astroimaging's Golden Age

We are denizens of a space age that began in the last century. Images of astronauts on a desolate Moon, rocks strewn across the rugged surface of Mars, and iconic views of planet Earth itself from a space-based perspective appear in our text books (fig. 4.1). Pictures of the cosmos from orbiting observatories and enormous telescopes on mountain tops provide even more distant vistas. In the images, astronomers read a story of origins in natal gas and dust clouds gathered in our Milky Way and a multitude of other galaxies flung across the distant Universe. The photographs represent a remarkable confluence of science and art. We seem to find the stories and pictures of vast, inhospitable environments and cosmic scales evocative on a fundamentally human level.

The appeal of astronomical subjects to artists has synergistically met with unprecedented access to photographic technology. The last decade in particular has seen an awakening in astronomical imaging, driven by the same advances in technology and economic forces that put cameras in cell phones. Archives of astronomical data, computational resources, and sensitive digital detectors that were once available only through large research institutions are now in the hands of individuals. Sometimes described as amateur astronomers, they explore the Universe in their spare time and with small telescopes from backyards and driveways. Unbeholden to grantors, employers, or research groups, they enjoy an increased freedom to experiment and make choices that enhance both a scientific and artistic process. The diversity of ideas that comes with increased access and multiple practitioners has engendered a golden age of astronomical imaging. That golden age has fueled this Starstruck exhibition.

Subjects pictured in Starstruck follow an impressive trail through the cosmos. Walking through the exhibition or flipping through these pages make you feel like an accomplished cosmic tourist!

Starting close to home, but still well within the realm of astrophotography, are Hans-Christian Schink's solarized trails of the moving Sun above the horizons of planet Earth (cat. nos. 67–90, catalogue, "Sun" section). His film-based technique produces images that appear unfamiliar and stark, but reflect on a consequence of something fundamental to the rhythm of daily life—the Earth's rotation on its axis. Choices of a digital photographic technique, palette, and location also reflect the rotation induced celestial motions in works like Babak Tafreshi's *Rio Moonset* (cat. 96, p. 174) reminiscent of Dada artist Marcel Duchamp's *Nude Descending a Staircase (No. 2)*, and Yuichi Takasaka's *Star Trails* (cat. 98, p. 157).

In fact, Starstruck works depicting the Sun, Moon, and other planets of the solar system cover a range still close to home compared to the exhibition's expansive arc of subjects. Neptune and large moon Triton are the most distant, but still only a mere four light-hours from the Sun. Dimly reflecting our star's light, their slender crescents are seen in Michael Benson's elegant reprocessing of Voyager 2 image data (cat. 1, p. 175), acquired as the spacecraft flew by the gas giant planet in 1989. By then, Voyager 2 had been traveling outward for 12 years following its 1977 launch. The crescent phases could only be captured after the spacecraft had crossed Neptune's orbit and pointed its camera back toward the planet's night side, normally hidden from the perspective of telescopes anchored closer to the Sun. An artist wanting to portray photographically the outer planets as crescents has no choice but to rely on probe data (fig. 4.5).

Traveling only a few light-years into interstellar space will take the Voyager spacecraft a few thousand centuries. But interstellar clouds of gas and dust on scales of tens to hundreds of light-

Fig. 4.1. *Volcanic Bumpy Boulder on Mars*, Mars Exploration Rover Mission, Cornell, JPL, NASA.

The plucky Spirit rover delighted the world with numerous images of the Martian landscape. Among them was this scene it captured of an unusually textured half-meter high rock, probably created by a volcano.

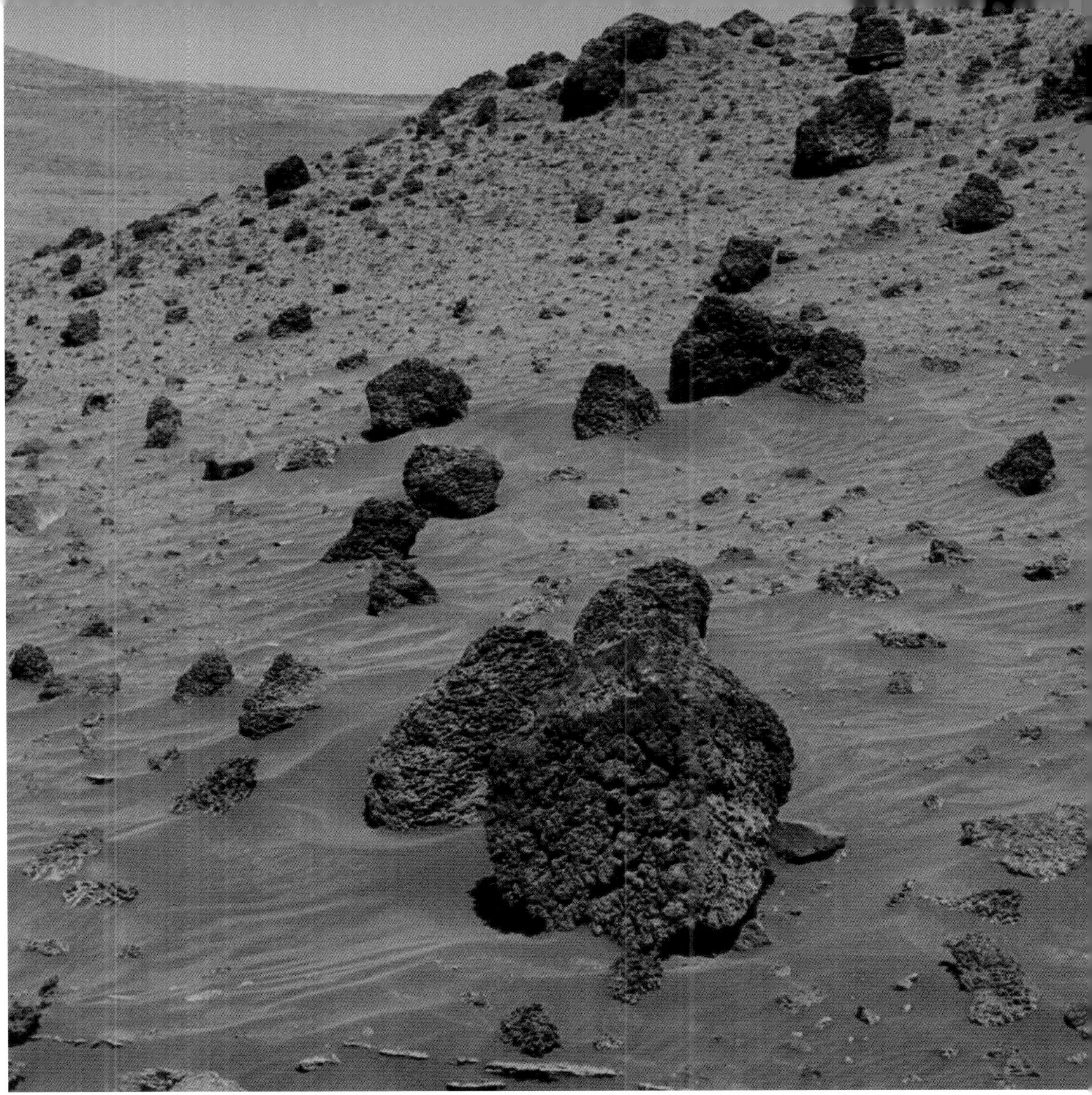

years and more are explored in Starstruck images, plumbed in cumulative exposures lasting hours. The deep space nebulae are tenuous and faint, by and large invisible to human eyes. If you could park your starship next to them and look out the window it would not help. Imaging them relies on a telescope and camera as an effective scientific instrument, gathering more light and revealing structure that your eyes can never see.

Works like Ken Crawford's depiction of nebula IC 5067 (cat. 16, p. 117) can also be seen as the result of an artistic process, creating a composition with balance, line, contrast, and color. But the imagery presents a rich scientific content. The framing captures significant details of star forming pillars of cold gas and dust. The tricolor palette chosen to display the data from narrow band filters highlights emission lines from atoms in the nebula, tracing temperatures and densities in the glowing atomic gas.

Deep space nebulae are tenuous and faint, by and large invisible to human eyes.

In fact, the tricolor palette Crawford uses has become known as the Hubble palette. The scheme for mapping atomic emissions recorded by narrowband filters to red, green, and blue colors does not mimic human color vision, arguably producing a false-color image, but has been made popular in images of star forming regions from the Hubble Space Telescope. While astrophotographers are free to make idiosyncratic choices while processing their image data, the Hubble palette is a favorite of many for compositions involving emission nebulae, probably because the palette seems vetted in the scientific community via Hubble images.

Applied to emission nebulae, the Hubble palette tends to produce aesthetically pleasing complimentary orange and blue hues that outline the depth and contours of structures in an apparently orthodox fashion (fig. 4.2). But astrophotography's golden age has encouraged amateur astronomers' further explorations driven by aesthetics, evidenced, for example, by Alan Friedman's *Blue Sun* (cat. 27, p. 77) and Ignacio Diaz-Bobillo's *Light-Drops* (cat. nos. 20–23, pp. 118–9, 126–7). Alternating palettes in *Light Drops* directly compare the range of visual perceptions possible using the same narrow band image data. Still, some of this exhibition's artists, such as Linda Connor (cat. nos. 9, 12, pp. 116, 78), Thomas Ruff (cat. nos. 64–6, pp. 178–80), and Michael Benson (cat. nos. 1–2, pp. 175, 155), seem the most willing to offer scientific image data, in this case appropriated from archives, for aesthetic appreciation alone, embracing the scientific content's contribution to the aesthetics of their images. And Hans-Christian Schink (cat. nos. 67–90, The Sun section, p. 81 ff.) and Sharon Harper (cat. nos. 37–38, pp. 170–3) reference a scientific process with their meticulous titles of original images.

Scanning along the stars, clusters, and nebulae of our own Milky Way galaxy, the narrow field of view afforded by most deep telescopic images can be restricting, akin to looking at the scene through a drinking straw. The nature of the image data offers a path to a broader view for artists who can devote time to acquiring deep images of adjacent fields to be digitally processed and combined into wide mosaics spanning many degrees on the sky.

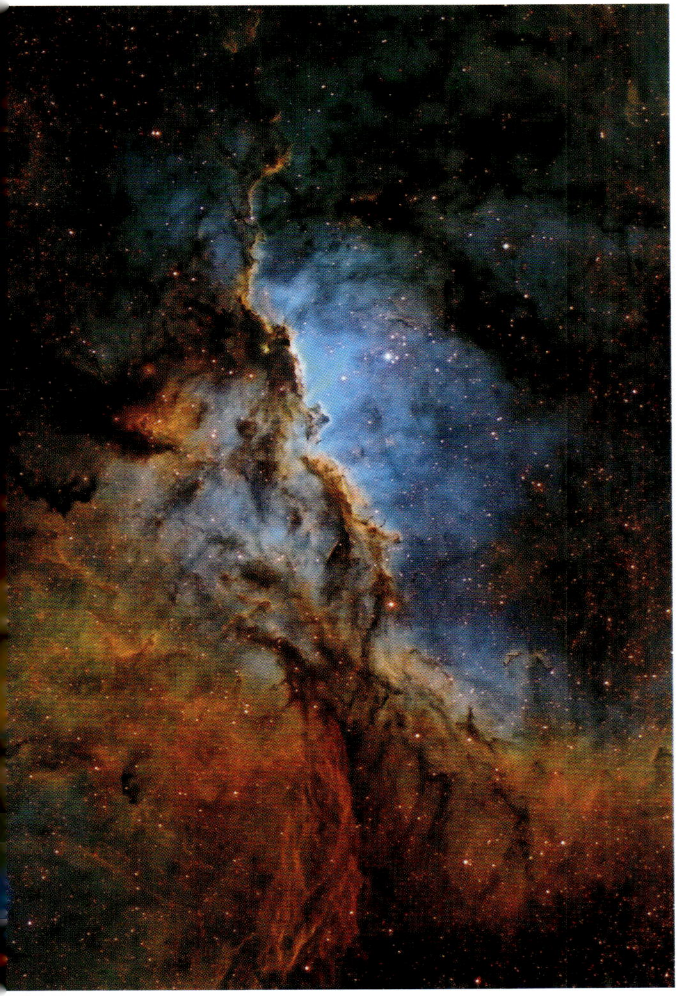

Fig. 4.2. Piotrek Sadowski, *Shaping NGC 6188*, 2010. Image courtesy of the artist. astrofotografia.com.pl

In Starstruck, the ultimate mosaic is Nick Risinger's monumental *Photopic Sky Survey* (cat. 59, pp. 186–7), which encompasses the entire night sky accessible from planet Earth and placing our Milky Way galaxy on a single canvas. Stitched together from over 37,000 images it is the result of substantial planning, a year-long effort under dark skies of the American west and the western Cape of South Africa, and a season processing the data and composing the image. The preparation and technical skill involved in this endeavor is reminiscent of that displayed by pioneer photographers working with large format cameras, glass plate negatives, and darkrooms transported by mule and wagon into remote wilderness areas.

In Risinger's all sky mosaic, obscuring dust clouds along the galactic plane are seen in silhouette against the Milky Way's stars and nebulae, but galaxies of stars beyond the Milky Way may also be spotted in the cosmic vista. At the beginning of the last century, though, the existence of galaxies beyond the Milky Way was not widely accepted. The turning point of that debate was Edwin Hubble's determination of the distance to the Andromeda Galaxy based on photographic observations of variable stars. [See Eric Wollman's "Starstruck Astronomy," p. 49 ff., for an expanded description of this discovery.—Editor]

Now known to be some 2.5 million light-years distant, the Andromeda Galaxy is the subject of one of Robert Gendler's (cat. 31, p. 147) Starstruck contributions. A champion example of astronomical imaging's golden age, the digital picture is a mosaic of ninety hours of image data acquired at an amateur observatory in New Mexico. Incredibly, in pixel dimensions it surpasses images of external galaxies from the Hubble Space Telescope. Gendler's work deftly follows the closely wound spiral arms of the massive galaxy, showing them to be laced with dust clouds, star forming nebulae, and star clusters, an echo of the structure of the Milky Way seen from an external perspective.

Gendler's galaxy portrait depicts the closest large spiral galaxy to the Milky Way and the largest member of the local group of galaxies. But Starstruck contributors portray even more distant galaxies in their compositions, some relying on a principle well known to modern astronomers, namely that galaxies are not randomly strewn through the Universe. Galaxies come in groups. The best place to find a galaxy is just next to another galaxy. Warren Keller's contribution frames a group of three bright spiral galaxies that amateurs call the Leo Triplet (cat. 43, p. 146). All are approximately 30 million light-years distant and offer drastically different profiles of flattened galactic disks seen tilted edge-on and nearly face-on. The scene is a naturally occurring cubist-style illustration of different perspectives viewed simultaneously.

The skill involved is reminiscent of pioneer photographers working with cameras, glass negatives, and darkrooms transported by mule into remote wilderness.

Drawn to cosmic train wrecks, Steven Mazlin and Jack Harvey collaborate on one of the most distant Starstruck subjects, the Antennae Galaxies (cat. 50, p. 153). The two galaxies, more formally catalogued as NGC 4038 and NGC 4039, are seen as they crashed together some sixty million light-years away (and thus sixty million years ago) toward the southern constellation Corvus. The tightly framed view highlights the maelstrom of star formation triggered by the collision of gas and dust clouds in the colliding galaxies. Although the scenario might seem wildly catastrophic, galaxy collisions and their eventual mergers are understood to be a normal part of the process of galaxy evolution. Other well-known systems of colliding galaxies are well within the reach of artists armed with gear accessible to amateur astronomers. The cosmic snap shot of the system chosen by Mazlin and Harvey captures a stage in the inevitable process. A similar fate awaits the Andromeda Galaxy and our own Milky Way in the far future.

Even as we travel, starstruck by this cosmic tour, I will remind us that that we are in the here and now, a special moment in astrophotography's history. The golden age that astroimaging now inhabits is an internet era that accelerates the infusion of astronomy images into popular culture. Today's astrophotographs are even more immediately accessible than the iconic Apollo era scenes from the lunar surface (fig. 4.3) or planet Earth poised above the limb of the Moon. The association of astronomical images with hip aesthetics and cutting-edge technology is a powerful combination recognized even in the corporate world by companies like Apple, which

Fig. 4.3. Charles Conrad, *Apollo 12: Self-Portrait.* Image courtesy NASA.

In November of 1969, Apollo 12 astronaut-photographer Charles "Pete" Conrad recorded this stark masterpiece while documenting colleague Alan Bean's lunar soil collection activities on the Oceanus Procellarum (largest of the lunar maria). The stark, harsh environment of the Moon's Ocean of Storms is echoed in Bean's helmet, a perfectly composed reflection of Conrad and the lunar horizon.

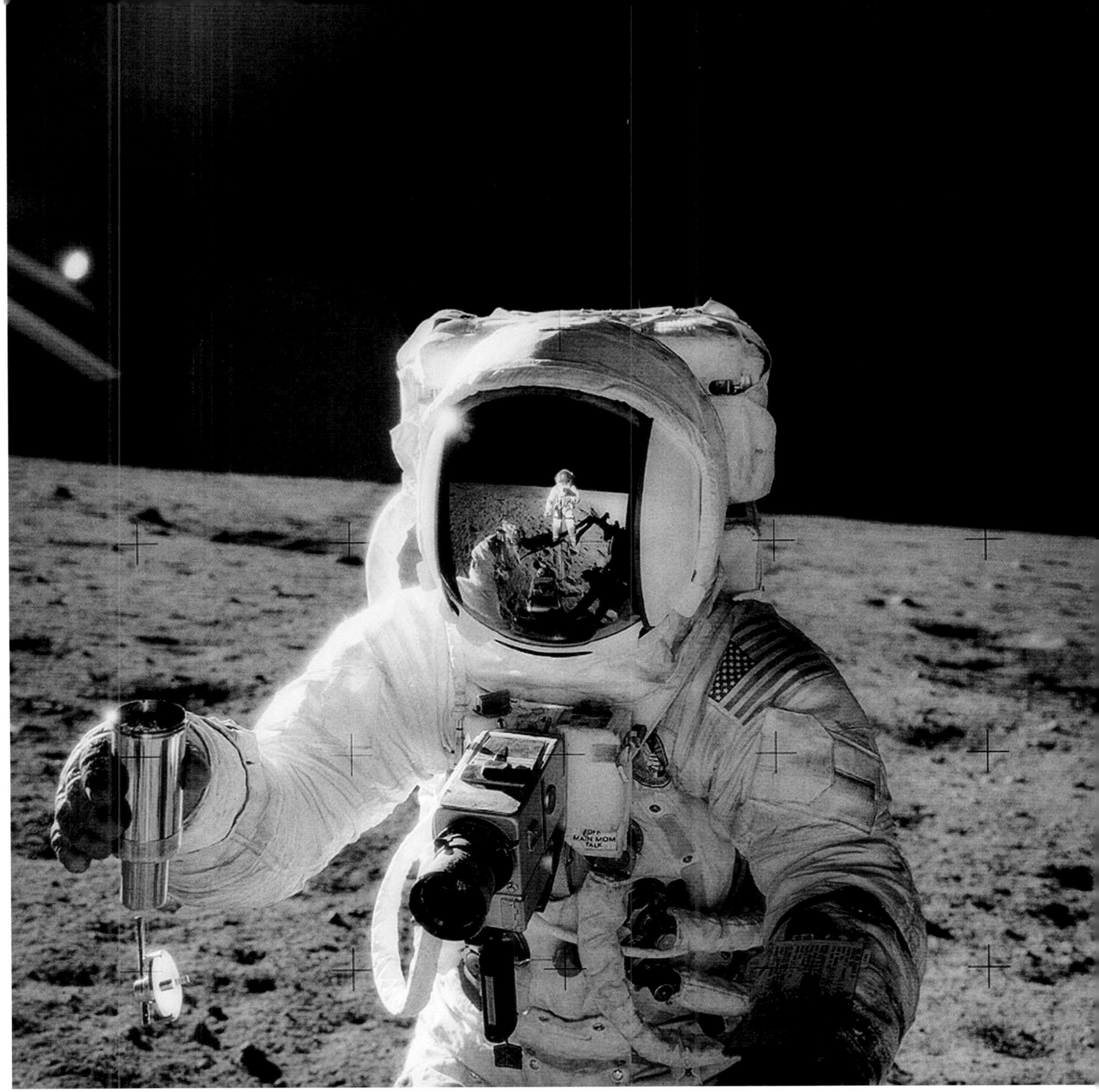

currently displays an image of the Andromeda Galaxy (aka M31) on their website's depiction of their computer monitors. And the demand for more astrophotographs grows, with Web sites like Astronomy Picture of the Day recording one million page views per day.

Astroimaging's current golden age has now brought about something else worth noting—exhibitions at traditional art venues like influential galleries and museums. These are important. They add credibility to the genre, allow it to be understood within the context of not the vastness of space, but the vastness of human endeavor that is the purview of art history. These exhibitions also raise the bar of excellence among many artists working outside the art mainstream by forcing them to create printed images which are more demanding on their skills than are images intended to be viewed on today's computer screens. By bringing works by amateur astronomers together with images created by mainstream artists represented in the world's great museums, exhibitions like Starstruck ultimately create a new bend in the art mainstream, redefining it so that it encompasses work made by artists who, until now, may have thought of themselves as outsiders.

It is no surprise that art historians and curators have recognized the rich genre of astrophotography. After all, who *wouldn't* find these images compelling? (fig. 4.4) Strolling through the Starstruck exhibition or the pages of this catalogue as a cosmic tourist, you are examining representations of the Universe light-minutes, light-hours, and ultimately millions of light-years in the past. Ironically, you may find yourself contemplating the future, inspired by a confluence of human endeavors in art and science and a golden age of astronomical imaging.

Fig. 4.4. *Earthrise*, 1968, Apollo 8.
Image courtesy NASA.

In December of 1968, the Apollo 8 crew of Frank Borman, James Lovell, and William Anders flew from the Earth to the Moon and back again. Among other impressive firsts the mission accomplished was the first photograph of Earth from deep space. As the Apollo 8 command module rounded the far side of the Moon, the crew could look toward the lunar horizon and see the Earth appear to rise, due to their spacecraft's orbital motion. The famous picture that resulted, of a distant blue Earth above the Moon's limb, was a marvelous gift to the world.

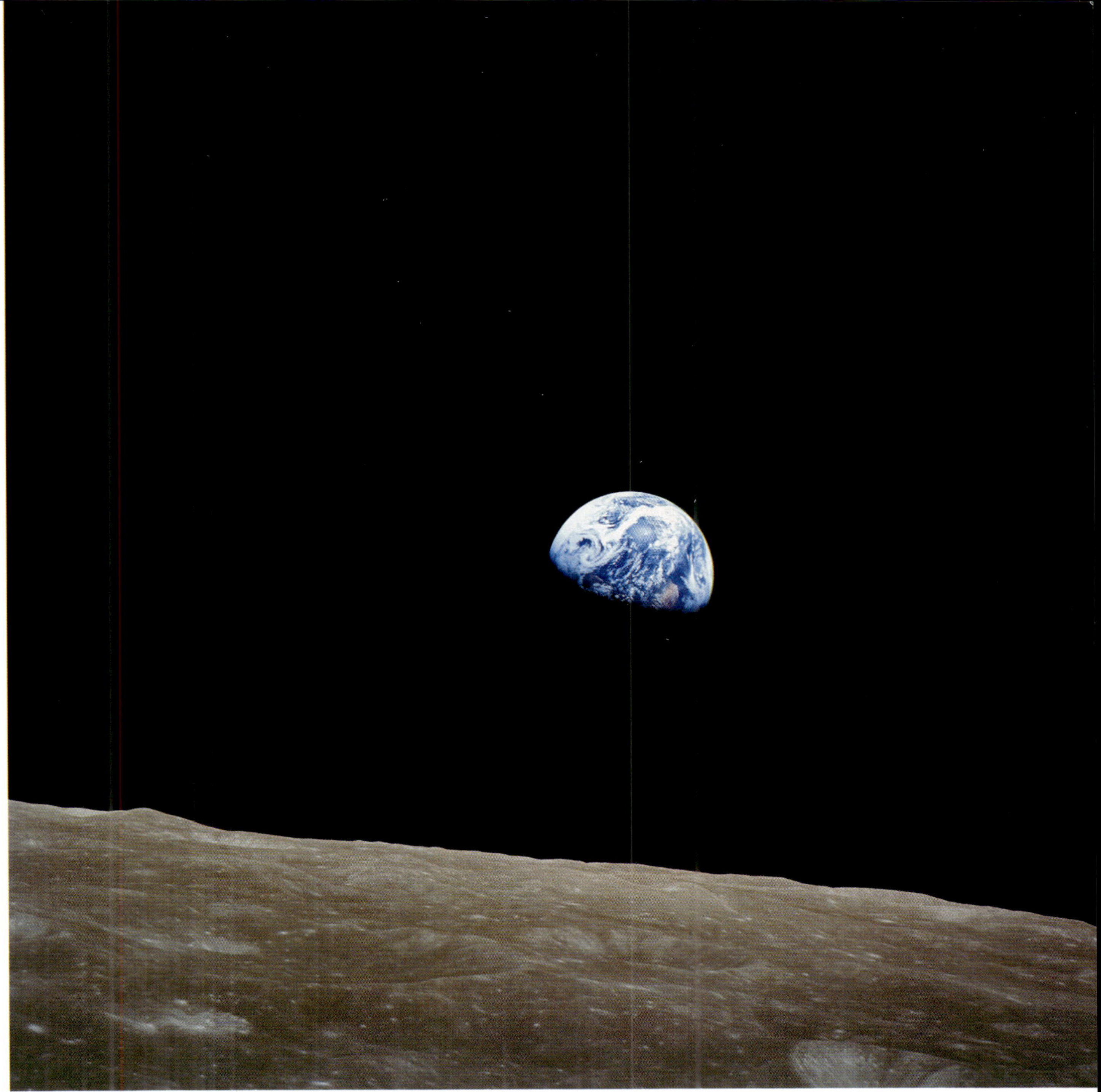

Fig. 4.5. *In the Shadow of Saturn*, Cassini Imaging Team, SSI, JPL, ESA, NASA.

In the shadow of Saturn, unexpected wonders appear. The robotic Cassini spacecraft now orbiting Saturn drifted in giant planet's shadow for about 12 hours in 2006 and looked back toward the eclipsed Sun. Cassini saw a view unlike any other. First, the night side of Saturn is seen to be partly lit by light reflected from its own majestic ring system. Next, the rings themselves appear dark when silhouetted against Saturn, but quite bright when viewed away from Saturn, slightly scattering sunlight, in this exaggerated color image. Saturn's rings light up so much that new rings were discovered, although they are hard to see in the image. Seen in spectacular detail, however, is Saturn's E ring, the ring created by the newly discovered ice-fountains of the moon Enceladus and the outermost ring visible above. Far in the distance, at the left, just above the bright main rings, is the almost ignorable pale blue dot of Earth.

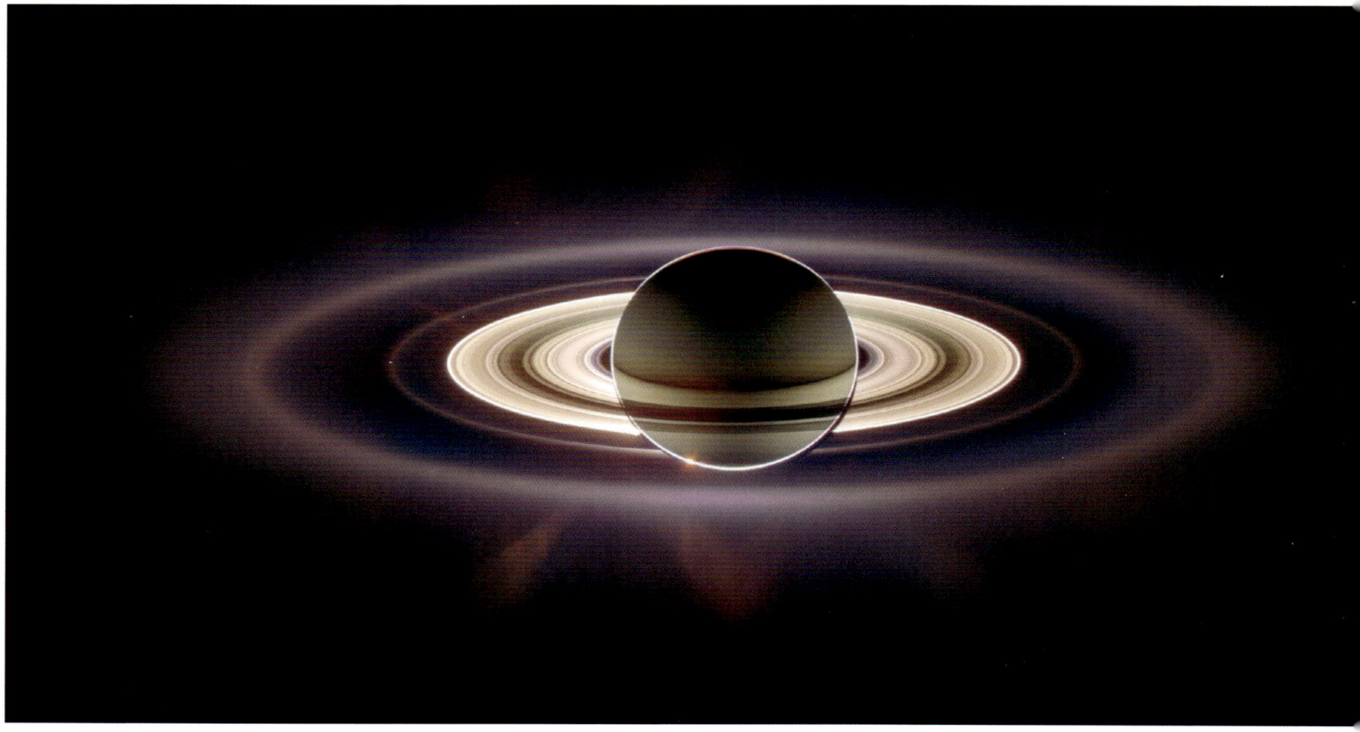

ERIC WOLLMAN
Professor of Physics, Bates College

Starstruck Astronomy

Aesthetics and knowledge are curiously entwined, with every field of experience being enriched by understanding. A gastronome who develops some skill in the kitchen or garden becomes aware of delightful new subtleties at the table just as a sports fan finds exciting new layers of intricacy in a game upon delving into its statistics. The same goes for astrophotography. While Starstruck's images are sensually captivating, they become even more powerful as one develops a sense of the scale of the space and time involved, an understanding of the structures and energies depicted, and a comprehension of our connection to all of it. Eric Wollman has spent his career doing just that. If you follow him through this brief introduction to astronomy, I think you will find that the insight you gain amplifies the mysteries presented by Starstruck's artists.

—Editor

NEARLY EVERYTHING YOU SEE in the *Photopic Sky Survey*, a ten-foot wide panorama by Nick Risinger reproduced in this catalogue at about 15½ percent (cat. 59, pp. 186–7), is constituent matter of the Milky Way Galaxy. The Milky Way Galaxy is a giant collection of a few hundred billion stars mixed with dusty gas, held together by gravity and flattened by rotation into a circular disk. The Sun is one of these stars. The disk is remarkably thin. Its thickness is only about 1⁄50 of its diameter. The imaginary plane that slices the disk into two equal halves, the way you slice an English muffin, is called the mid-plane of the Galaxy. The Sun is located very near the mid-plane and roughly half way between the center and the outer edge of the disk. The Milky Way is a spiral galaxy. If you could rise up out of the disk and look down on it, you would see a pattern of bright "arms" spiraling out from the central region. The arms are bright because they are regions of active star formation triggered by a gravity-induced spiral-shaped density wave that travels around the disk.

People use the word "awesome" rather casually these days. But to say that the Galaxy[1] is awesome in size hardly captures the reality. In one second, light travels a distance about equal to seven times around the Earth. The Earth orbits the Sun at a distance of about eight light minutes—the distance light travels in eight minutes. The stars closest to the Sun are a few light years away. The disk is about two thousand light years thick, and it takes light about one hundred thousand years (about three trillion seconds!) to traverse the diameter of the galactic disk. At seven Earth circumferences per second, this makes the diameter of the Galaxy's disk about twenty-one trillion times the circumference of the Earth!

To connect what you are looking at in the image with what you see outside on a clear night, you must imagine wrapping the Risinger panorama around you, joining the luminous band into a continuous circle and then bending the panorama inward above and below the band to form a spherical shell. Thus you turn the flat rectangular panorama into a celestial sphere —the imaginary spherical shell on which the stars and all other celestial objects appear to lie.[2] The luminous band running through the panorama becomes the band in the sky we call the Milky Way, a name that predates by millennia the modern idea of a galaxy. It runs completely around the celestial sphere. A line through the luminous band, circling around you and slicing the band in half, represents the mid-plane of the Galactic disk seen edge-on. Because the Earth is very near the mid-plane, this line divides the Celestial Sphere into two hemispheres.

From our location near the mid-plane, we step outside at night and see individual stars in all directions. These are the brighter stars that are sprinkled fairly uniformly throughout the *Photopic Sky Survey*. Most of these stars are relatively close to us, at distances ranging from a few light years to several hundred light years. You can see particularly luminous stars (stars with extraordinarily large light power output) out to distances of several thousand light years.

As noted above, the disk is approximately two thousand light years thick, and we are near the mid-plane. Therefore, in directions perpendicular to the mid-plane, the collections of stars extends only about a thousand light years. In directions in and near the mid-plane, you are looking through the disk edge-on. In these directions the stars are spread out to much greater distances. All around the disk, the enormous number of very distant (and therefore faint) stars blur together to produce the white luminous haze that stretches across Risinger's panorama and gives the Milky Way its ancient name.

Facing the center of the *Photopic Sky Survey*, you are facing the center of the Galaxy. That is, you are looking toward the center of the disk from your position halfway to the outer edge. The Milky Way band bulges out in this region, corresponding to what is called the nuclear bulge of the Galaxy—a concentration of stars around the Galactic center. This central concentration of stars is more spherical than disk-like.

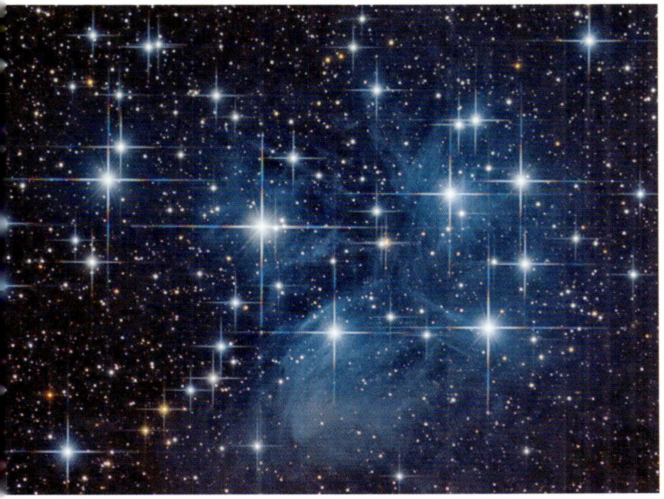
Catalogue 41. Warren Keller, *Seven Sisters*, 2011. [detail]

You may recognize the Pleiades star cluster about 9/10 of an inch in from the left edge of the panorama and about 1¼ inches below the mid-plane. You can find the hunter Orion just below the mid-plane near the right edge of the panorama. A diffuse semicircle of red light loops around Orion's belt and sword and terminates at the bright whitish star Rigel. The famous red supergiant Betelgeuse occupies a position symmetrically opposite Rigel on the other side of the belt. Again facing the center of the Galaxy and wrapping the panorama around you, the Pleiades and Orion end up behind you. From our location in the Milky Way, you must turn around and face away from the Galactic center to look at the Pleiades and Orion.

The orientation of the Milky Way band in the sky outside at night is determined by the orientation of the Earth's spin axis relative to the Galactic disk and also by the date, time of day, and location on Earth. The North Pole of the Earth points in a direction 123 degrees counterclockwise (to the left, looking at the panorama) from the Galactic center and 27 degrees up out of the disk. In the catalogue reproduction of the panorama, this direction is about 2½ inches from the left edge and 1¼ inches above the mid-plane. There you find the star Polaris in Ursa Minor, about one-third of the way between the Galactic plane and the Big Dipper of Ursa Major. On midsummer evenings at mid-northern latitudes, the Milky Way stretches across the sky roughly northeast to southwest. This half of the complete Milky Way band corresponds to the 9¼ inch section of the Risinger panorama beginning about 1 55/64 inches from the left edge and ending about 11 inches from the left edge. This section includes the nuclear bulge, visible in the summer Milky Way close to the southern horizon. The last 7½ inches or so on the right connected to the first 1¾ inches on the left passes overhead roughly southeast to northwest on midwinter evenings, with Orion dominating the night sky.

Even the briefest discussion of the Galaxy as a whole must mention the extraordinary fact that what you see is much less than what you get. The disk of a typical spiral galaxy is surrounded by an invisible halo of mass called, because we cannot see it, dark matter. We deduce the presence of this dark matter from its gravitational influence on matter we can see. The mass of the dark matter halo is at least several times the mass of the stellar disk; and although these dark halos were discovered over three decades ago, the identity of the dark matter remains a mystery.

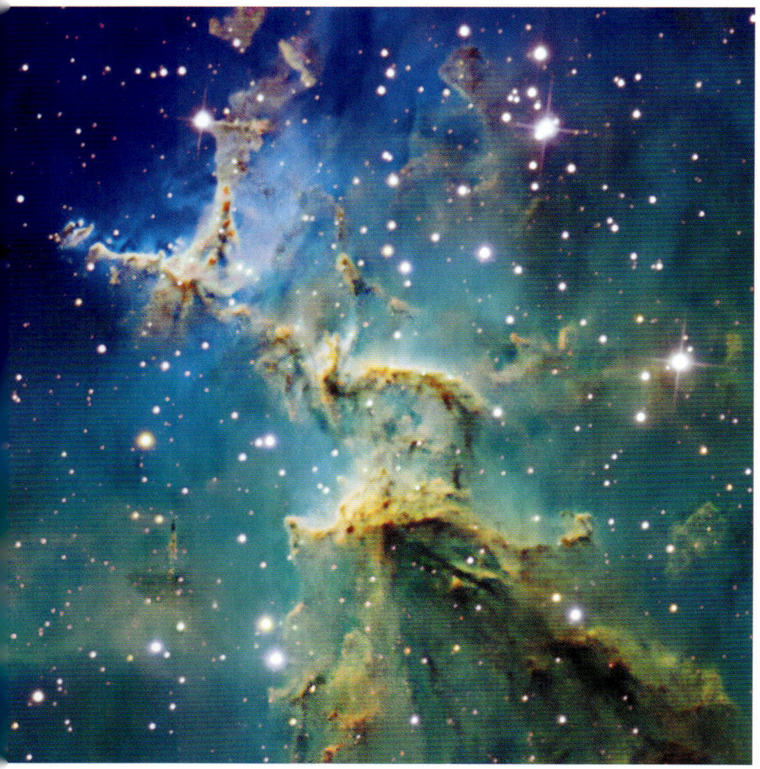

Catalogue 34. Don Goldman, *Cosmic Dust.* [detail]

Star Birth and the Interstellar Medium

The diffuse dusty gas in which the stars are immersed is called the interstellar medium. The gas is mostly hydrogen along with some helium and small amounts of heavier elements. The interstellar dust is exactly what you understand "dust" to be. Each dust grain is a microscopic speck of rock, like the stuff that collects on tables and shelves in your home. Both the density and temperature of the interstellar medium vary considerably from place to place. Reddish blotches in the panorama are clouds of gas hot enough to glow. As part of their quantum-mechanical behavior, atoms of a hot dilute gas emit a set of pure colors. Each pure color corresponds to a specific light wavelength, which in turn corresponds to one of the set of specific frequencies at which the electronic cloud of an atom vibrates. The set of colors is different for each element but the same for all atoms of the same element. This is how the light from a distant astronomical object tells us about its composition. The most prominent color radiated by hydrogen atoms is in the red region of the spectrum, which is why hot hydrogen-rich interstellar gas glows with a reddish color.[3] In hot gas at higher densities, such as the surface of a star, the pure atomic vibrations are disturbed by collisions among atoms. The result is more chaotic electronic motion, radiating a mix of all colors and producing the whitish light we see from stars.

Interstellar gas itself is quite transparent at visible wavelengths, but the dust severely limits our view. The dark areas throughout the luminous band in the Risinger panorama are regions where dust blocks the view of the distant stars. Looking within the disk, our eyes see little that is beyond a few thousand light years. The center of the Galaxy is completely hidden from view. To explore distant parts of the Milky Way disk, astronomers use modern detectors sensitive to electromagnetic radiation at other wavelengths. For example, infrared and radio waves pass through the dust much more easily that visible light. Because the disk is thin, we get a relatively clear view out of the disk in directions perpendicular to the mid-plane. These directions are best for studying the universe beyond our Galaxy.

For astrophotographers, much of the action is in the dusty gas of the Galactic disk, because this is where stars begin and end their lives. Both events create visually spectacular effects. Stars are born inside cold and dark clouds of the interstellar medium. These clouds are called molecular clouds, because at their low temperatures the atoms combine into molecules. The low temperature also means that the gas pressure is low, so that a sufficiently dense sub-region of a molecular cloud will collapse under its own gravity. As it collapses, it fragments into many smaller clumps. Ultimately, these smaller clumps become individual stars, and the collapsing sub-region becomes a cluster of stars. The bottom line is that stars are born in clusters inside cold clouds of dusty gas.

A collapsing clump of gas actually becomes a star when its self-gravitational compression elevates the core temperature sufficiently to trigger nuclear fusion of hydrogen into helium. This nuclear fusion releases energy at an enormous rate, sufficient to stabilize the star against further contraction. The energy ultimately escapes from the star as light radiated from its surface. The total light power radiated is called the luminosity of the star. The luminosity of our Sun, an average star, is about 4×10^{26} watts. That is a 4 followed by 26 zeros. Or put differently, four hundred trillion trillion watts! Not your average household light bulb.

Much of the action is in the dusty gas of the Galactic disk, because this is where stars begin and end their lives. Both events create visually spectacular effects.

Stars within a new cluster range in mass, and the nuclear power output of a star depends strongly on its mass. Most stars form with masses less than that of the Sun. These are low-luminosity stars. A small fraction of the stars of a newborn cluster are significantly more massive than the Sun. These stars are very luminous. The luminosity of a newborn star with a mass twenty times the Sun's mass is about a thousand times the luminosity of the Sun. Most of the light from a new cluster of stars comes from the few most massive stars.

While on the topic of luminosity, it is worth mentioning that gazing at the stars at night does not directly tell you which are highly luminous and which are not. When you look at a star, what registers is what we call its brightness. Brightness is related to how much light enters the pupil of your eye. A low-luminosity star appears bright if it is not too far away. A very luminous star is faint if it is very far away. Compare for example Sirius and Rigel. Sirius is the very bright star about 1¼ inches to the left of Orion. Of these two stars, Sirius is the brighter and is in fact the brightest star in the sky. Rigel, on the other hand, is over three thousand times more luminous than Sirius. But Rigel is 860 light years away whereas Sirius is only 8.6 light years away. The decisive fact is that brightness decreases as the square of the distance. So Rigel, being 100 times farther away than Sirius, is 10,000 times fainter than it would be if it were side-by-side with Sirius at a distance of only 8.6 light years. This issue of brightness versus luminosity will reenter the discussion later on, in connection with Edwin Hubble's discoveries in the 1920s.

Returning to properties of stars, another difference between newly born high-mass and low-mass stars is surface temperature. The Sun's surface temperature is about 6,000 Kelvin. High mass is associated with high surface temperature (above 10,000 Kelvin), and low mass with low temperature (only a few thousand Kelvin). This difference is obvious to the eye. Hot stars glow bluish-white; cool stars glow reddish-yellow. Since the massive stars dominate the light from a newborn cluster, the cluster light is bluish-white. See, for example, Warren Keller's image of the Pleiades star cluster (cat. 41, p. 133). The haze in the image is light from the bright bluish stars scattered by surrounding dust.

Backlit cold and opaque clouds appear as lumpy and tentacled dark forms, some amusing, some sinister.

When a new star cluster forms inside a large interstellar gas cloud, the light from the new stars heats the surrounding gas. So brand new star clusters are embedded in clouds of glowing gas. If the cluster forms deep in the interior of a cold cloud, the cluster and its surrounding cloud of hot gas are hidden from view by dust (although not hidden from the probing eye of a telescope equipped with an infrared detector). But if the cluster is near the edge of an opaque cloud, and that edge happens to be on our side of the cloud, then we can see the new cluster along with the hot gas surrounding it. The most famous example of this phenomenon is the Great Nebula in Orion, visible to the unaided eye as the middle of the three stars of Orion's sword. Even a small telescope reveals that this middle "star" is not a star but is instead a glowing gas cloud. With a modest size telescope you can see the more luminous of the new-born stars within the cloud. In the Risinger panorama, this object is obviously not a star. Robert Gendler's image (cat. 29, p. 131) presents the Orion Nebula in beautiful detail.

The Orion Nebula is a hot region of star formation on the near side of a giant complex of molecular clouds. The whole complex spans several hundred light years, from north of Orion's belt to south of his sword. Several other star-forming regions lie within the complex. The red hydrogen emission visible in the Risinger panorama gives an idea of the extent of the complex.

Fig. 5.1. Walter Koprolin, *Horsehead and Flame Nebulae.* Image courtesy of the artist.

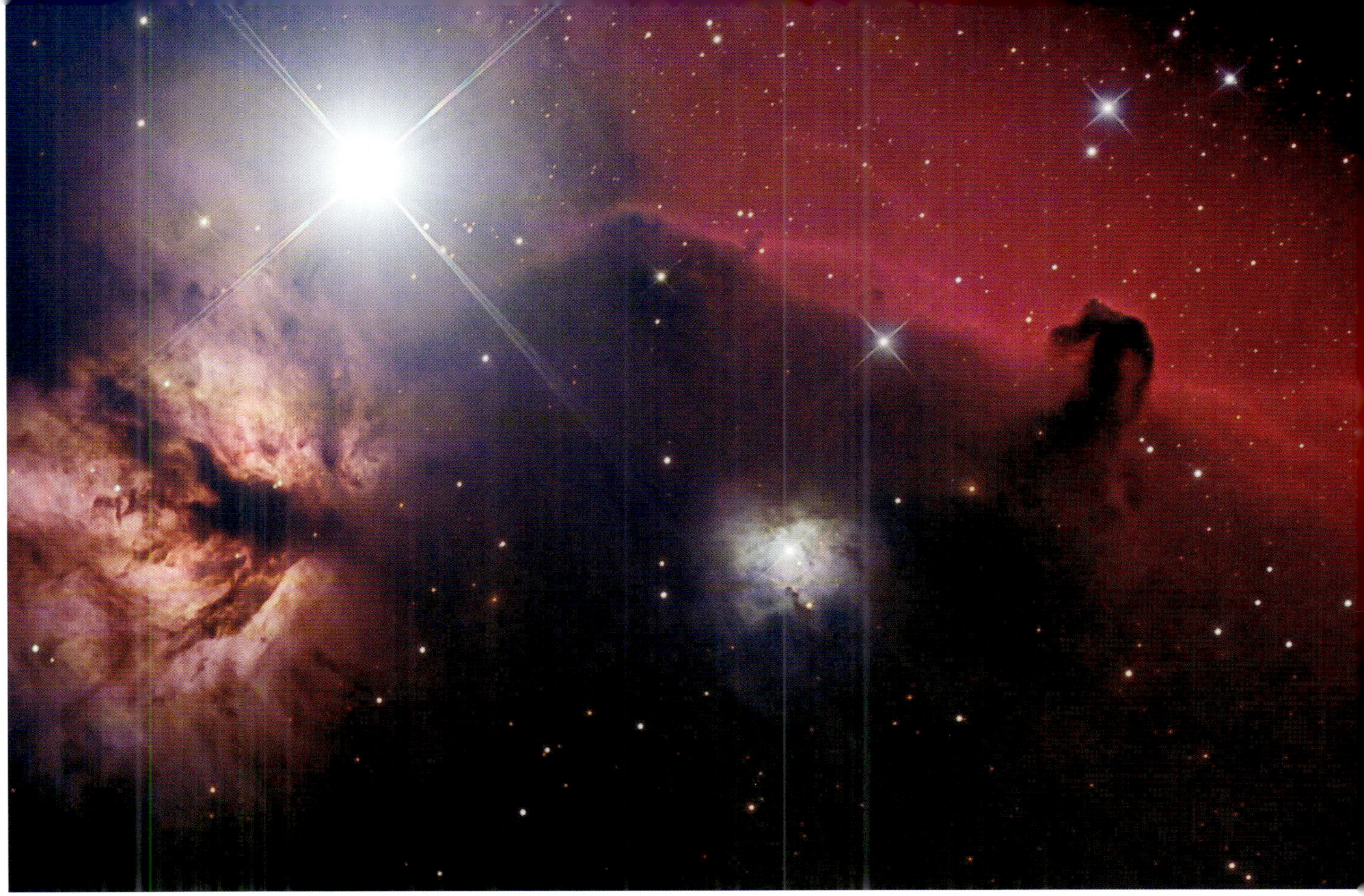

The Orion complex is not unique. Giant molecular clouds are spread throughout the disk of the Milky Way. Within them, gravity is giving birth to new star clusters, and intense radiation from newborn massive stars is heating up and lighting up the surrounding. The radiation is powerful enough to eat into nearby dark dusty gas, pushing it around and creating weird and fantastic shapes. Backlit cold and opaque clouds appear as lumpy and tentacled dark forms, some amusing, some sinister. Backlit dust creates, for example, the horsehead near the star Alnitak at the east end of Orion's belt (Fig. 5.1). Front-lit dust clouds, called reflection nebulae, become pale specters such as the Witch Head Nebula in the lower left corner of John Davis' *Sword to Witch* (cat. 19, p. 134) This image also includes Deneb and the Orion Nebula. The Orion Nebula, the Horsehead Nebula, and the Witch Head Nebula are all within the Orion cloud complex. Similar objects abound throughout the astrophotographer's paradise that is the Milky Way.

Radiation from the stars of a cluster, along with other environmental effects, eventually disperses the residual gas and dust of the cloud from which the cluster formed. We then have a clear view of the cluster, if it is nearby. However, most of the stars you see in the night sky are spread around fairly randomly, not bunched into clusters. If stars are born in clusters, why don't we see lots of star clusters? The answer is that on the Galactic timescale, clusters do not live very long. They eventually dissolve, with individual stars taking separate paths. Several processes are involved. Stars within a cluster can get thrown free from the cluster's gravity by close gravitational encounters with other stars in the cluster. Also, an entire cluster can get pulled apart by a close encounter with some other large mass in the Galactic disk, such as a large molecular cloud. The opposite does not happen: stars in the disk do not spontaneously form clusters. The result is that at any given time, most of the stars we see are not in clusters.

Globular clusters played a crucial role in the discovery of the layout of the Galaxy and our location in it.

There is a special class of amazing and beautiful star clusters called globular clusters. A globular cluster is a collection of typically one hundred thousand to one million stars contained in a spherical region a few tens of light years in radius. The brightest and largest of the Milky Way's globular clusters is NGC (New General Catalog) 5139 in the southern constellation Centaurus. This globular cluster, also known as Omega Centauri, is the subject of David Malin's work (cat. 47, p. 135). You can see it in the *Photopic Sky Survey* as a white dot, somewhat larger than a star, about 2½ inches to the right of center and about ⅝ inch above the mid-plane. The Hercules globular cluster in the northern sky is difficult to pick out in the panorama but is a prime object for small and medium-size telescopes. There are about 150 globular clusters in the Milky Way Galaxy. They did not form in the Galactic disk, however. They are composed of very old stars and probably formed as the Galaxy itself was forming. They are concentrated in an extended spherical region around the Galactic center. As seen from our position halfway out toward the edge of the disk, the globular clusters occupy a large and approximately circular area of the sky centered on the nuclear bulge.

The globular clusters played a crucial role in the discovery of the layout of the Galaxy and our location in it. At the beginning of the twentieth century, astronomers thought that the Sun was somewhere near the center of the Galaxy and that the Galaxy was much smaller than we now know it to be. Dust caused this error. Because of dust, we see through the disk about the same distance in all directions, several thousand light years. However, a hundred years ago astronomers were not yet aware that dust blocks our view of more distant regions of the disk. Instead, they thought that the disk of stars extends several thousand light years in all directions and that we are somewhere near its center. In the second decade of the twentieth century, Harlow Shapley carried out an intensive study of the globular clusters, focusing especially on their distances and bringing to bear new developments in distance measurements. He found that the clusters form a roughly spherical assembly centered on a location far from us in the direction of the richest part of the Milky Way (what we now know to be the nuclear bulge). He then argued convincingly that the center of the distribution of the globular clusters must coincide with the center of the Galaxy. For removing us from the center of the Milky Way, Shapley has been likened to Copernicus, who, four centuries earlier, removed the Earth from the center of the Solar System.[4]

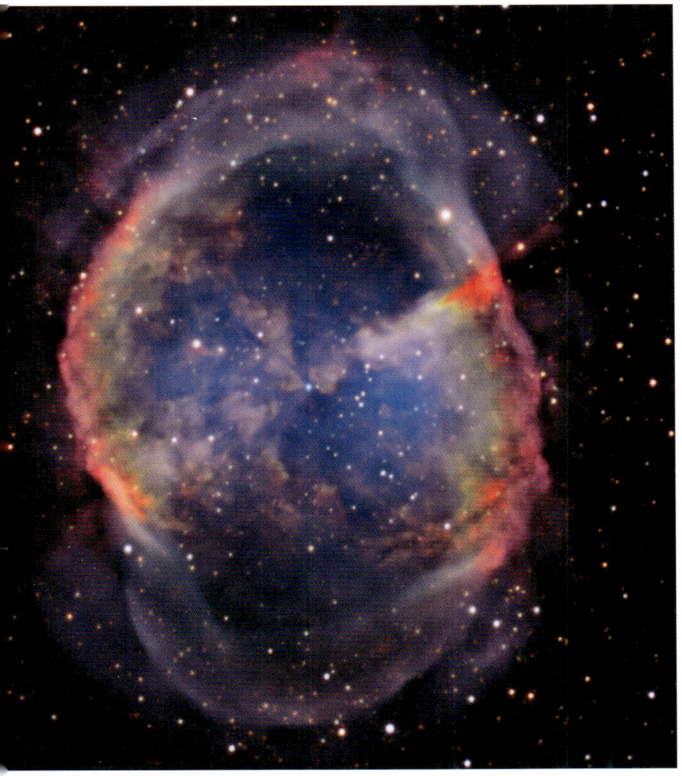

Catalogue 58. Martin Pugh, *M27 in Vulpecula (The Dumbbell Nebula)*. [detail]

Star Death and the Interstellar Medium

Star death also provides dramatic material for the astrophotographer. To better understand the relevant processes, it helps to learn a bit more about the life of a star. In the first and longest stage of a star's life, called the main sequence phase, the star fuses hydrogen into helium in its core. Eventually, the core runs out of hydrogen, and nuclear fusion stops, because the core is not hot enough for the next stage—fusion of helium into carbon. The core does not cool off, however, and the star's life is not yet over. When hydrogen fusion stops, the outward pressure of radiation drops; and instead of cooling, gravity further compresses the core, which elevates the temperature. When it gets hot enough, fusion of helium to carbon begins, and the star is again stabilized. But this time the luminosity is significantly higher, because the fusion reactions proceed at a faster rate in the now hotter and denser core. The increased flow of radiant energy from the core to the surface pushes on the outer portion of the star, which swells up. With its larger surface area, the surface of the star can radiate the increased luminosity at a lower temperature, giving the surface a reddish color. The star has become a red giant.

Stars go through a sequence of fusion stages, converting lighter elements to heavier elements. Each fusion stage generates enough power to stabilize the core of the star against gravitational contraction. At the end of the fusion stage, when the element currently being fused is used up, contraction resumes until the core temperature is high enough to trigger the next stage. Each successive fusion stage is shorter in duration and more furious in power output, as required to counteract the increasing self-gravity of the ever shrinking core.

How far a star gets through this fusion sequence depends on the mass of the star. A low mass star does not get past helium and carbon. A high mass star can make it all the way to iron, becoming a red supergiant along the way. Betelgeuse, the prominent red star of Orion, is a red supergiant. If the Solar System were placed inside Betelgeuse, with the Sun at its center, Mercury, Venus, Earth, Mars, and the asteroid belt would all be inside Betelgeuse.

Either way—low mass or high mass—a star ends its life throwing off its outer layer, leaving behind the core. For lower mass stars, the outer layer of the star is thrown off rather gently. The expanding shell is heated by ultraviolet radiation coming off the hot exposed core. The results are beautiful, and the Dumbbell Nebula (cat. 58, p. 140) is a fine example. The core left behind is called a white dwarf, with a mass comparable to the Sun and a size comparable to the Earth. The white dwarf is not generating nuclear energy. It is prevented from collapsing under its own gravity by the quantum-mechanical effect known as the exclusion principle: no two electrons can be in the same quantum energy state.

A high-mass star arrives at a more spectacular finale. Once nuclear fusion has converted the core to iron, a catastrophic end is unavoidable. Iron is the most stable nucleus, and no more nuclear energy is available to provide additional pressure to counter gravity. Even the electron exclusion principle is inadequate. As the core collapses, protons and electrons are driven together to form neutrons. The result is a neutron star, a ball of neutrons with a mass several times the mass of the Sun packed into a sphere the size of a small city. At such high density, the exclusion principle, which applies to neutrons as well as electrons, can halt further contraction if the neutron star mass is not too large. Meanwhile, the stellar mass outside the core falls inward onto the neutron star. The collapse of the core to form a neutron star and the collapse of outer layers onto the core together convert an enormous quantity of gravitational potential energy to heat. The result is a colossal explosion above the surface of the neutron star. This explosion, called a Type II supernova, blows off the stellar material outside the core.

You do not want to be too close to a supernova. For a few weeks, the explosion pours out radiant energy at a spectacular rate. At peak luminosity, the power output of the explosion can equal that of an entire galaxy. The outer portion of the star blasts into its surroundings, sweeping up surrounding gas. The expanding shell of gas, called the supernova remnant, can be visible to telescopes with sensitive detectors thousands of years after the explosion. If this shell of gas plows into a nearby interstellar cloud, the compression of the cloud can be sufficient to trigger a burst of star formation in the cloud. Thus star birth and star death sometimes share the stage.

A famous supernova remnant is the Crab Nebula in Taurus. The supernova which formed it was seen and recorded in the year 1054, although it was not understood to be a stellar explosion until the twentieth century. Older supernova remnants and shell fragments adorn the Milky Way, and include the Veil Nebula in Cygnus and the supernova remnant (cat. 32, p. 142) in the southern constellation Vela. Don Goldman's *Witch's Broom* (cat. 36, p. 143) is a segment of the Veil Nebula and The Pencil Nebula, seen in Ken Crawford's *NGC 2736*, (cat. 17, p. 144) is a small piece of the Vela supernova remnant.

The neutron star left behind is itself a fearsome object.

David Malin's *The Light Echo of Supernova 1987A* (cat. 45, p. 139) captures a special effect. As its designation indicates, this supernova was seen in 1987. It was the explosion of a star in the Large Magellanic Cloud, a small irregular galaxy near the Milky Way about which I will have more to say later. In this image, the bright light from the explosion itself has faded. However, the burst of light from the explosion travels out as an expanding shell, reflecting off surrounding dust as it expands. We see this reflected light, the light echo, as a ring around the position of the explosion. The echo arrives later than the light directly from the explosion because of the longer indirect path, first from the explosion to the dust and then from the dust to us. The two separate rings probably come from two separate dust clouds. Light echoes from supernovae thus probe the interstellar medium in the vicinity of the explosion.

The neutron star left behind is itself a fearsome object. The compression of the core and accretion of orbiting gas convert slow rotation into rapid spin. Moreover, the neutron star is a magnet of unimaginable strength. This powerful, rapidly spinning magnet beams out electromagnetic radiation like a rotating searchlight. Such objects are called pulsars. The Crab nebula, Veil nebula, and Vela supernova remnant all house pulsars.

If the core mass exceeds about six Solar masses, the exclusion principle for neutrons is inadequate to resist the extreme gravitational forces. The core collapses to a black hole, an object with so much mass in so little volume that not even light can escape its gravity.

To summarize, the interstellar medium is cycled continuously through stars. Gravity compresses clumps of gas into stars. Stars generate power by nuclear fusion deep in their interiors, converting hydrogen and helium into heavier elements. In death, stars blow off their outer layers, ejecting gas that rejoins the surrounding interstellar medium, gradually enriching it with newly formed heavier elements. This is an ongoing process, with interstellar gas cycling repeatedly through stars. About five billion years ago, roughly seven billion years after the Galaxy formed, a clump of gas collapsed under gravity to form the Sun and its planets. Eventually humans appeared on Earth. According to the current understanding, every atom of carbon, nitrogen, oxygen, and all heavier elements in your body was created in the nuclear furnace of some long-dead star.

Every atom of carbon, nitrogen, oxygen, and all heavier elements in your body was created in the nuclear furnace of some long-dead star.

The Extragalactic Realm

Finally, we venture into the extragalactic realm—that which lies beyond the Milky Way. About three inches from the left edge of the panorama and about 1¼ inches below the mid-plane, near the lower edge of the luminous band, a small fuzzy oval white object stands out against the dark background. The object subtends an angle of several degrees in the sky. Because of its location in the Andromeda constellation, it came to be called the Andromeda Nebula. (It is Number 31 in Charles Messier's eighteenth-century catalogue of diffuse objects, and subsequently has been called Messier 31 or M31.) It is the brightest and largest in angle of many such objects. Another, M33, is the smaller and less elongated oval about ⅝ inch to the left and ⅝ inch down from M31. With a large telescope, the number of these pale oval objects, more and less circular, is staggering.

Mid-eighteenth-century astronomers were well aware of this class of puzzling objects. Thinking about them led philosopher Emmanuel Kant to a daring proposal. Knowledgeable about Newtonian dynamics, Kant first argued that the Milky Way is a circular disk-like system of stars flattened by rotation. He then followed this hypothesis to a radical but logical conclusion: the fuzzy oval objects are similar systems. The fact that some are more circular and some more elongated is, according to Kant, simply due to the various angles at which we see them. A disk viewed face-on appears circular; a disk tilted away from our line of sight will appear oval-shaped. Kant's idea, published in 1755, is the first enunciation of the modern conception of the universe of galaxies.

The idea gained little traction. For one thing, eighteenth century technology provided no way to test the proposition. But astronomy was making progress. In the mid-1800s, William Parsons, Third Earl of Rosse, built the largest telescope of his time and with it discovered spiral structure in many of these nebulous objects. They came to be known as "spiral nebulae." In the 1860s, William Huggins attached a spectrometer to a telescope. His spectrum of M31, the Andromeda Nebula, showed light composed of the full range of colors, suggesting that the nebula is composed of stars. Through nineteenth century telescopes, astronomers could not see individual stars in M31, indicating that the nebula, if stars, must be at least moderately distant from the Sun. But is it within the Milky Way and much smaller that the Milky Way? Or is it far outside the Milky Way and perhaps similar in size to the Milky Way? Either possibility is consistent with its angular size.

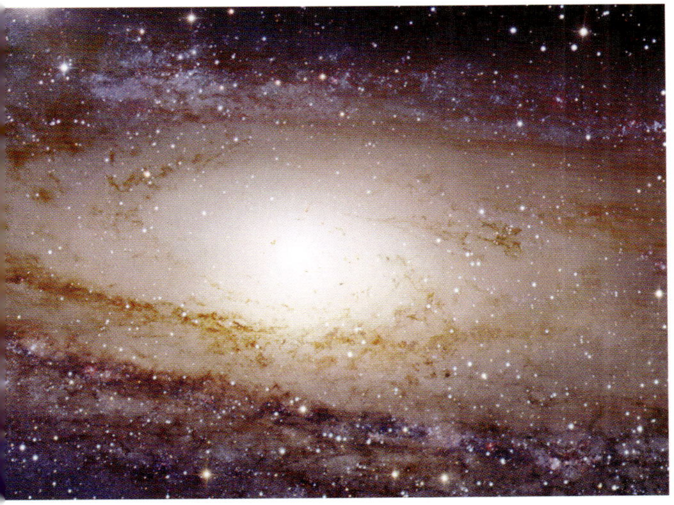

Catalogue 31. Robert Gendler, *The Andromeda Galaxy*, 2005. [detail]

In the 1920s, Edwin Hubble resolved the issue; and to understand how he did this requires a brief digression. In the Risinger panorama, about 3¾ inches to the right of the Galactic center and 1 55/64 inches below the mid-plane, you see two bright diffuse objects about a foot apart from each other. These are the Large Magellanic Cloud (LMC) and the Small Magellanic Cloud (SMC). First made known to Europeans by companions of Magellan on his circumnavigation of the Earth, they are prominent objects of the southern sky, not visible from mid-northern latitudes. By the late nineteenth century, these objects were known to be composed of stars and gas. By the mid-1920s, distance estimates placed them well outside the Milky Way disk. We now identify them as low-mass galaxies orbiting the Milky Way Galaxy. Current distance estimates are about 160,000 light years for the LMC and about 200,000 light years for the SMC.[5]

During the first decade of the twentieth century, Edward Pickering and Henrietta Leavitt of the Harvard College Observatory were studying variable stars in the Magellanic Clouds. These are stars that undergo variations in brightness. Of particular interest were members of the class of variable stars called Cepheid variables, which get brighter and dimmer in a regular and identifiably distinctive fashion. Analyzing brightness measurements of these variables in the SMC, Leavitt made what turned out to be a tremendously important discovery. She found a tight relationship between the period of variability (time to complete a cycle) and peak brightness of the SMC Cepheids: Cepheids with longer periods are brighter. She analyzed enough Cepheids to conclude also that different Cepheids with the same period have the same peak brightness. Because all the Cepheids are in the SMC, they are all at about the same distance. For two objects at the same distance to be equally bright, they must have the same luminosity. Therefore, Leavitt could say that two Cepheids with the same period have the same peak luminosity. Her results were published in 1908.

We now return to Edwin Hubble. Available to Hubble was the new 100-inch Hooker Telescope on Mount Wilson, completed and put into operation in 1917. With this telescope, Hubble was able to measure the brightness of individual Cepheid variable stars in both M31 and M33. With repeated measurements, he gathered information on peak brightness and period of variability for these stars. In 1926 he published his work on M33. The most important finding was that the peak brightness of a Cepheid in M33 is roughly 64 times fainter than the peak brightness of a Cepheid of the same period of variability (and therefore the same peak luminosity) in the SMC. Since the brightness of an object decreases as the square of the distance to the object, M33 must be about eight times farther from us than the SMC. The SMC was already known to be in the outskirts of the Milky Way. Therefore, M33 is a separate, distant, and large system of stars, similar to the Milky Way.

In 1929, Hubble published his analysis of M31, the Andromeda Nebula, placing it far beyond the Milky Way. The combined results left little doubt about the nature of the spiral nebulae. They are not relatively small objects in or near the Milky Way. They are separate star systems, galaxies like the Milky Way. Robert Gendler's *The Andromeda Galaxy* (cat. 31, p. 147) is seen through a field of Milky Way stars, the former a thousand times more distant than the latter. More distant galaxies appear in the work of Adam Block (cat. 4, p. 148), Warren Keller (cat. 43, p. 146), and Steven Mazlin and Jack Harvey (cat. 50, p. 153).

Before Hubble's work on the spiral nebulae, the prevailing (but not unanimous) view was that the Milky Way is a unique giant star collection isolated in space and that all other objects are parts of or satellites of the Milky Way. Hubble laid that view to rest, replacing it with the picture enunciated by Kant 170 years earlier. Our Milky Way Galaxy is a giant star system. But on the cosmic scale it is a mere speck, one of hundreds of billions of similar star systems spread throughout the vast space accessible to modern telescopes. The universe, according to our current understanding, has no center and no edge, and we do not occupy a special position in it.

In 1929, the year in which he published his study of M31, Hubble published an even more dramatic discovery. In addition to distances, Hubble, with the assistance of his colleague Milton Humason, had been measuring the Doppler shift of light from spiral nebulae to determine their line-of-sight velocities. By 1929 he had discovered that these nebulae, now understood to be galaxies, are receding from the Milky Way at speeds proportional to their distances. Hubble had discovered that the universe is expanding.

The universe has no center and no edge, and we do not occupy a special position in it.

Hubble's discovery of the layout of the universe and Einstein's creation of the general theory of relativity, the theoretical foundation for describing cosmic dynamics, together launched a revolution in cosmology, the study of the universe as a whole. The pace of observational work accelerated throughout the twentieth century and now into the twenty-first century, with new technology opening new windows of the electromagnetic spectrum and with steady improvements in detector sensitivity, imaging capability, and spectroscopy.

You sometimes hear people say that it's all wrapped up; all the good stuff has been discovered; or there is not much left to do and little new to see. Don't believe it. The images of Starstruck, which speak of modern science, are also windows into the unknown, a realm as vast as the universe itself.

Notes

1. By convention, "Galaxy" with a capital G refers to the Milky Way Galaxy.

2. Because of distortions caused by representing a spherical surface on a rectangular plane, the panorama does not include the entire sky. Some regions away from the mid-plane are omitted.

3. The panorama combines images taken with several different wavelength filters, one of which transmits only the wavelength of the red hydrogen emission and accentuates this emission. The unaided eye is not sufficiently color-sensitive to see this reddish hue.

4. The heliocentric model was first proposed by the Greek astronomer and mathematician Aristarchos in the third century BCE, a fact acknowledged by Copernicus.

5. Robert Gendler's image of the Andromeda Galaxy shows that it also has two small satellite galaxies. One is below and to the right of the center of Andromeda. The other is smaller and to the left of Andromeda's center.

Selected Bibliography

Astronomy [magazine]. Waukesha, Wis.: Kalmbach Publishing Co.

Benson, Michael. *Beyond: A Solar System Voyage.* New York: Abrams for Young Readers, 2009.

—— *Beyond: Visions of the Interplanetary Probes.* New York: Harry N. Abrams, 2003.

—— *Far Out: a Space-time Chronicle.* New York: Harry N. Abrams, 2009.

Bonnell, Jerry T., and Robert J. Nemiroff. *Astronomy: 365 Days.* New York: Harry N. Abrams, 2006.

Brashear, Ronald, with Daniel Lewis and Owen Gingrich. *Star Struck: One Thousand Years of the Art and Science of Astronomy.* Seattle, Wash.: University of Washington Press, 2001.

Brewster, Sir David. *A Treatise on Optics.* Philadelphia: Lea & Blanchard, 1847.

Connor, Linda. *Odyssey: The Photographs of Linda Connor.* San Francisco: Chronicle Books, 2008.

Cornell, Daniel. *Alfred Stieglitz and the Equivalent: Reinventing the Nature of Photography.* New Haven, Conn.: Yale University Art Gallery, 1999.

Covington, Michael A. *Digital SLR Astrophotography.* Cambridge, U.K.: Cambridge University Press, 2007.

Devorkin, David. *Hubble: Imaging Space and Time.* Washington, D.C.: Smithsonian National Air and Space Museum, in association with National Geographic, 2008.

Fitzpatrick, Chris. "Post-Pastoral." *Camerawork* 36.1 (Spring/Summer 2009): 20–5.

Folberg, Neil, and Timothy Ferris. *Celestial Nights: Visions of an Ancient Land: Photographs from Israel and the Sinai.* New York: Aperture Foundation, 2001.

Galilei, Galileo. *Sidereus Nuncius or The Sidereal Messenger.* Translated, with introduction, conclusion, and notes by Albert Van Helden. Chicago: University of Chicago Press, 1989.

Gendler, Robert. *Capturing the Stars: Astrophotography by the Masters.* Minneapolis, Minn.: MBI Publishing, 2009.

—— *A Year in the Life of the Universe: a Seasonal Guide to Viewing the Cosmos.* St. Paul, Minn.: Voyageur, 2006.

Gendler, Robert, Lars Lindberg Christensen, and David Malin. *Treasures of the Southern Sky.* New York: Springer, 2011.

Harper, Sharon. *From Above and Below.* Sante Fe, N. Mex.: Radius Books, 2012.

Jenkins, Steven. "A Conversation with Richard Misrach." *Artweek* 28 (July 1997): 21–2.

Kier, Ruben. *The 100 Best Astrophotography Targets: A Monthly Guide for CCD Imaging with Amateur Telescopes.* New York: Springer, 2009.

Lacroux, Jean, and Christian Legrand. *Discover the Moon.* Cambridge, U.K.: Cambridge University Press, 2003.

Leydier, Richard, and L. S. Torgoff. "David Malin." *Art Press* 308 (January 2005): 81–2.

Light, Michael, and Andrew Chaikin. *Full Moon*. New York: Alfred A. Knopf, 1999.

Malin, David. *Ancient Light: a Portrait of the Universe*. London: Phaidon, 2009.

—— *The Invisible Universe*. Boston: Little, Brown and Company, 1999.

Malin, David, and Paul Murdin. *Colours of the Stars*. Cambridge, U.K.: Cambridge University Press, 1984.

Naef, Weston J. *The J. Paul Getty Museum Handbook of the Photographs Collection*. Malibu, Calif.: The J. Paul Getty Museum, 1995.

Nemiroff, Robert J., and Jerry T. Bonnell. *The Universe 365 Days*. London: Thames & Hudson, 2003.

Paglen, Trevor, and Ute Reiss. *Trevor Paglen: A Compendium of Secrets*. Bielefeld, Germany: Kerber, 2010.

Paglen, Trevor, and Tom Vanderbilt. "1000 Words: Trevor Paglen Talks about 'The Other Night Sky.' " 2007. *Artforum International* 47.7 (March 2009): 224–9.

Pang, Alex. "Technology, Aesthetics, and the development of Astrophotography at the Lick Observatory." In *Inscribing Science*, edited by Timothy Lenoir, Stanford, Calif.: Stanford University Press, 1998, pp. 223–48.

Parker, Greg, and Noel Carboni. *Star Vistas: A Collection of Fine Art Astrophotography*. New York: Springer, 2009.

Penpase, Brian E. *The Power of the Stars: How Celestial Observations Have Shaped Civilization*. New York: Springer, 2010.

Pocock, Philip, and Thomas Ruff, interview. "thomas ruff." *Journal of Contemporary Art* (n.d.). http://www.jca-online.com/ruff.html Accessed April 24, 2012.

Proctor, Minna. "Sharon Harper." *Bomb* 111 (Spring 2010): 68–9.

Roth, Evelyn. "Roger Ressmeyer: A Space Photography Specialist." *American Photographer* 22 (April 1989): 60–7.

Schink, Hans-Christian. Hans-Christian Schink: *1hr.* Ostfildern, Germany: Hatje Cantz, 2011.

Seip, Stefan, and Elke Schulz. *Digital Astrophotography: A Guide to Capturing the Cosmos.* Santa Barbara, Calif.: Rocky Nook, 2008.

Sky & Telescope [magazine] Cambridge, Mass.: Sky & Telescope Media.

Smyth, Diane. "Invisible." *British Journal of Photography* 157 (August 2010): 19.

Solnit, Rebecca. *The Sky Book/Richard Misrach.* Santa Fe, N. Mex.: Arena Editions, 2000.

Story, Alfred T. *The Story of Photography.* London: George Newnes Limited, 1898.

Sutton, Thomas A., and George Dawson. *A Dictionary of Photography.* London: Sampson Low, Son & Marston, 1867.

Travis, David B. "A Mind Above the Clouds: Alfred Stieglitz's Equivalents Photographs." In *At the Edge of the Light: Thoughts on Photography and Photographers, on Talent and Genius.* Jaffrey, N.H.: David R. Godine, 2003.

"Viewfinder" *Green Places* 78 (October 2011): 2–3.

Exhibitions

Amsterdam 2010

Asser, Saskia, and Huib Henrichs. *First Light: Photography & Astronomy.* Amsterdam: Huis Marseilles Museum of Photography, 2010.

Berlin 2010

Blood to Ghosts. Martin Klosterfelde Gallery.

Chicago 2012

Exposure: Matt Keegan, Katie Paterson, Heather Rasmussen. Art Institute of Chicago.

Kansas City 2011

Heavens: Photographs of the Sky & Cosmos. Nelson-Atkins Museum of Art.

Malibu, Calif. 2011

In Focus: The Sky. The J. Paul Getty Museum.

New York 2009–10

Kuhr, Jonathan, ed. *Visions of the Cosmos: From the Milky Ocean to an Evolving Universe.* New York: Rubin Museum of Art, 2009.

New York 2010

Sharon Harper. One Month, Weather Permitting. Rick Wester Fine Art.

New York 2011

Michael Benson. Beyond. Hasted Kreautler Gallery.

Oldenburg, Germany 2011

Wild Sky. Edith Russ Site for Media Art.

Philadelphia 2001

Take Five: The American Landscape. Nexus Gallery in conjunction with the Center for the Photographic Image.

Philadelphia 2007

Moon Studies and Star Scratches. The Print Center.

Richmond, Australia 2008

David Stephenson. Star Drawings. John Buckley Gallery.

San Francisco–Amsterdam–London–New York–Sydney, Australia–Göteborg, Sweden 1999–2006

Michael Light. Full Moon.

Vienna 2011

Space: About a Dream. Kunsthalle Wien.

WEB SITES

Astronomy Picture of the Day, NASA. Eds., Robert J. Nemiroff and Jerry T. Bonnell. apod.nasa.gov/apod

International Dark-Sky Association. darksky.org

World at Night, The (TWAN). Ed., Babak A. Tafreshi. twanight.org/newTWAN

FILMS

Man Who Colours Stars, The. Kent, U.K.: York Films, 2005.

City Dark, The. Produced and directed by Ian Cheney. Brooklyn, N.Y.: Wicked Delicate Films, 2011.

“Katie Paterson: When art meets astronomy.” Haunch of Venison, London (April 4, 2011). http://haunchofvenison.com/films/katie_patersonwhen_art_meets_astronomy/ 04:09. Accessed April 24, 2012.

Catalogue

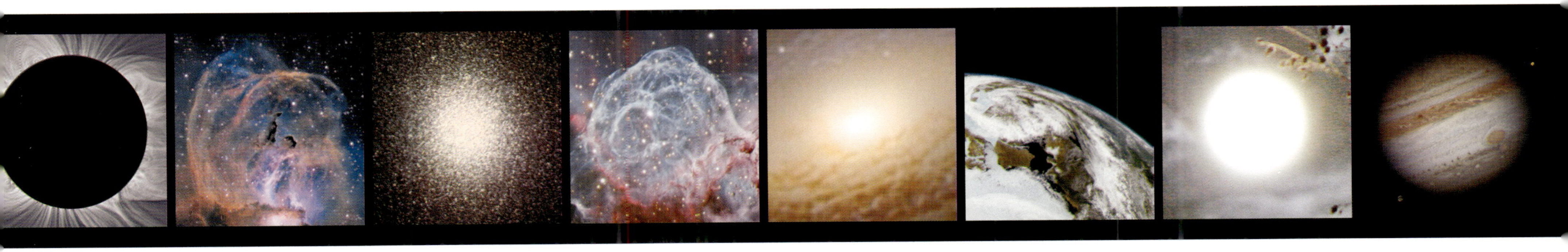

Astronomical Subject Guide

Listed by catalogue number

*Not shown due to map projection distortions combined with artist's tilting of Galactic plane.
—Editor

Abbreviations refer to scientific catalogues

vdB van den Bergh

IC Index Catalogue

M Messier Catalogue

NGC *New General Catalogue of Nebulae and Clusters of Stars*

Artists

Starstruck is an international exhibition including artists from eleven countries and representing every continent except Africa and Antarctica. Unless artists are from the United States, their nationality is listed. All artists are living, except for Alfred Stieglitz. An illustrated listing of works by artist with their catalogue numbers and page references is provided in the Checklist on page 189 ff.

—Editor

BENSON, Michael
BLOCK, Adam D.
CONNOR, Linda
CORMIER, James
CRAWFORD, Ken
DAVIS, John A.
DIAZ BOBILLO, Ignacio *(Argentina)*
DI CICCO, Dennis
FRIEDMAN, Alan
GENDLER, Robert
GERRITSEN, Jacob
GOLDMAN, Don
HARPER, Sharon
HARVEY, Jack
KELLER, Warren
KOPROLIN, Walter *(Austria)*
MALIN, David *(Australia)*
MAZLIN, Steve
NICKELSON, Jim
ØYE, Geir *(Norway)*
PASACHOFF, Jay
PEACH, Damian *(United Kingdom)*
PUGH, Martin *(Australia)*
RISINGER, Nick
ROGERS, Kirk
ROUX, Jean Paul *(France)*
ROWELL, Tony
RUFF, Thomas *(Germany)*
SCHINK, Hans-Christian *(Germany)*
STETSON, John
STIEGLITZ, Alfred
TAFRESHI, Babak A. *(Iran)*
TAKASAKA, Yuichi *(Canada)*
WALKER, Sean
WANG, Wei-Hao *(Taiwan)*
WOODS, Jacqueline

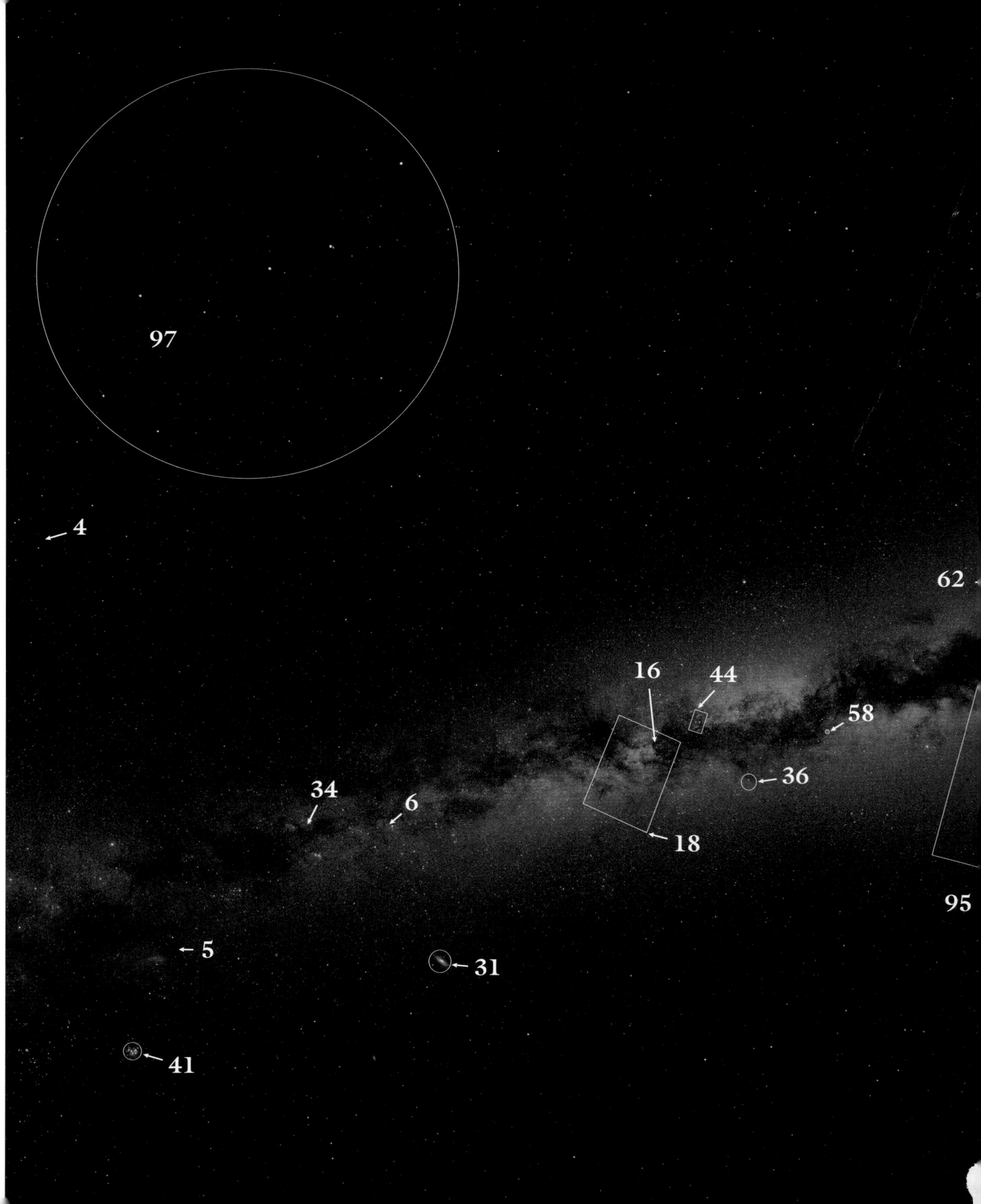

59. The celestial sphere. [background image]

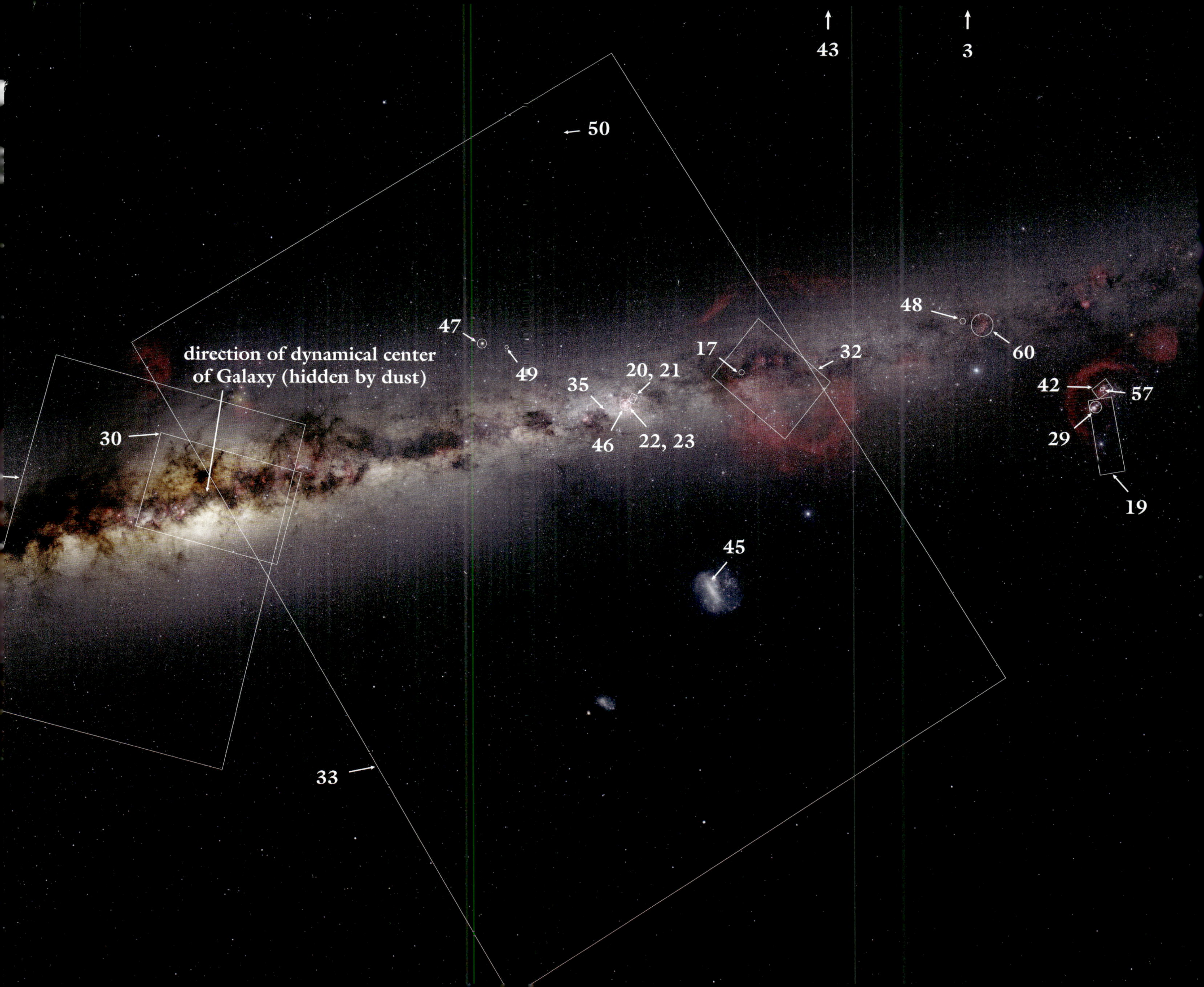
43
3
50
47
49
48
60
17
32
direction of dynamical center
of Galaxy (hidden by dust)
20, 21
35
46
22, 23
42
57
29
19
30
45
33

FOLD OUT THIS FLAP to find a map of Starstruck's images keyed to Nick Risinger's *Photopic Sky Survey*, selected for this purpose because, as Eric Wollman explains in his essay, it is a map projection of the celestial sphere—the imaginary shell upon which celestial objects appear to lie. This map is intended to assist the viewer in understanding the location and relative apparent size of the various subjects depicted. Numbers on the map refer to catalogue numbers in the Astronomical Subject Guide at left.

To visualize the panorama as the celestial sphere, imagine wrapping it around you, joining the starry band into a continuous circle and then bending the panorama inward above and below the band to form a spherical shell. Because of distortions caused by representing a spherical surface on a rectangular plane and the artist's tilting of the image, some regions away from the mid-plane are omitted, as are solar system bodies such as the Sun, Moon, and planets.

—Editor

The Sun

Images are shown grouped by subject. Catalogue numbers refer to the Checklist organized alphabetically by artist.

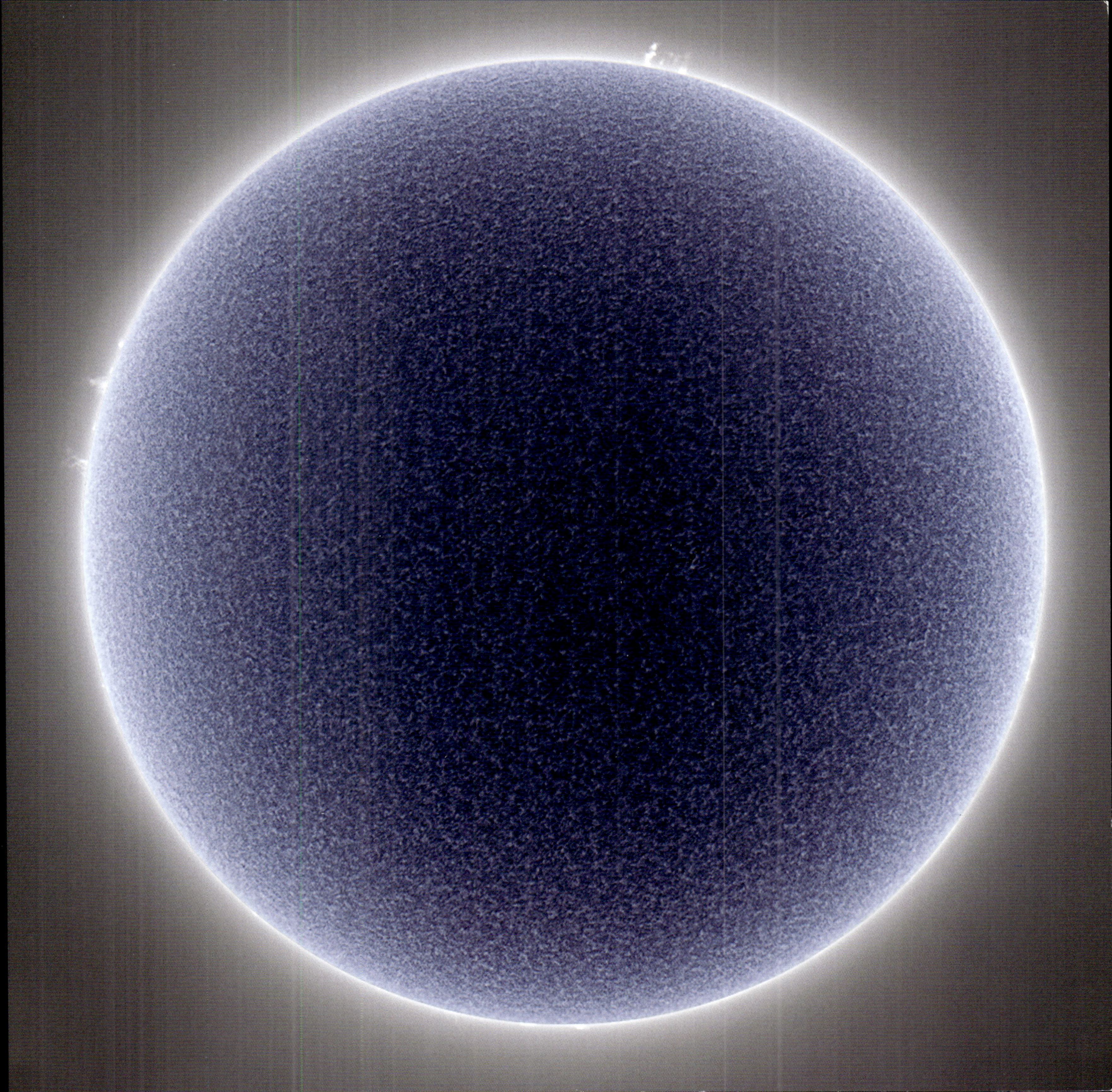

27. Alan FRIEDMAN, *The Blue Sun*.

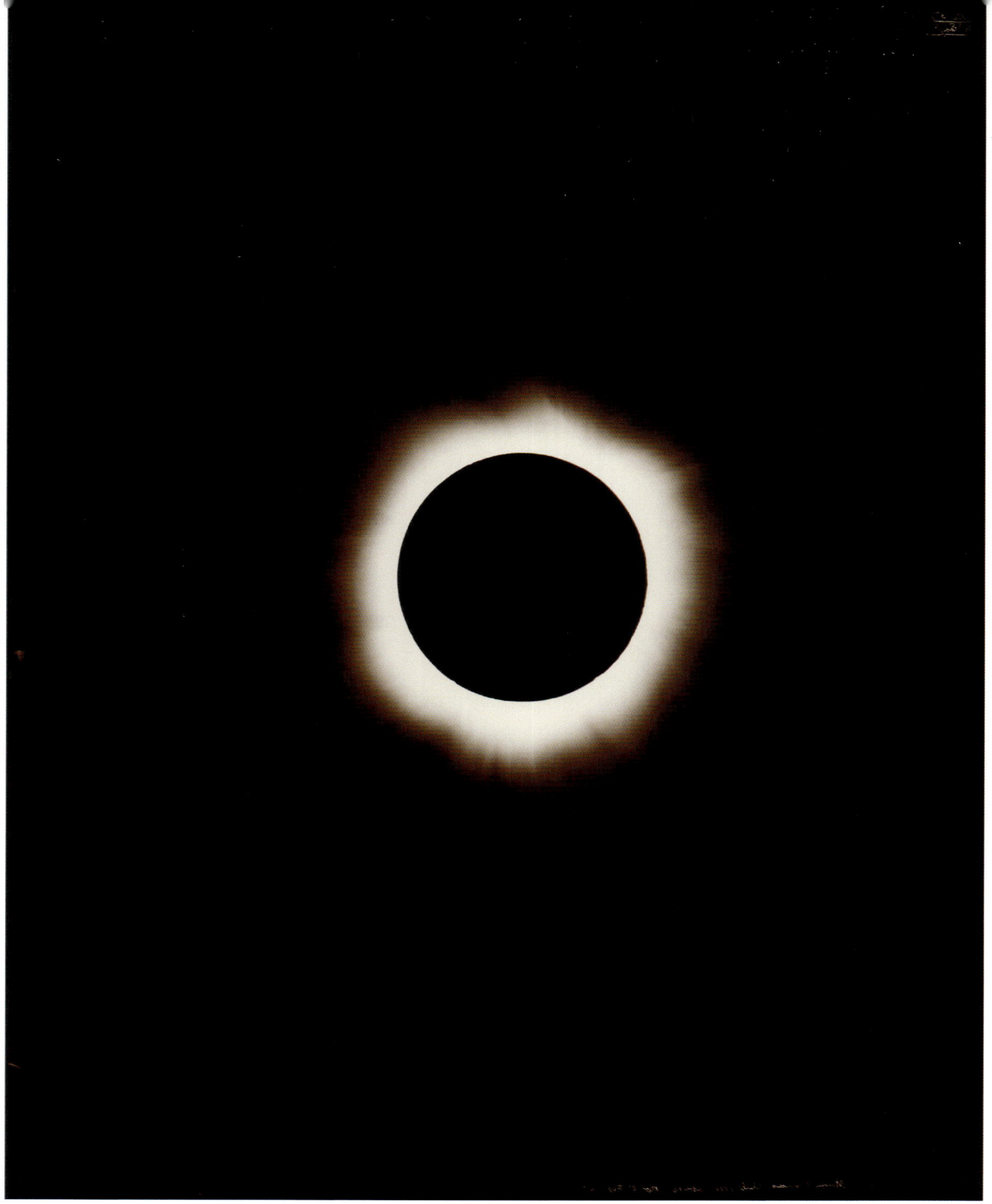

12. Linda CONNOR, *Solar Eclipse, Chile, April 16, 1893,* 1998. Printed from Lick Observatory Archive negative. Original photo by John Schaeberle. © UC Regents/Lick Observatory.

54. Jay PASACHOFF, *Easter Island Eclipse, July 11,* 2010.

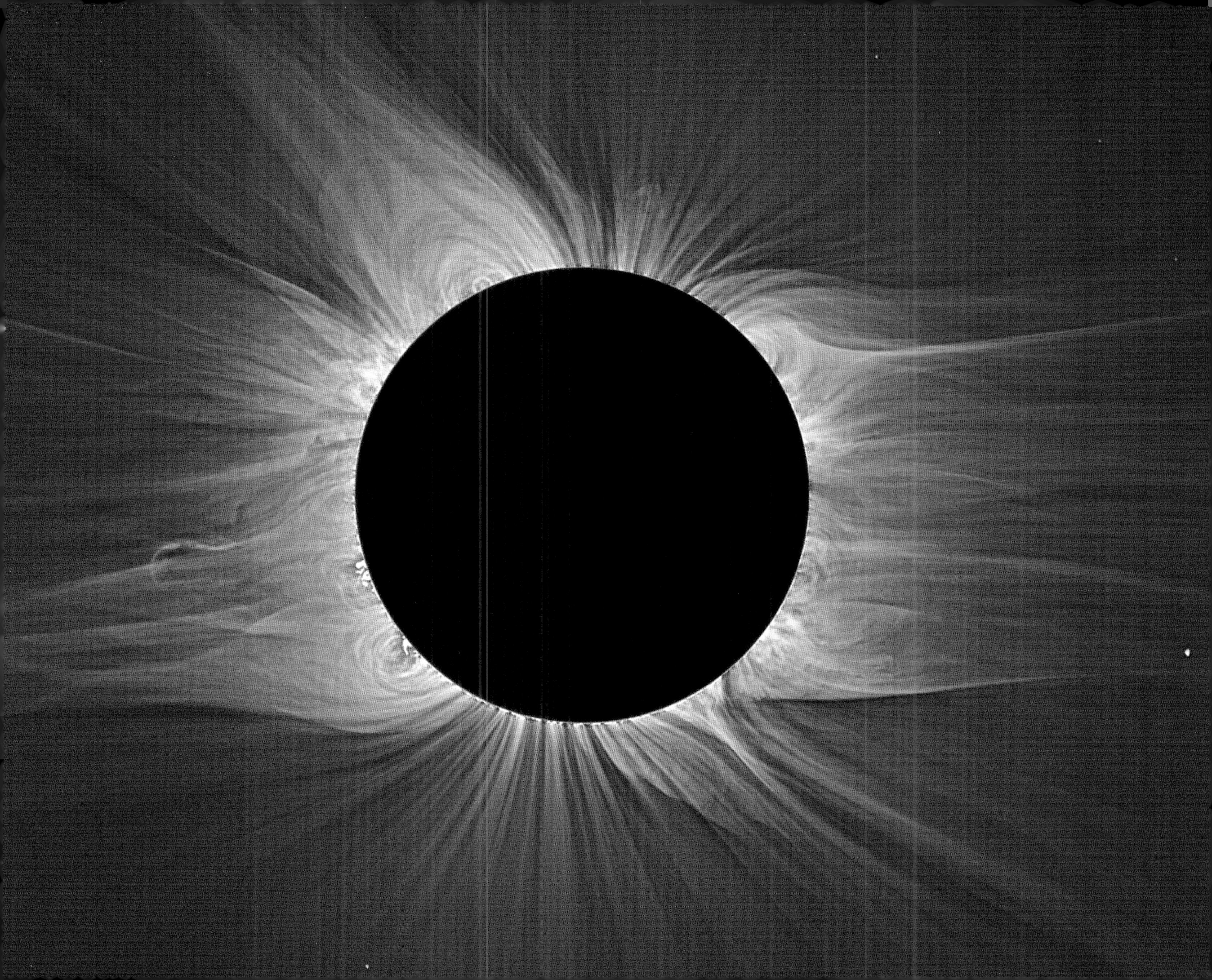

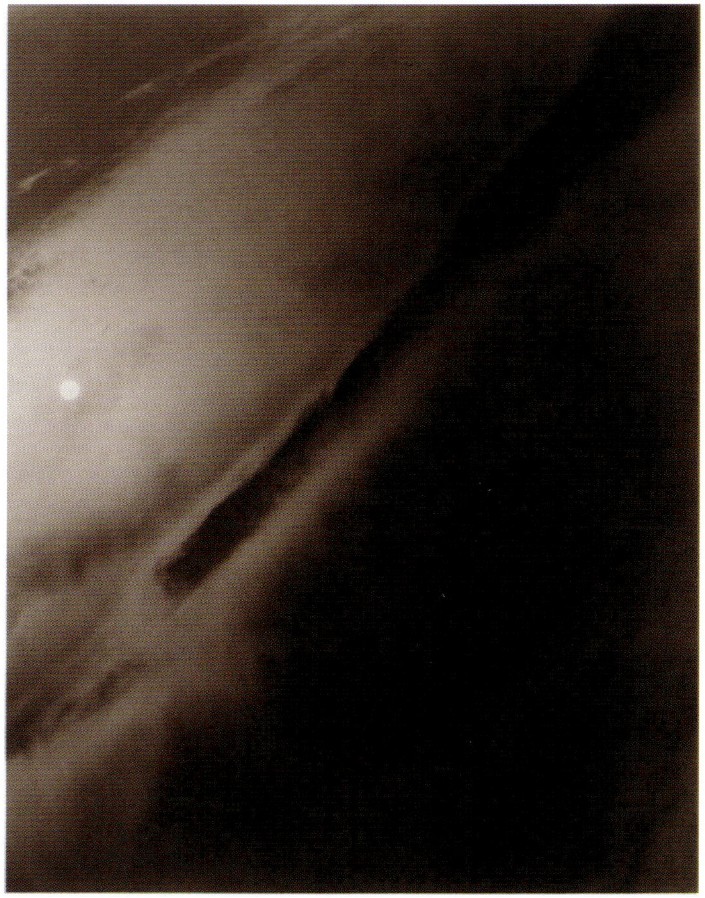

94. Alfred STIEGLITZ, *Equivalent*. National Gallery of Art, Washington, D.C., Alfred Stieglitz Collection.

81. Hans-Christian SCHINK, *1hr* series, *2/21/2010, 7:00–8:00 p.m., S 38°49.042' E 174°34.976'*. Courtesy Wilson Centre for Photography.

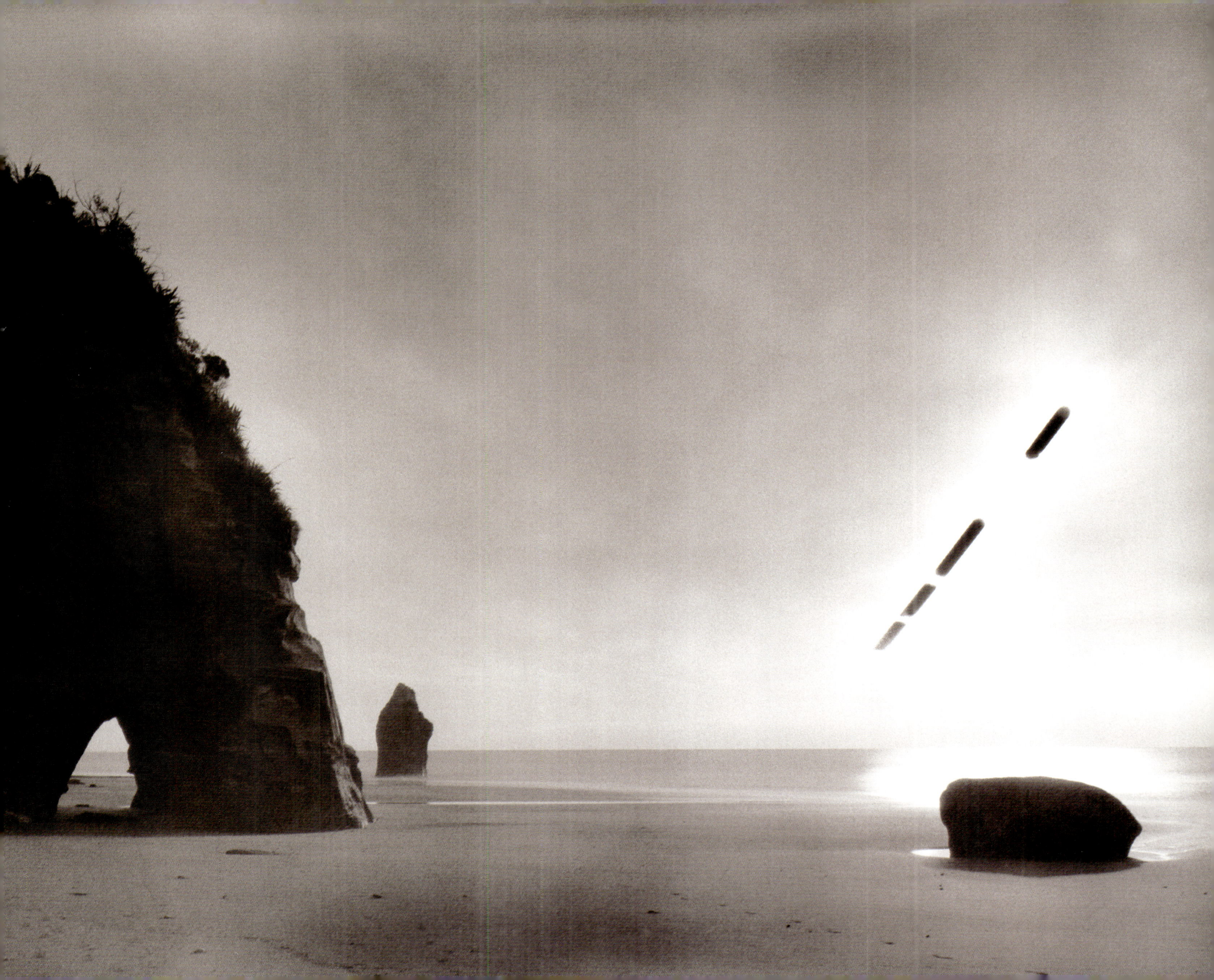

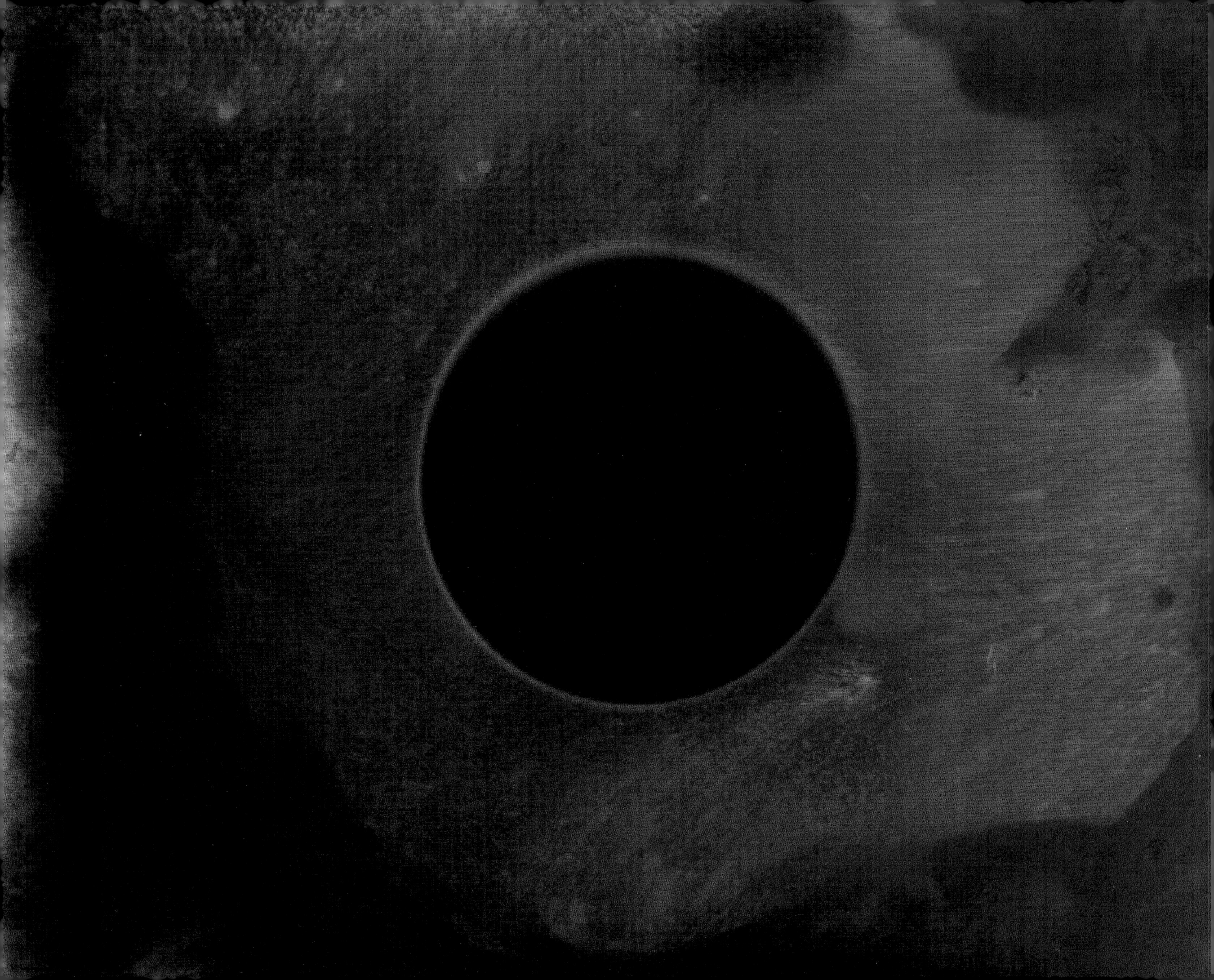

93. Alfred STIEGLITZ, *Equivalent, Set C2, No. 2*. National Gallery of Art, Washington, D.C., Alfred Stieglitz Collection.

104. Jacqueline WOODS, *Black Sun #30*, 2011.

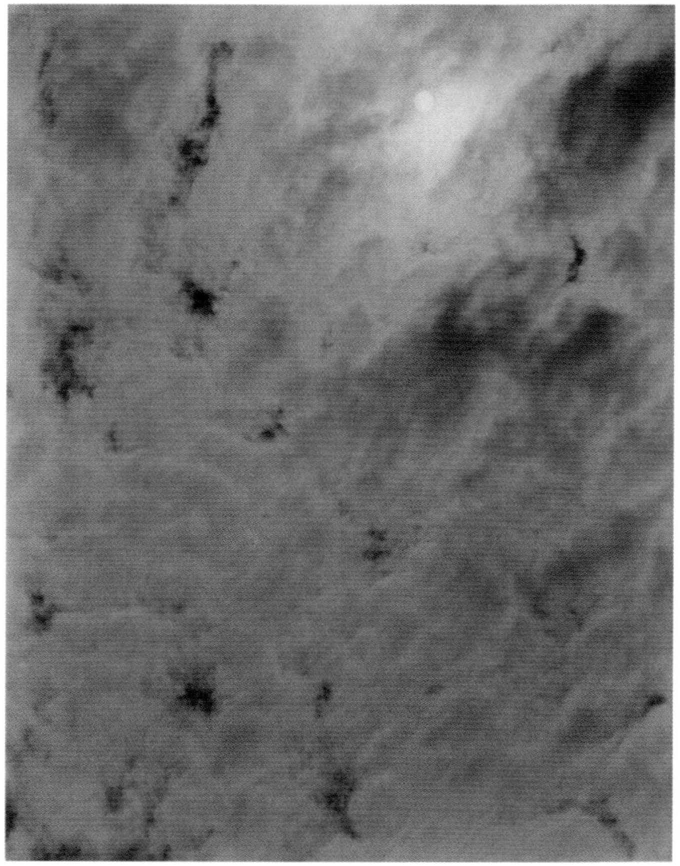

92. Alfred STIEGLITZ, *Song of the Sky*. National Gallery of Art, Washington, D.C., Alfred Stieglitz Collection.

76. Hans-Christian SCHINK, *1hr* series, *5/12/2006, 6:44–7:44 a.m., N 24°43.562' E 055°57.061'*. Courtesy Wilson Centre for Photography.

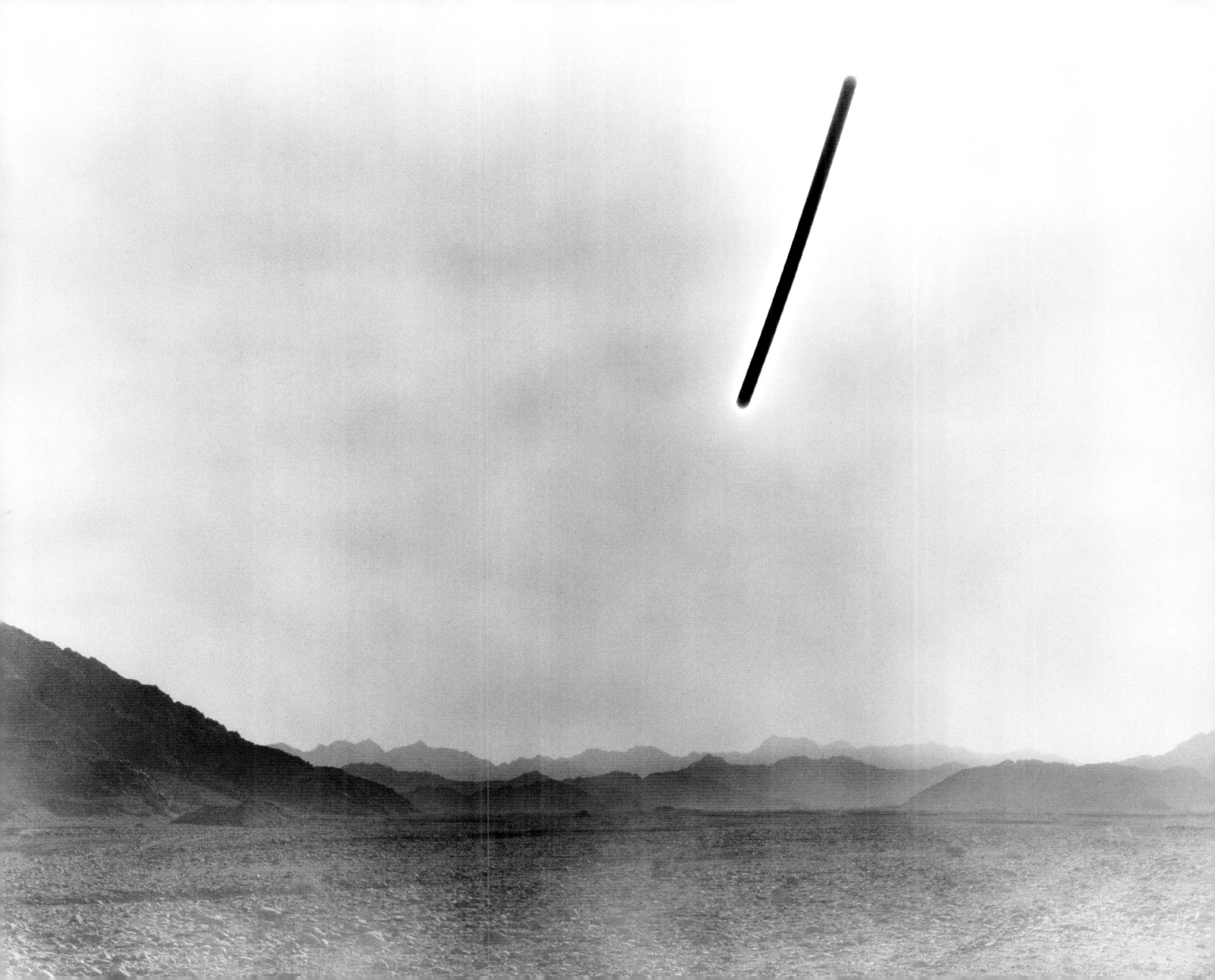

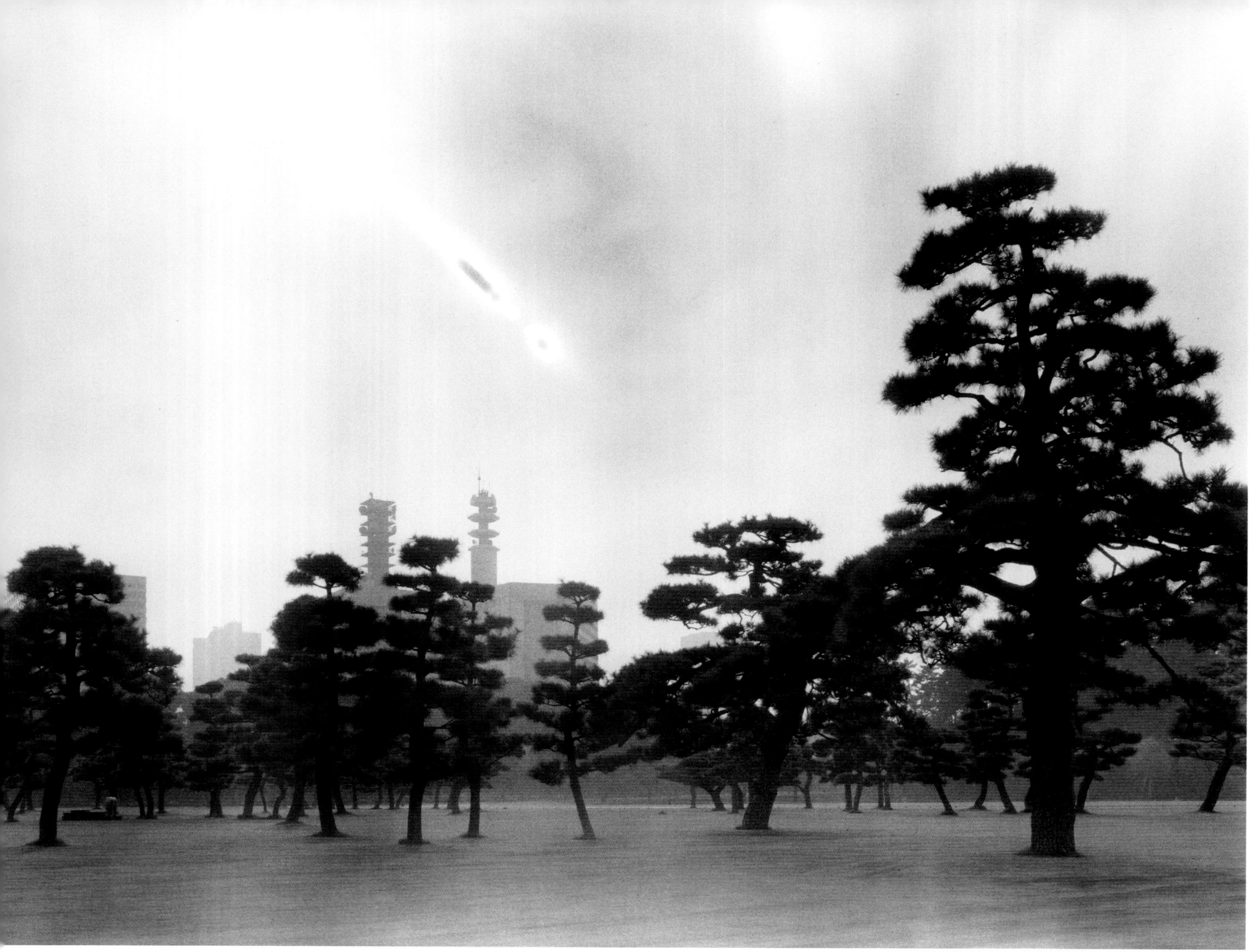

72. Hans-Christian SCHINK, *1hr* series, *10/28/2005, 2:03–3:03 p.m., N 35°40.460' E 139°45.243'*; and

84. *4/10/2009, 4:11–5:11 p.m., S 26°28.034' E 018°16.142'*. Courtesy Wilson Centre for Photography.

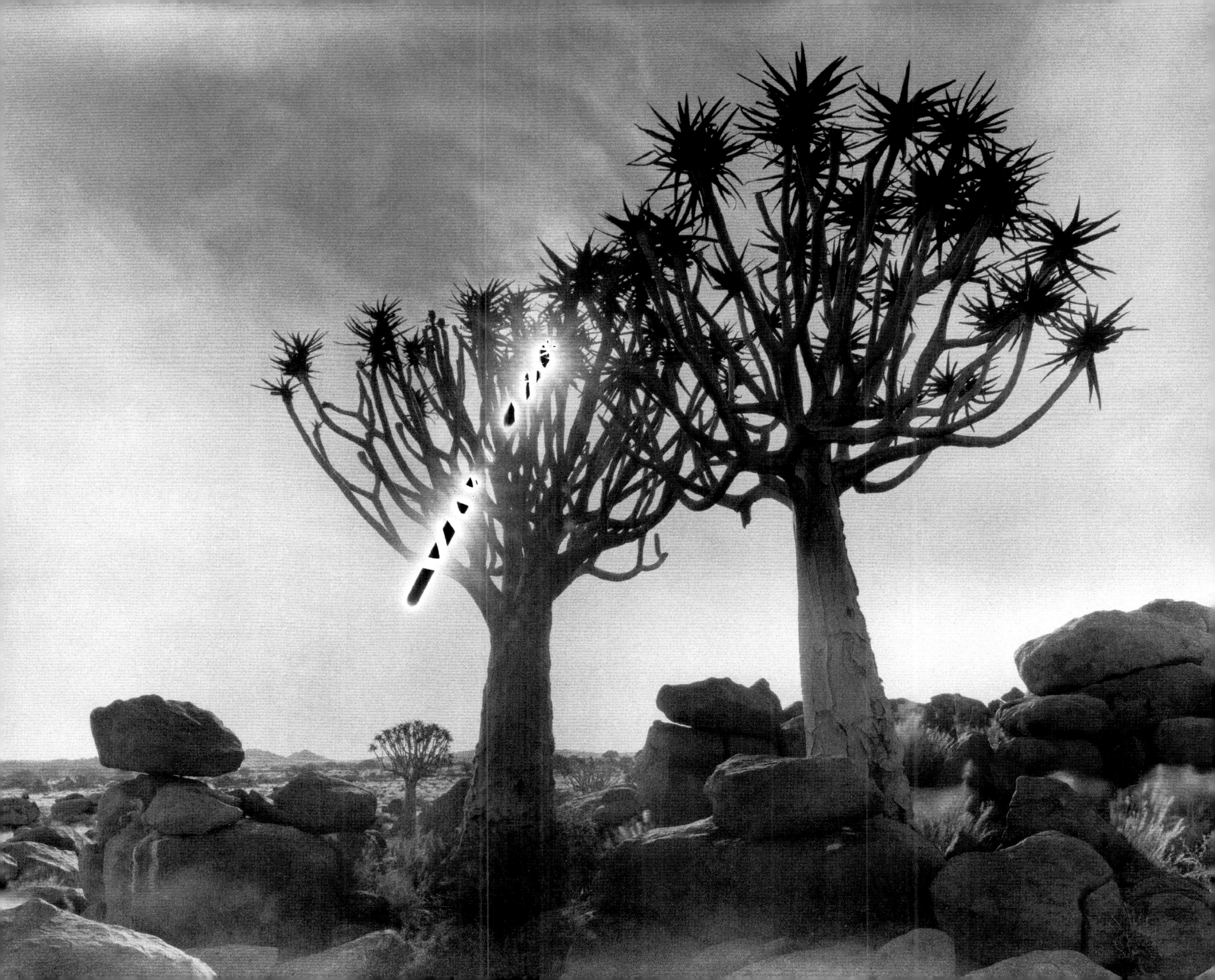

101. Jacqueline WOODS, *Black Sun #4*, 2011.

8. Linda CONNOR, *Open Dome, Caravansary, Turkey*, 1992.

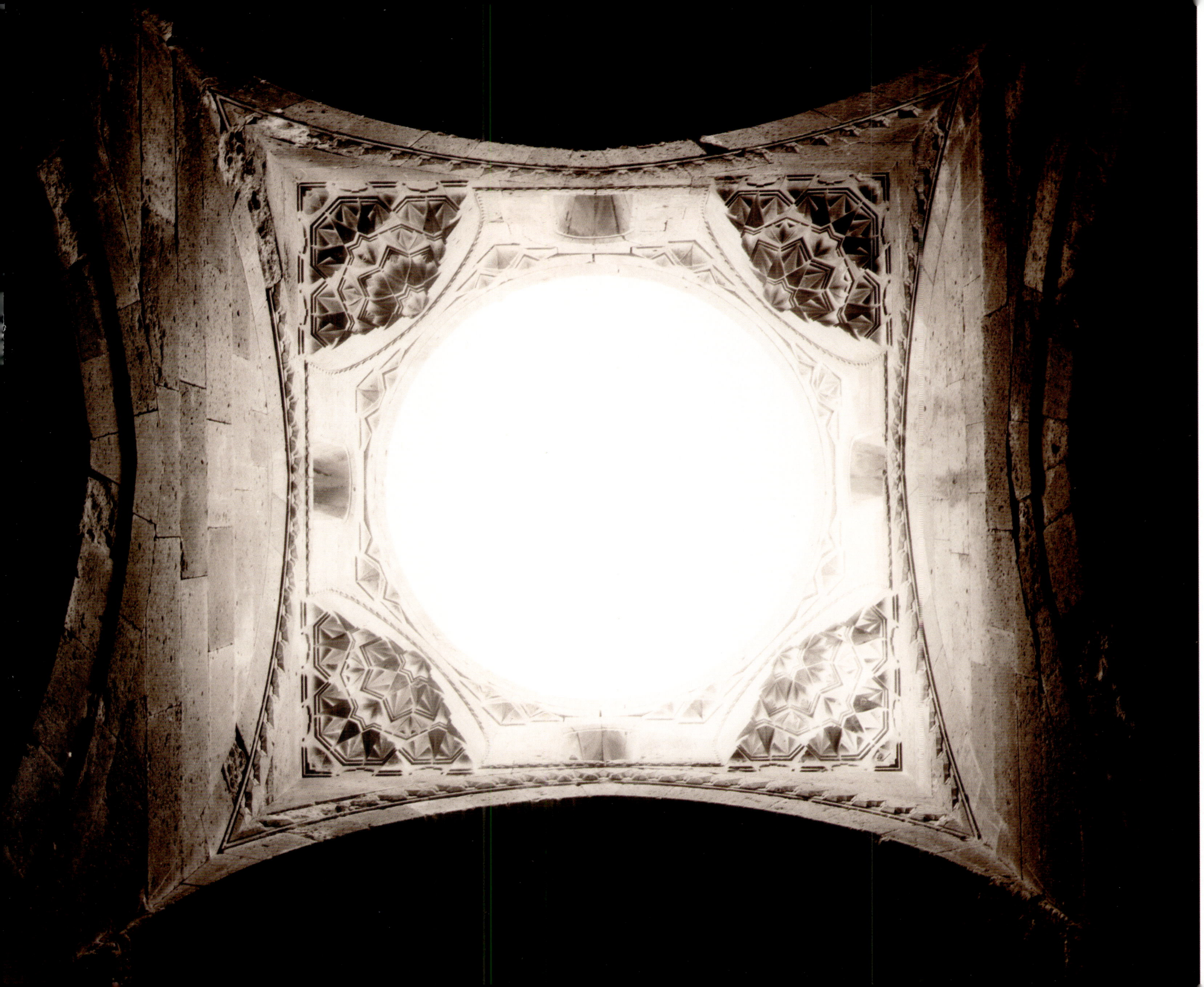

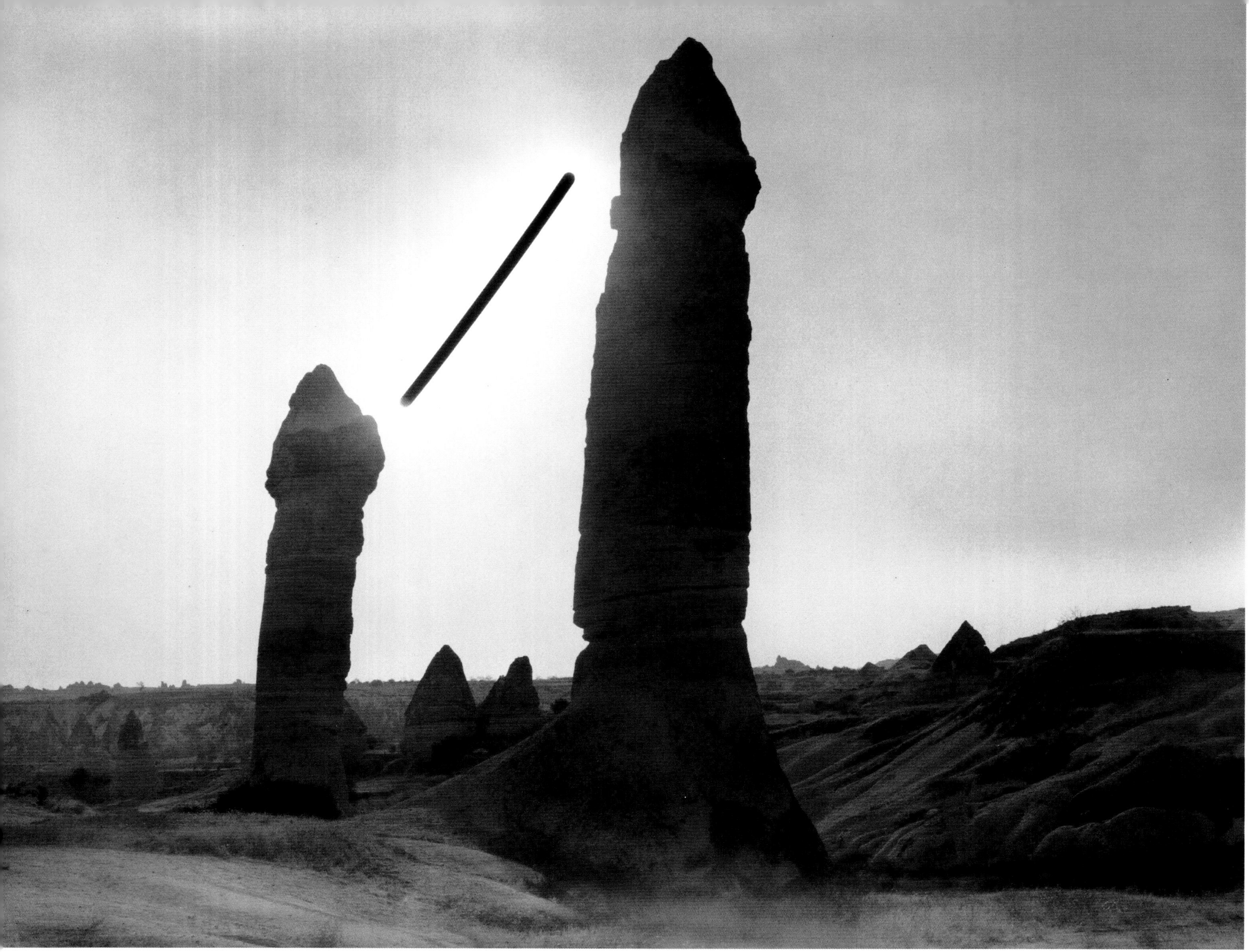

71. Hans-Christian SCHINK, *1hr* series, *7/27/2008, 7:07–8:07 a.m., N 38°38.269' E 034°49.963'*; and

78. *1/09/2008, 4:38–5:38 p.m., N 21°48.913' E 006°30.297'*. Courtesy Wilson Centre for Photography.

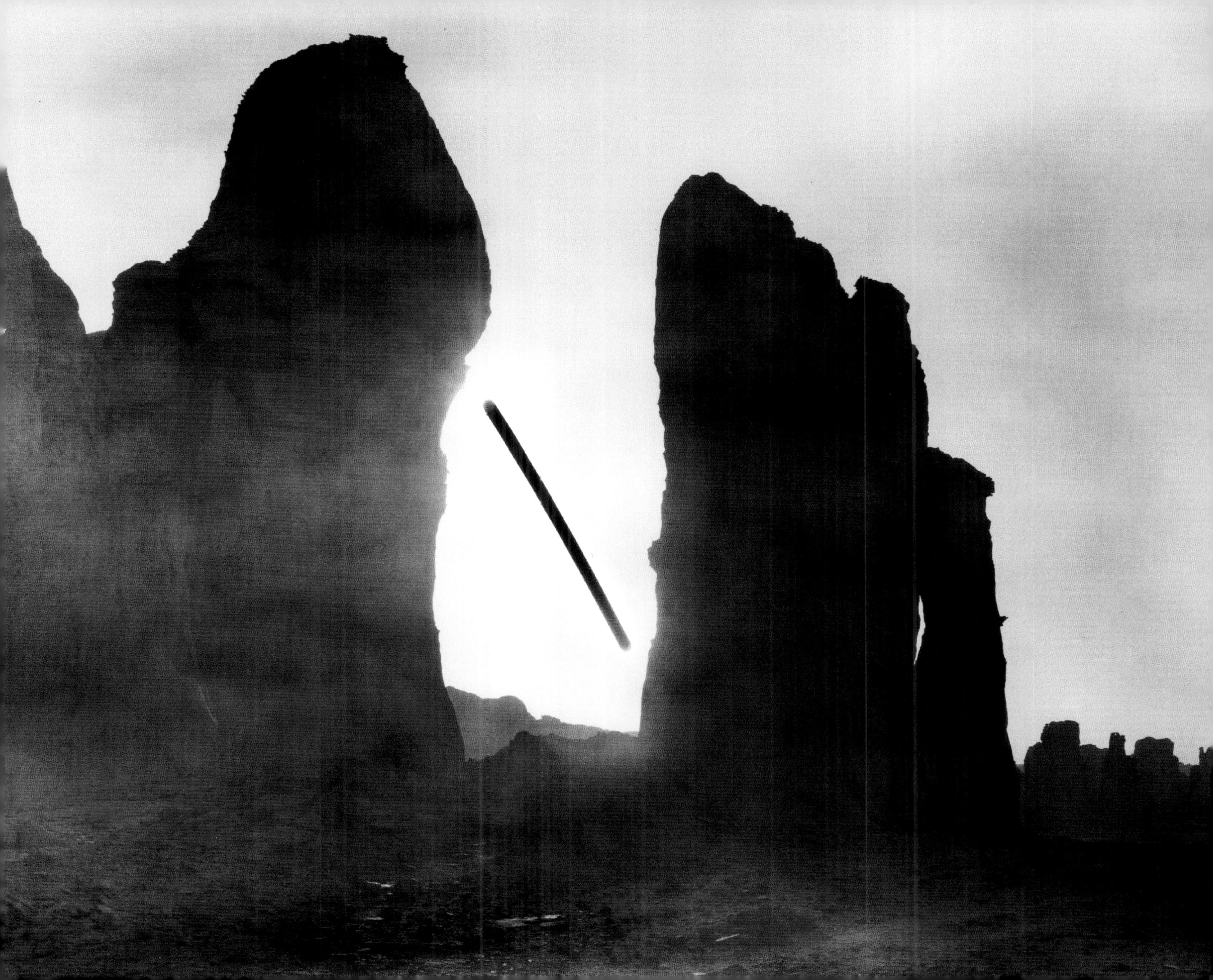

106. Jacqueline WOODS, *Black Sun #37*, 2011.

90. Hans-Christian SCHINK, *1hr* series, *1/05/2010, 5:46–6:46 p.m., S 06°26.486' E 039°27.776'*. Courtesy Wilson Centre for Photography.

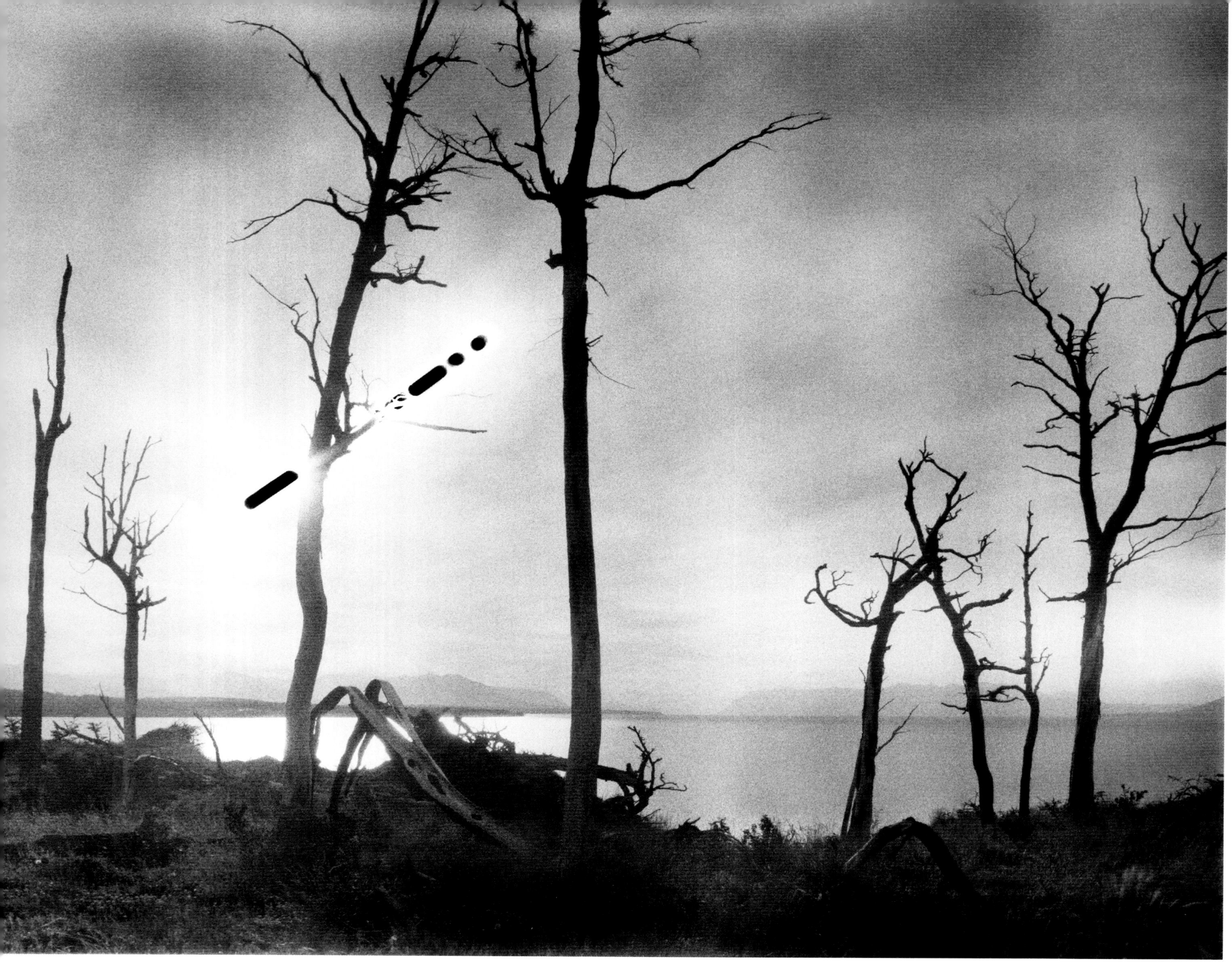

80. Hans-Christian SCHINK, *1hr* series, *2/04/2010, 7:22–8:22 p.m., S 54°35.873' W 067°22.541'*; and

85. *6/01/2008, 9:18–10:18 am, S 26°03.817' W 065°54.723'*. Courtesy Wilson Centre for Photography.

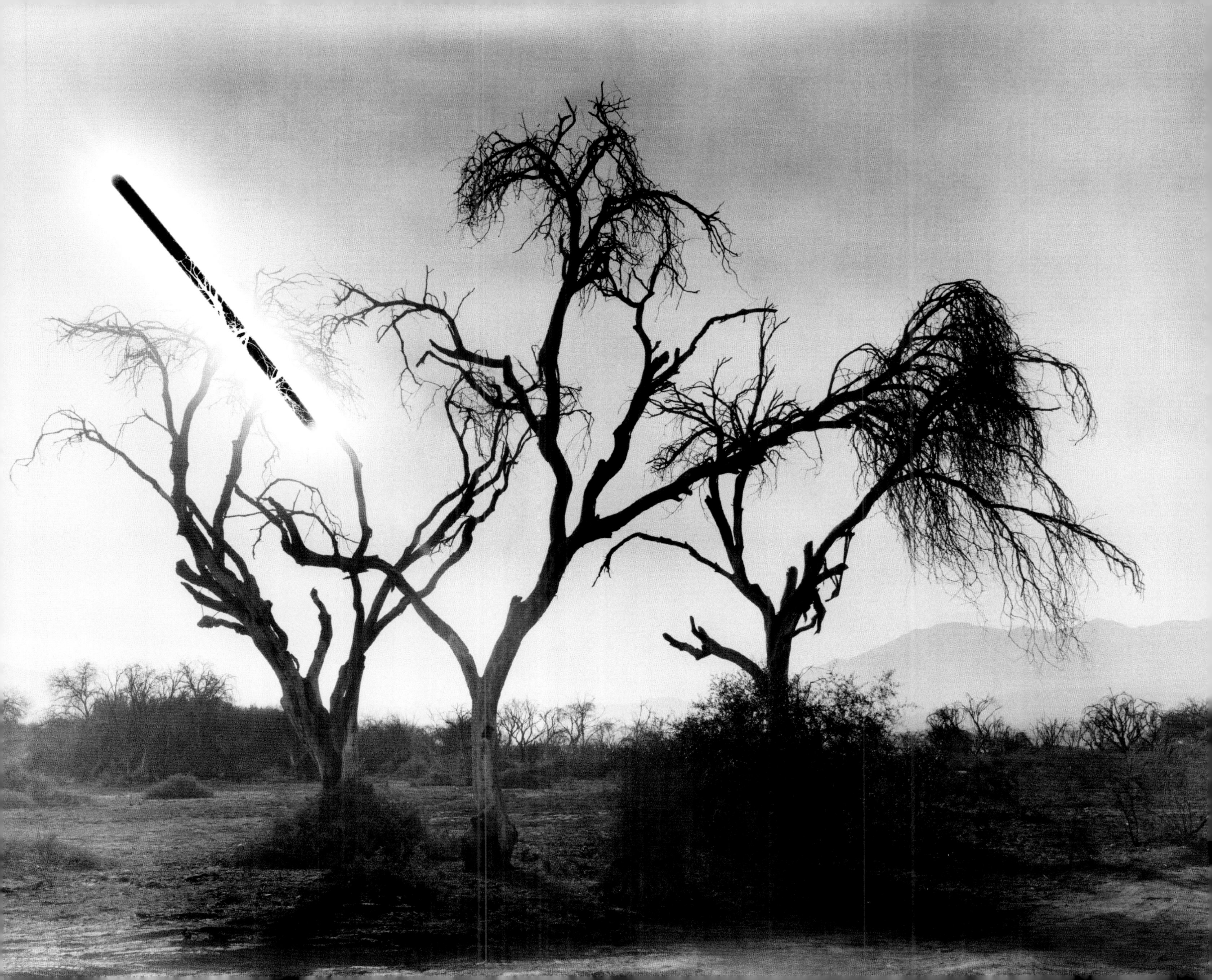

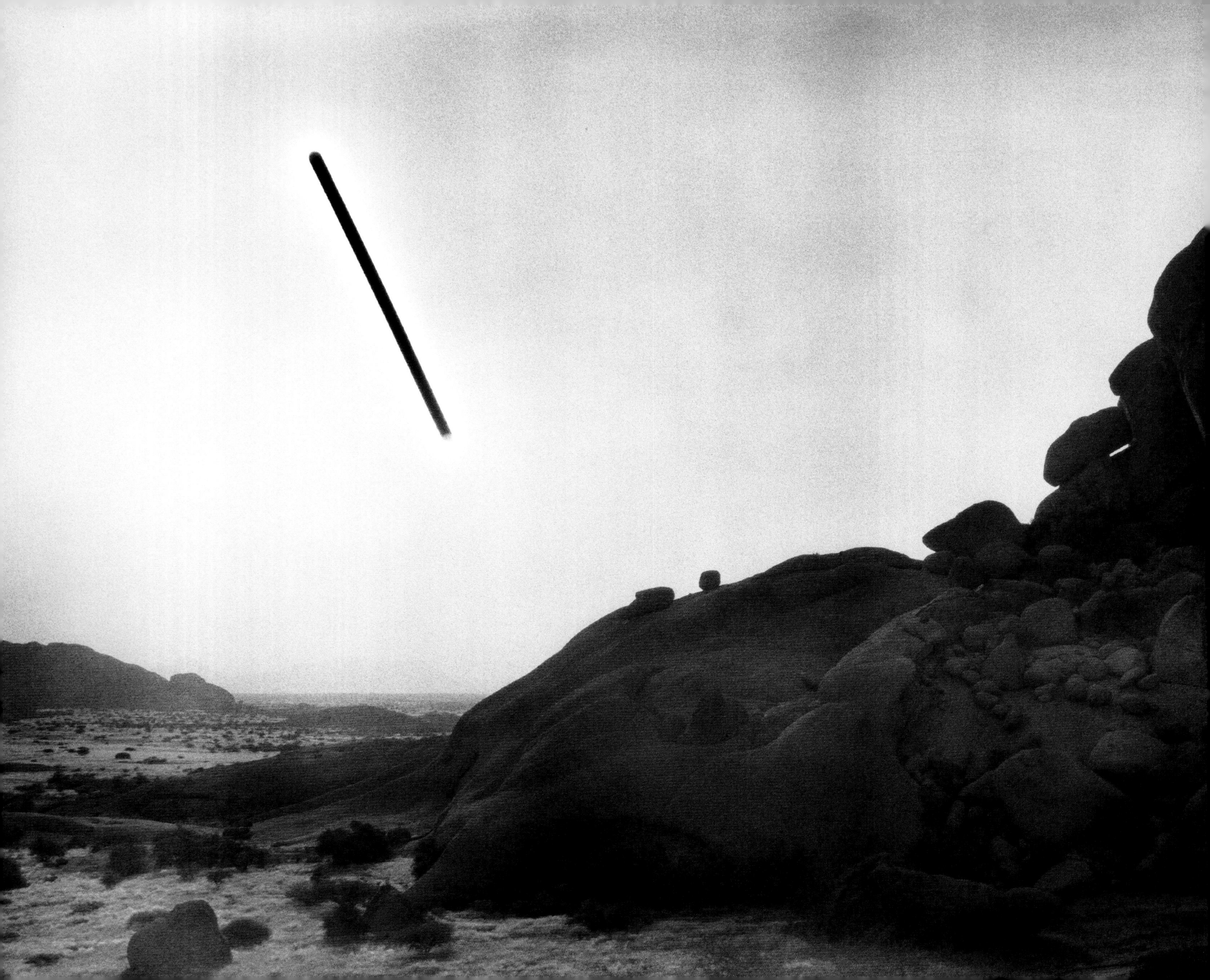

103. Jacqueline WOODS, *Black Sun #16*, 2011.

86. Hans-Christian SCHINK, *1hr* series, *4/15/2009, 7:02–8:02 a.m., S 21°50.398' E 015°11.352'*. Courtesy Wilson Centre for Photography.

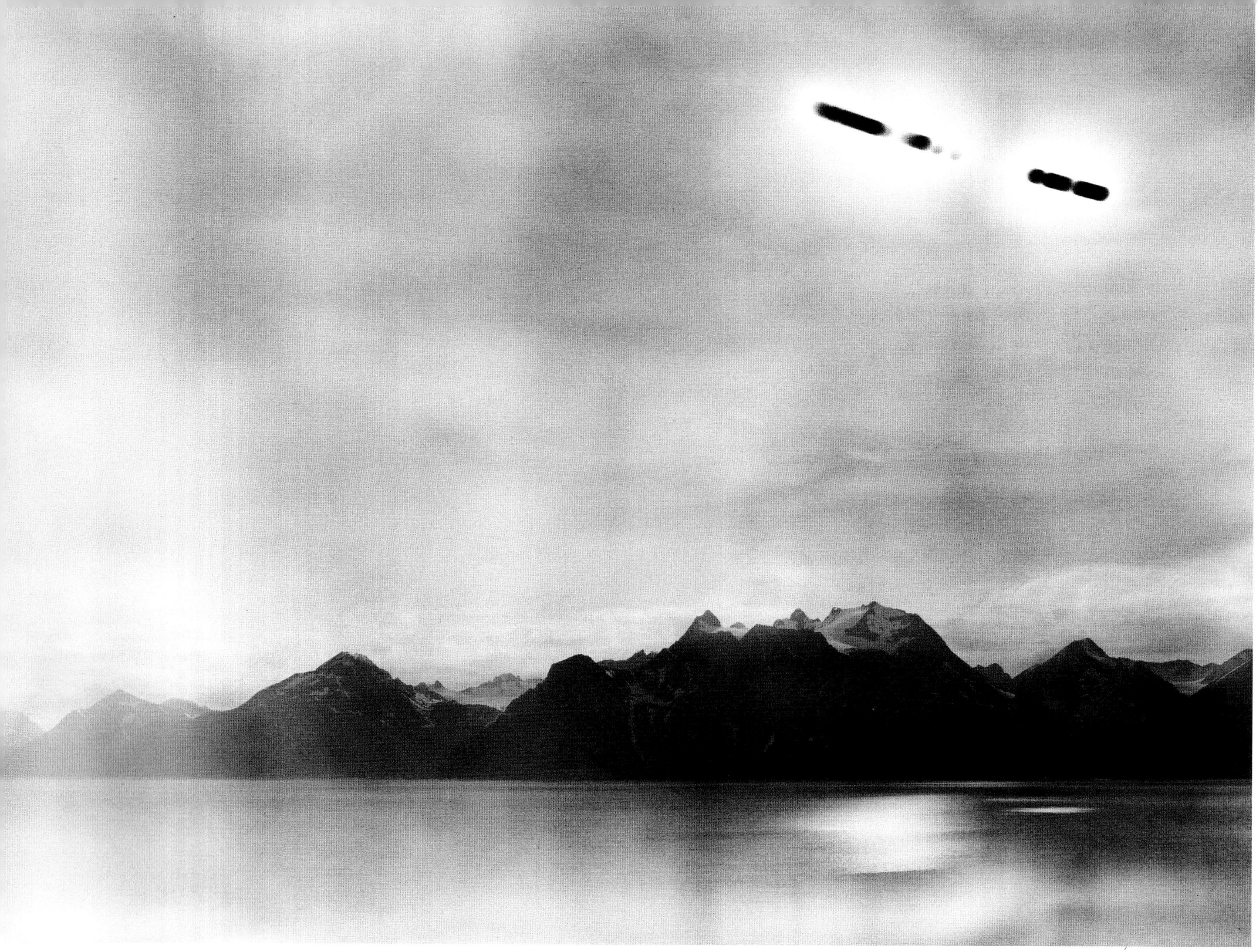

68. Hans-Christian SCHINK, *1hr* series, *7/11/2007, 5:27–6:27 p.m., N 69°45.199' E 020°29.497'*; and

88. *1/19/2010, 6:42–7:42 a.m., S 20°22.016' E 057°42.293'*. Courtesy Wilson Centre for Photography.

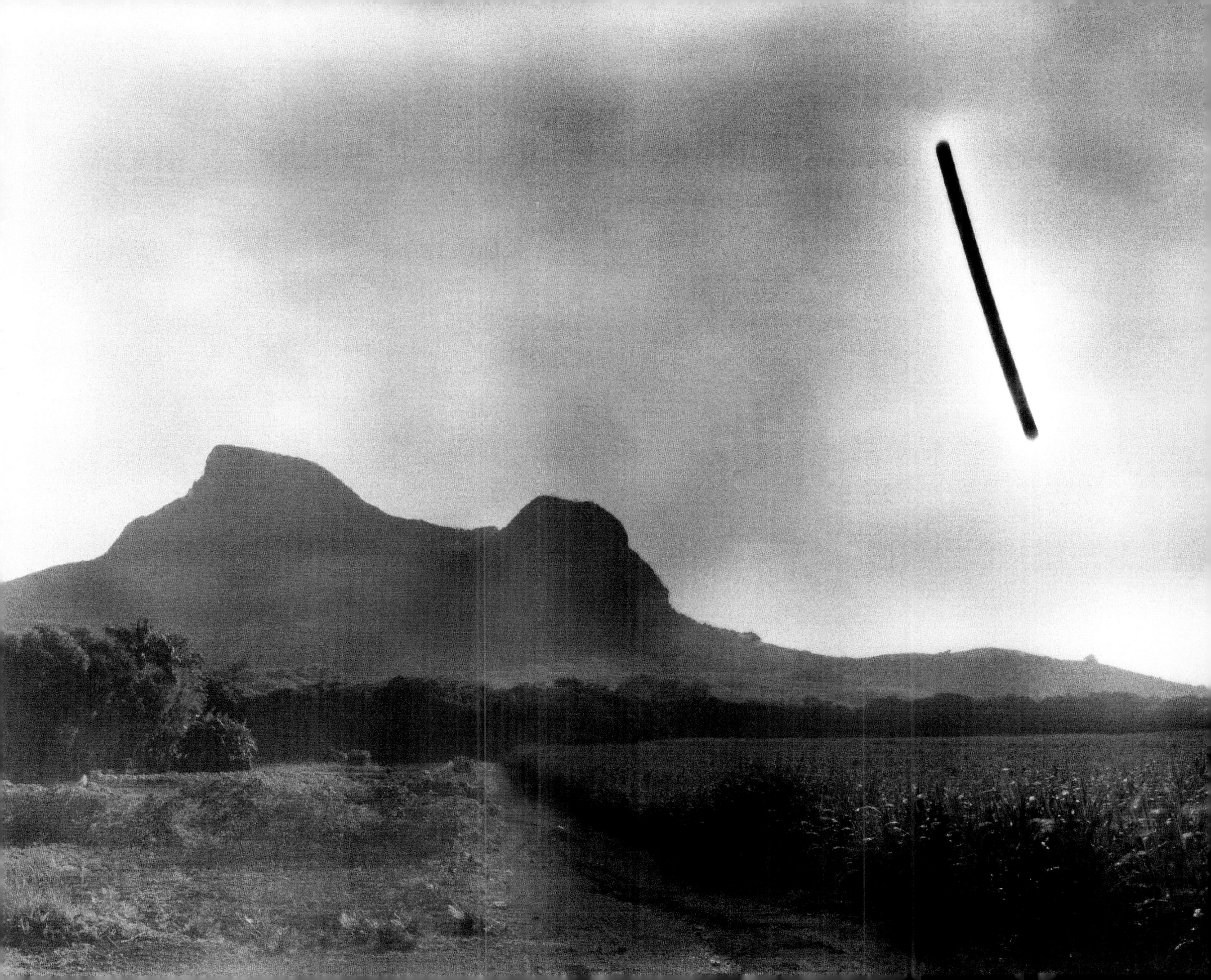

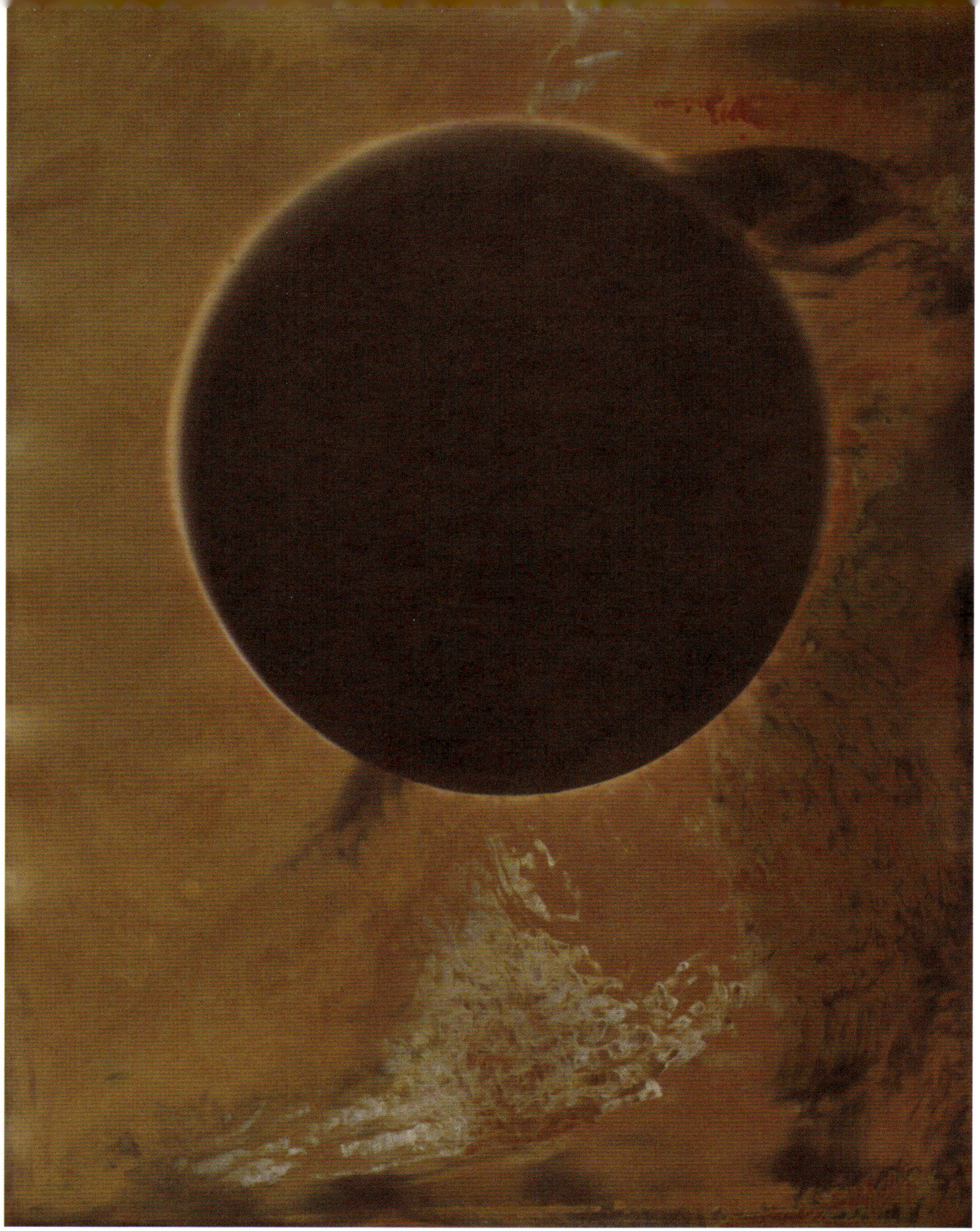

105. Jacqueline WOODS, *Black Sun #33*, 2011.

87. Hans-Christian SCHINK, *1hr* series, *4/12/2009, 4:11–5:11 p.m., S 21°47.094' E 015°39.829'*. Courtesy Wilson Centre for Photography.

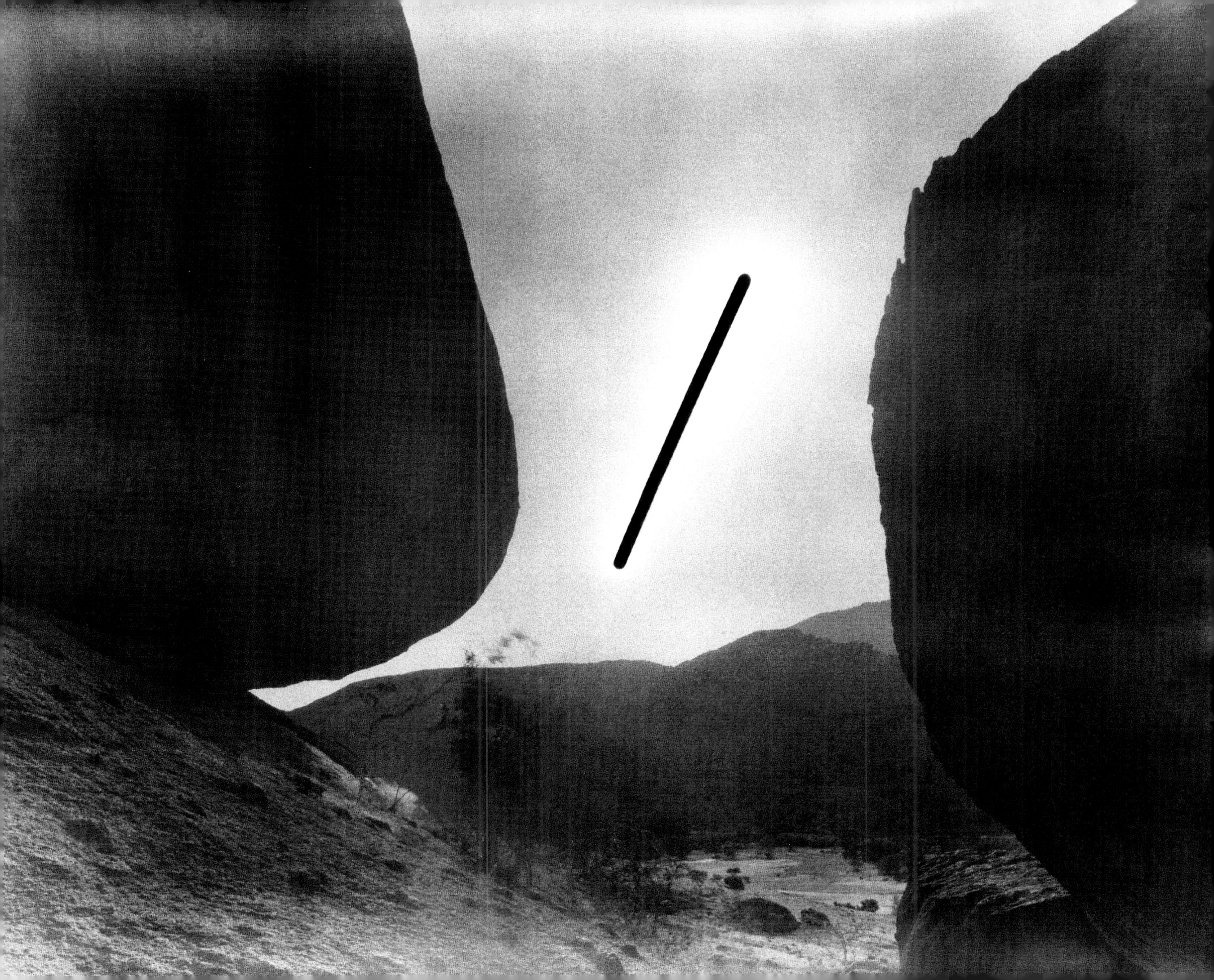

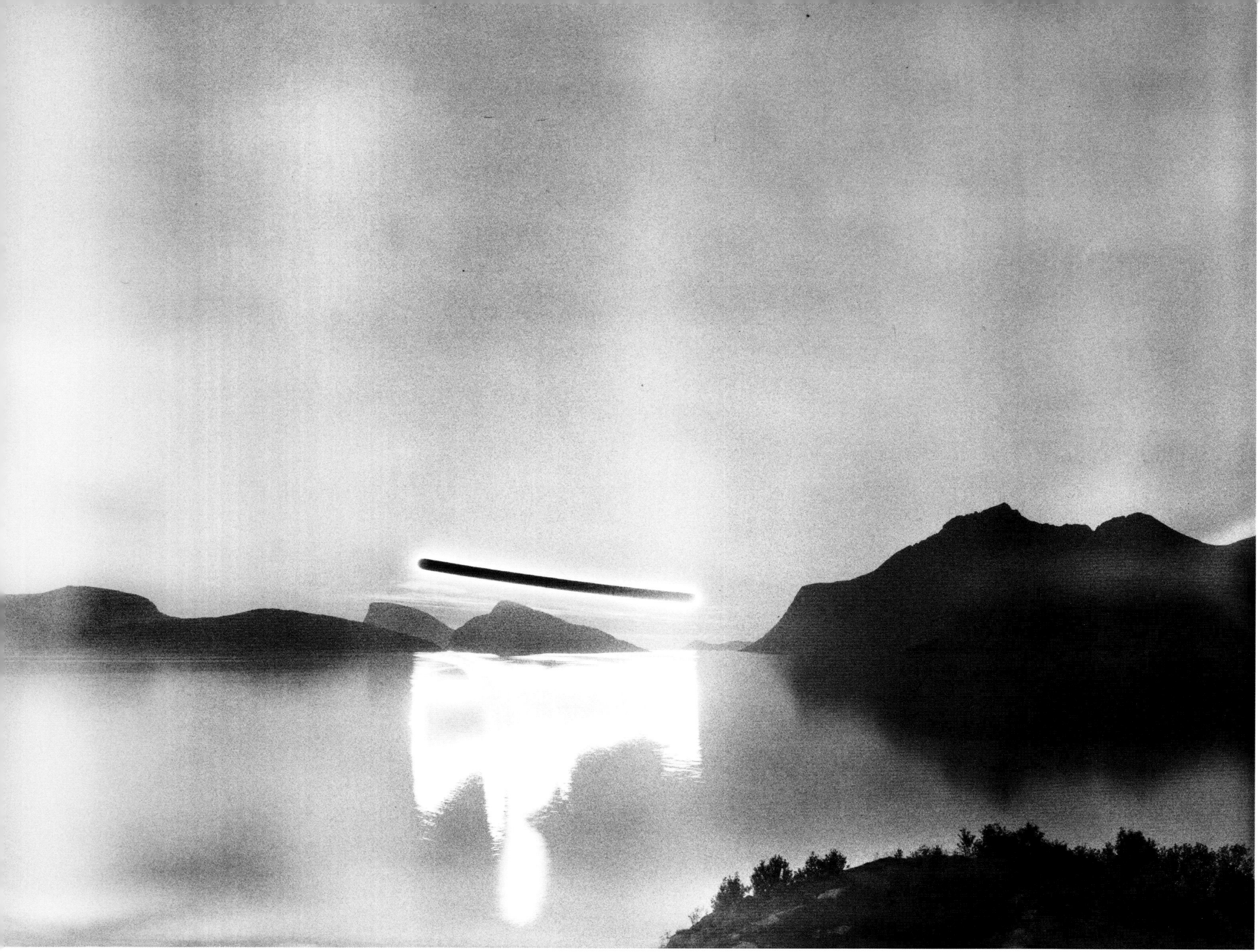

69. Hans-Christian SCHINK, *1hr* series, *7/14–15/2007, 11:28 p.m.–0:28 a.m., N 69° 37.661' E 018°13.470'*; and

67. *9/17/2006, 8:45–9:45 a.m., N 78°13.370' E 015°40.024'.* Courtesy Wilson Centre for Photography.

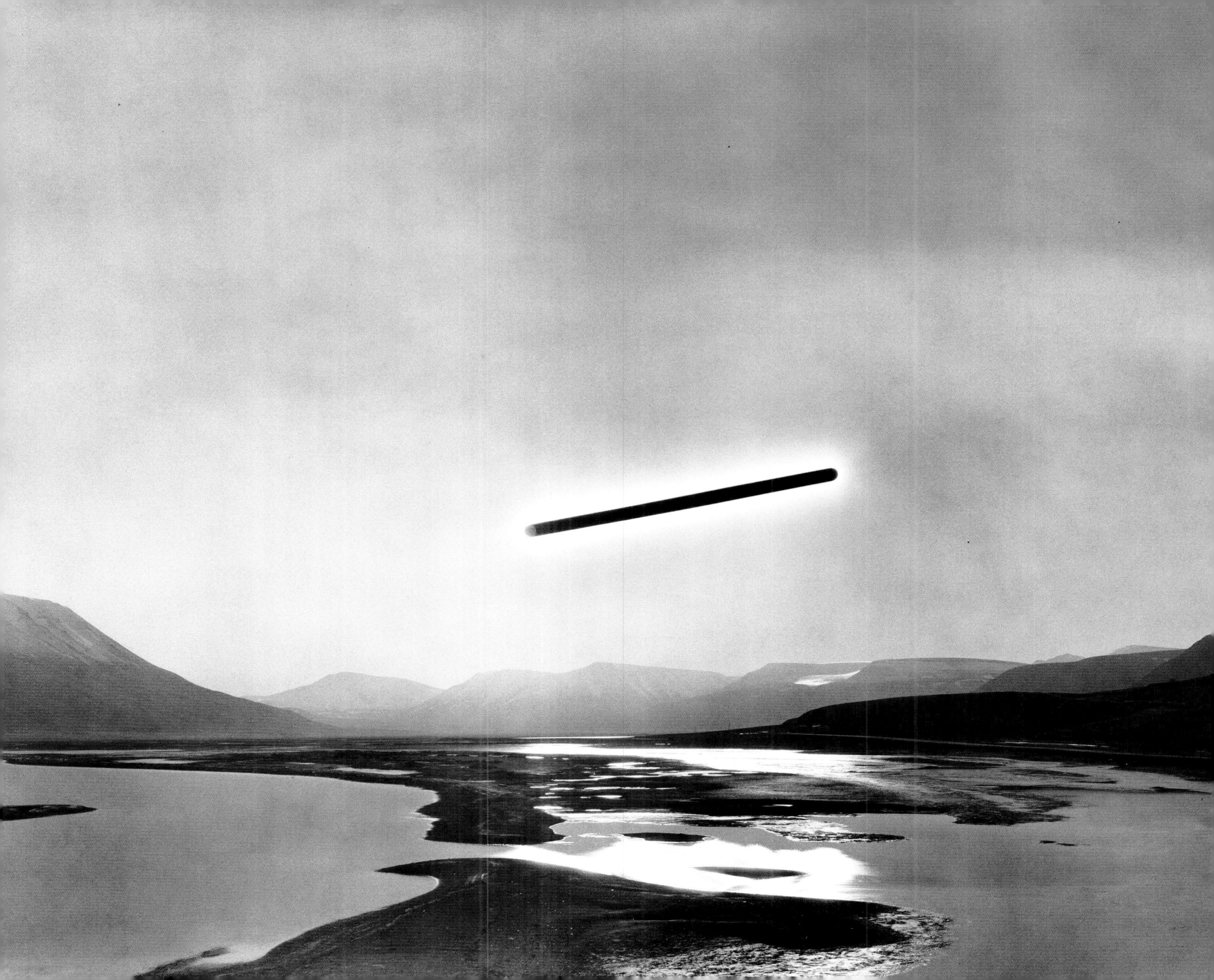

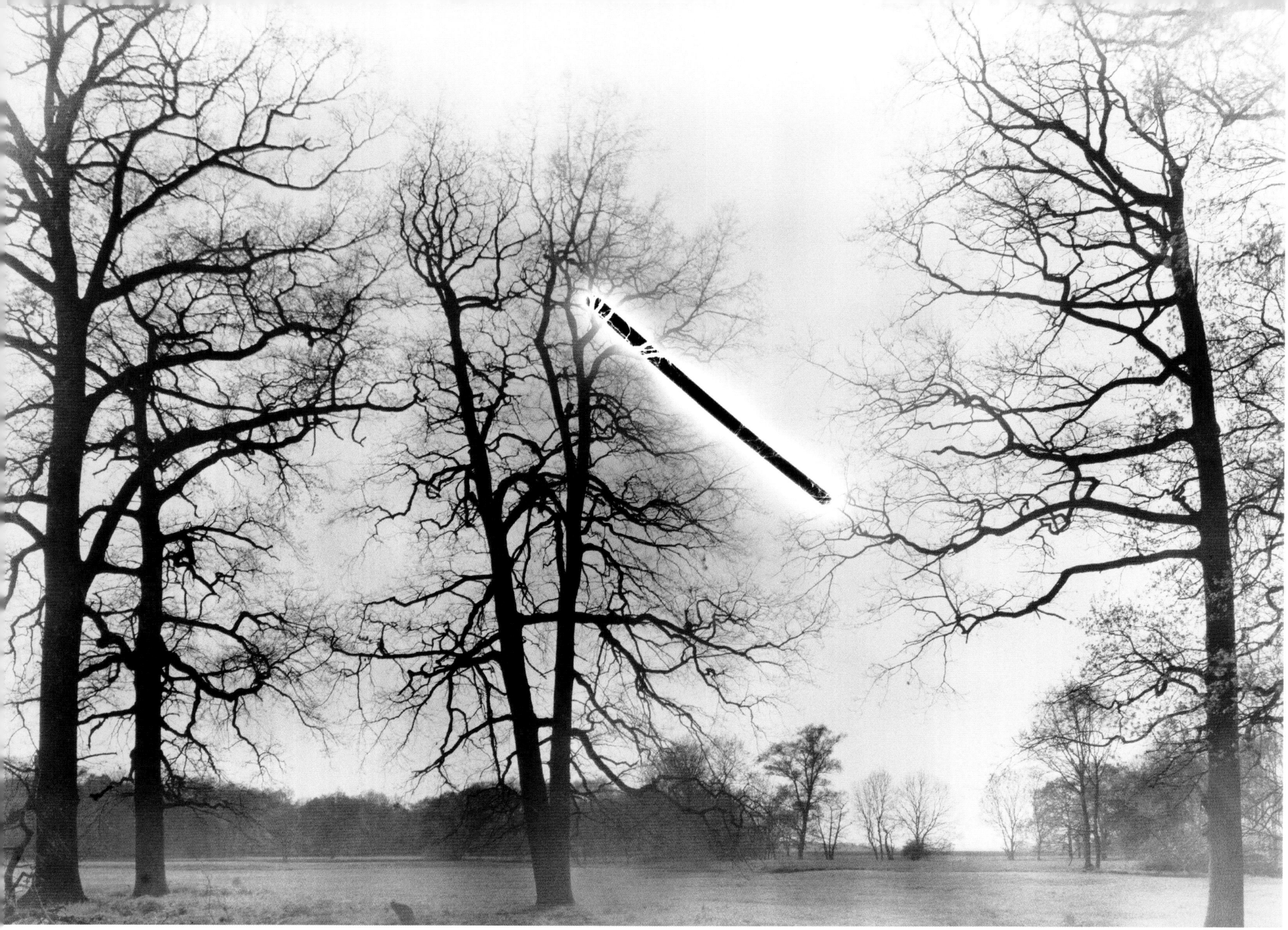

70. Hans-Christian SCHINK, *1hr series*, *5/03/2006, 6:04–7:04 p.m., N 51°22.126' E 012°09.310'*. Courtesy Wilson Centre for Photography.

25. Dennis DI CICCO, *Solar Analemma*, 1978–79.

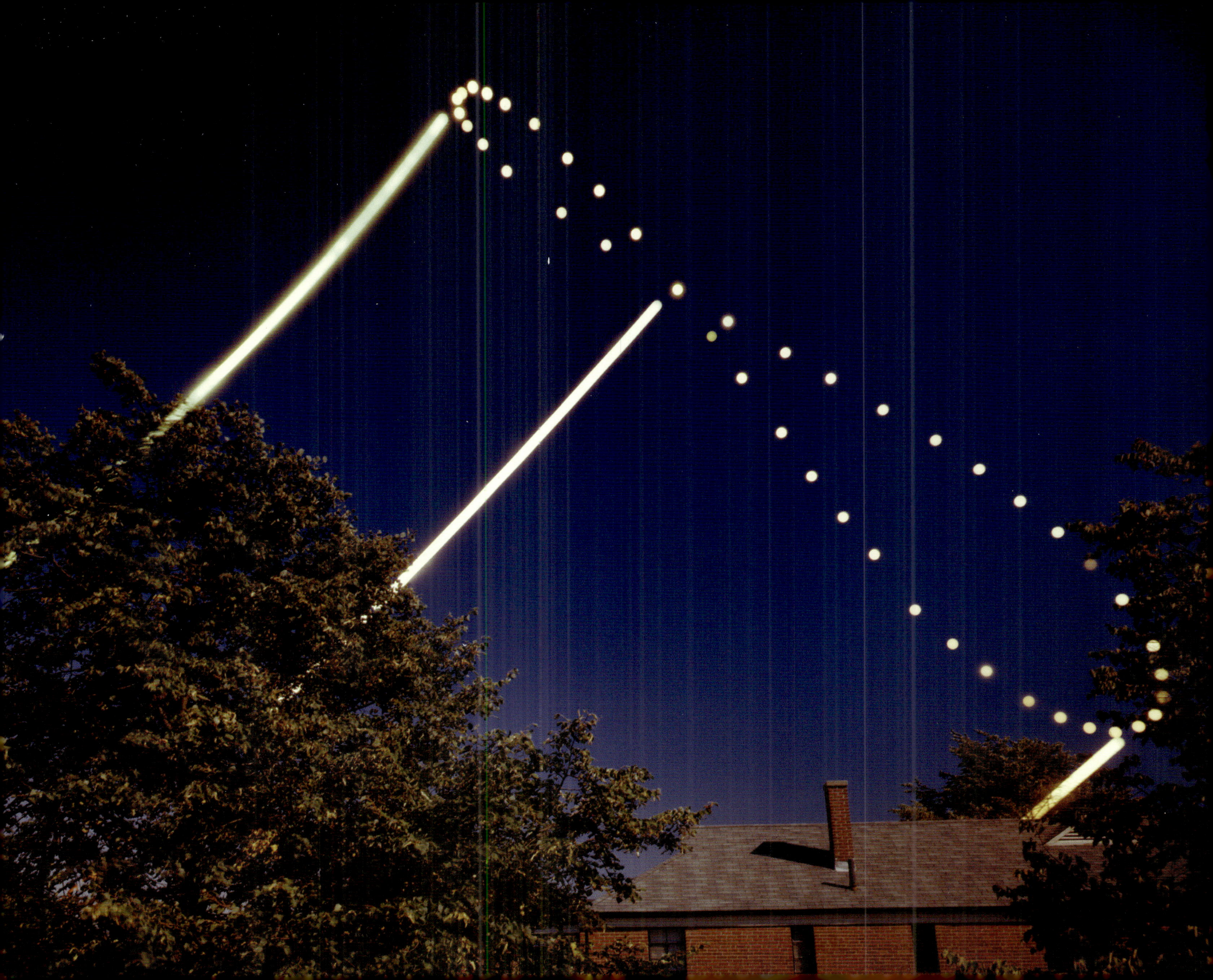

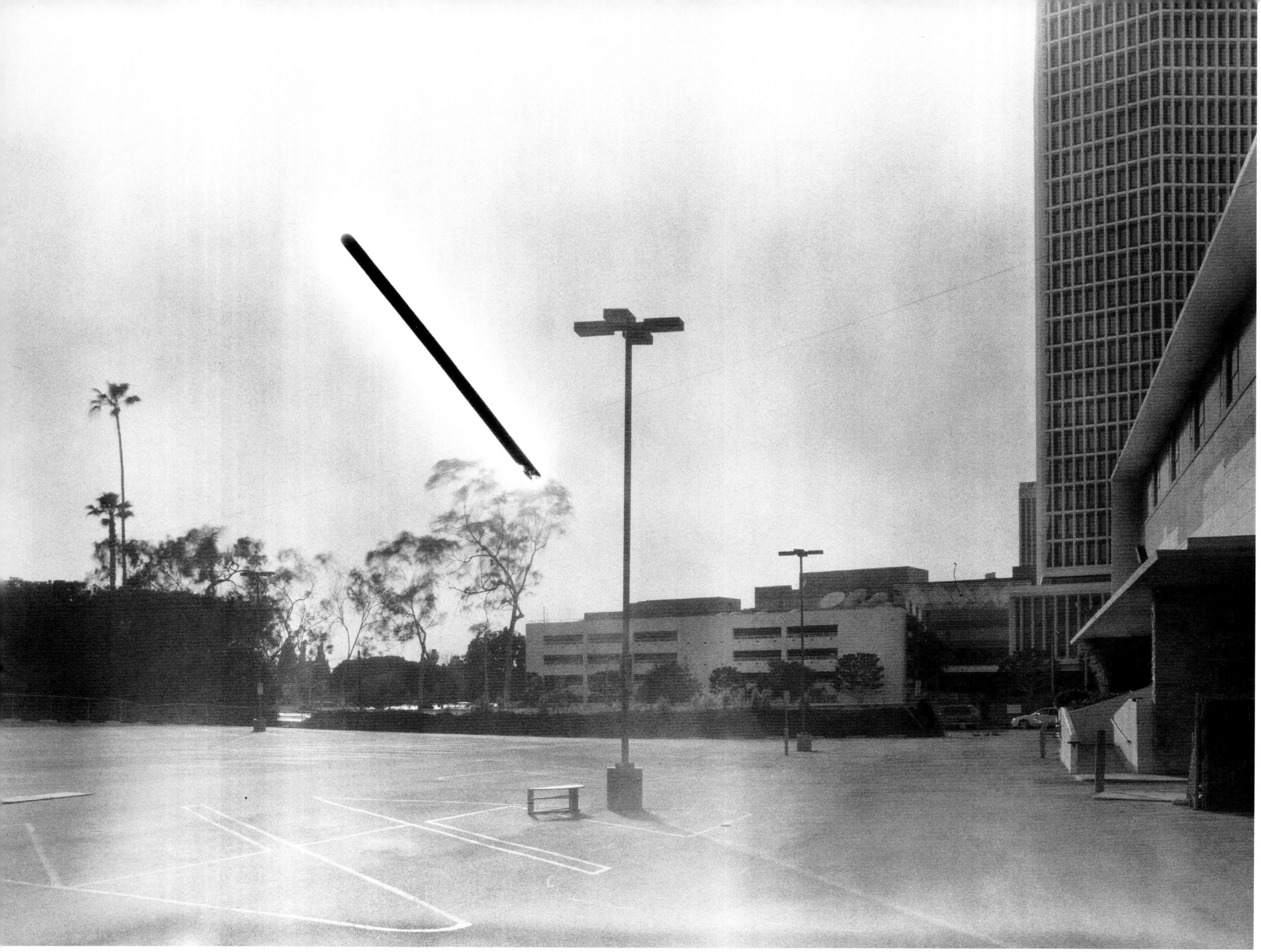

74. Hans-Christian SCHINK, *1hr* series, *2/23/2006, 4:04–5:04 p.m., N 34°03.712' W 118°20.979'*; and

73. *2/22/2006, 7:32–8:32 am, N 34°30.850' W 117°12.266'*. Courtesy Wilson Centre for Photography.

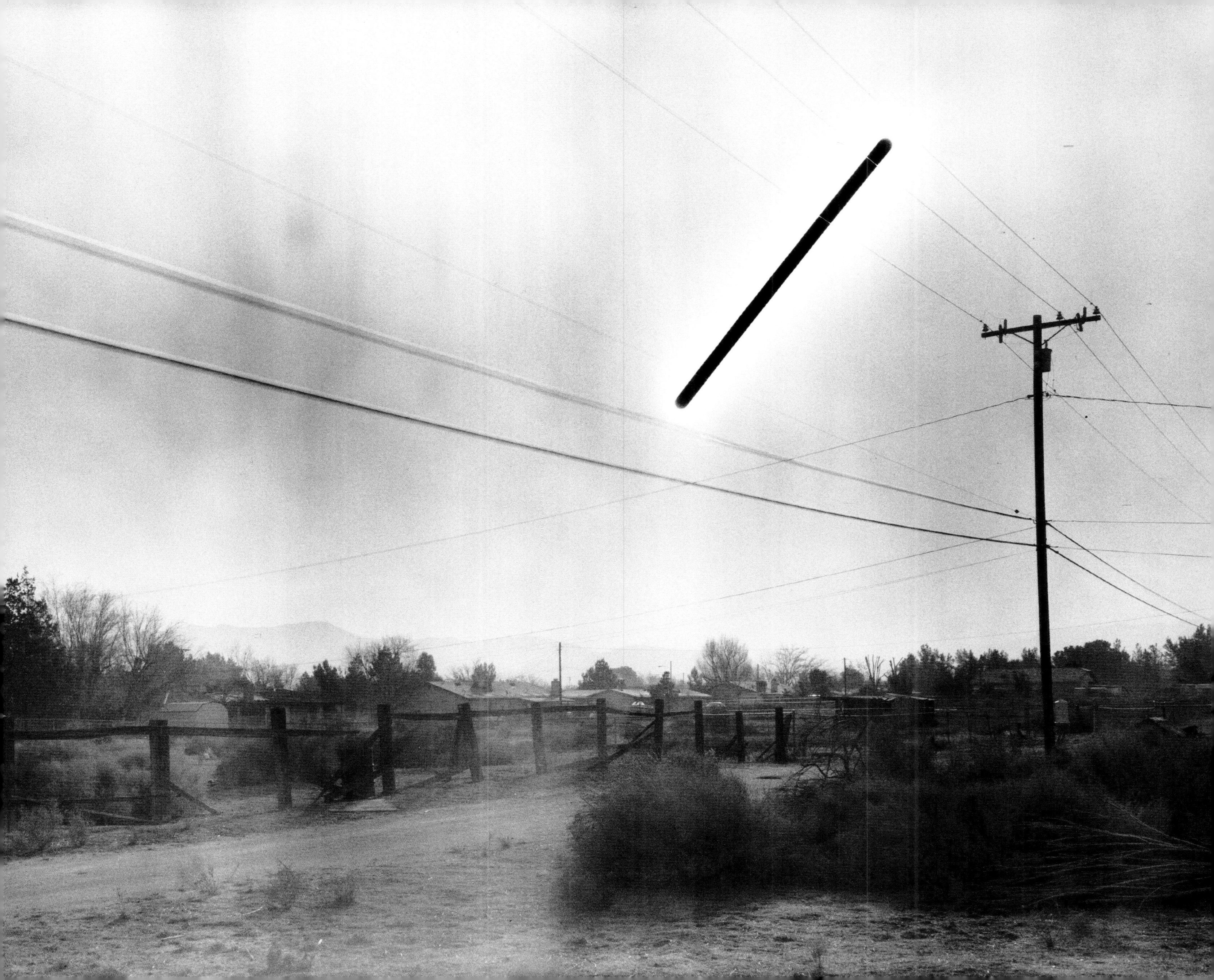

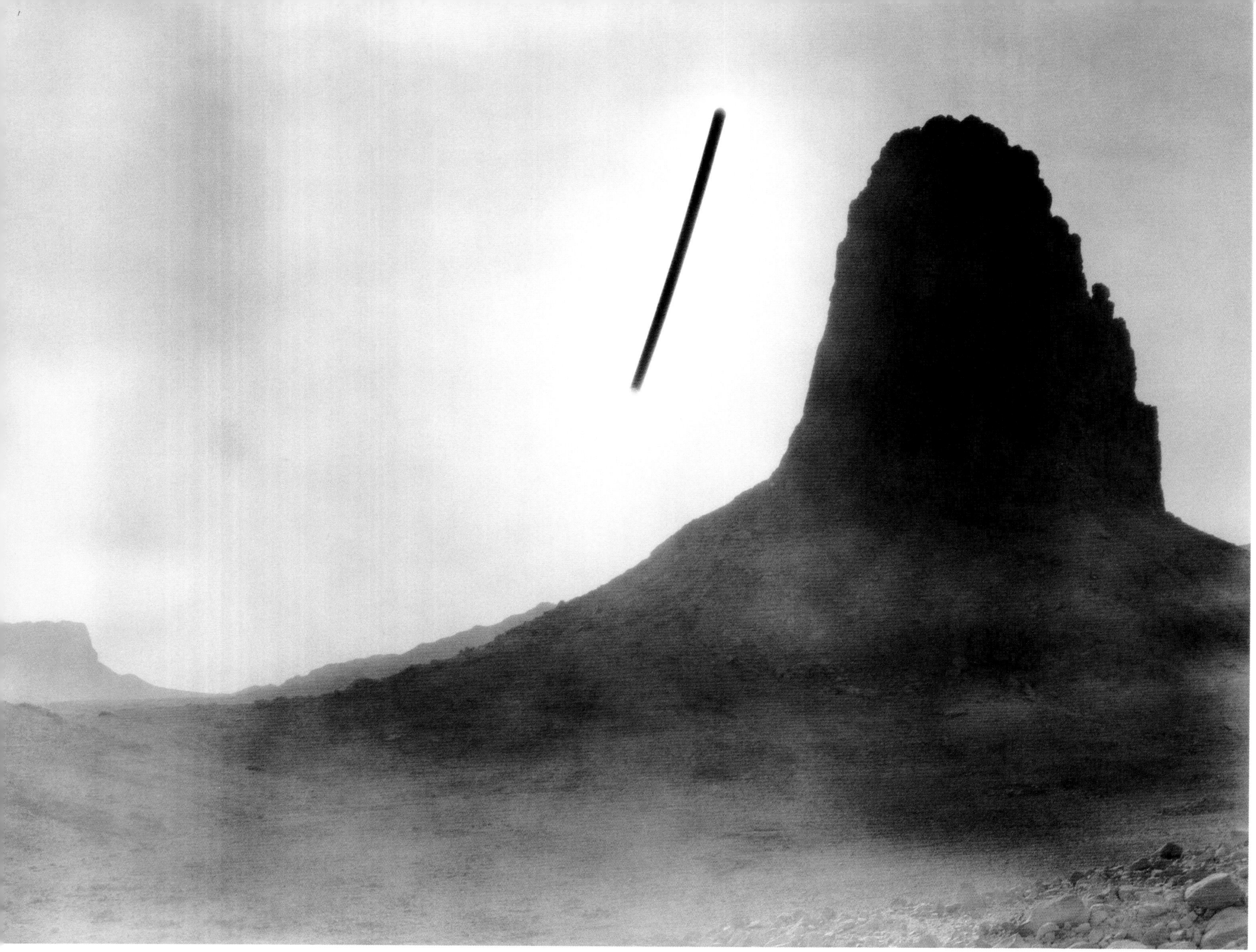

77. Hans-Christian SCHINK, *1hr* series, *6/20/2007, 7:02–8:02 a.m., N 22°54.402' E 005°40.083'*; and

83. *2/26/2010, 7:54–8:54 a.m., S 36°49.622' E 175°47.340'*. Courtesy Wilson Centre for Photography.

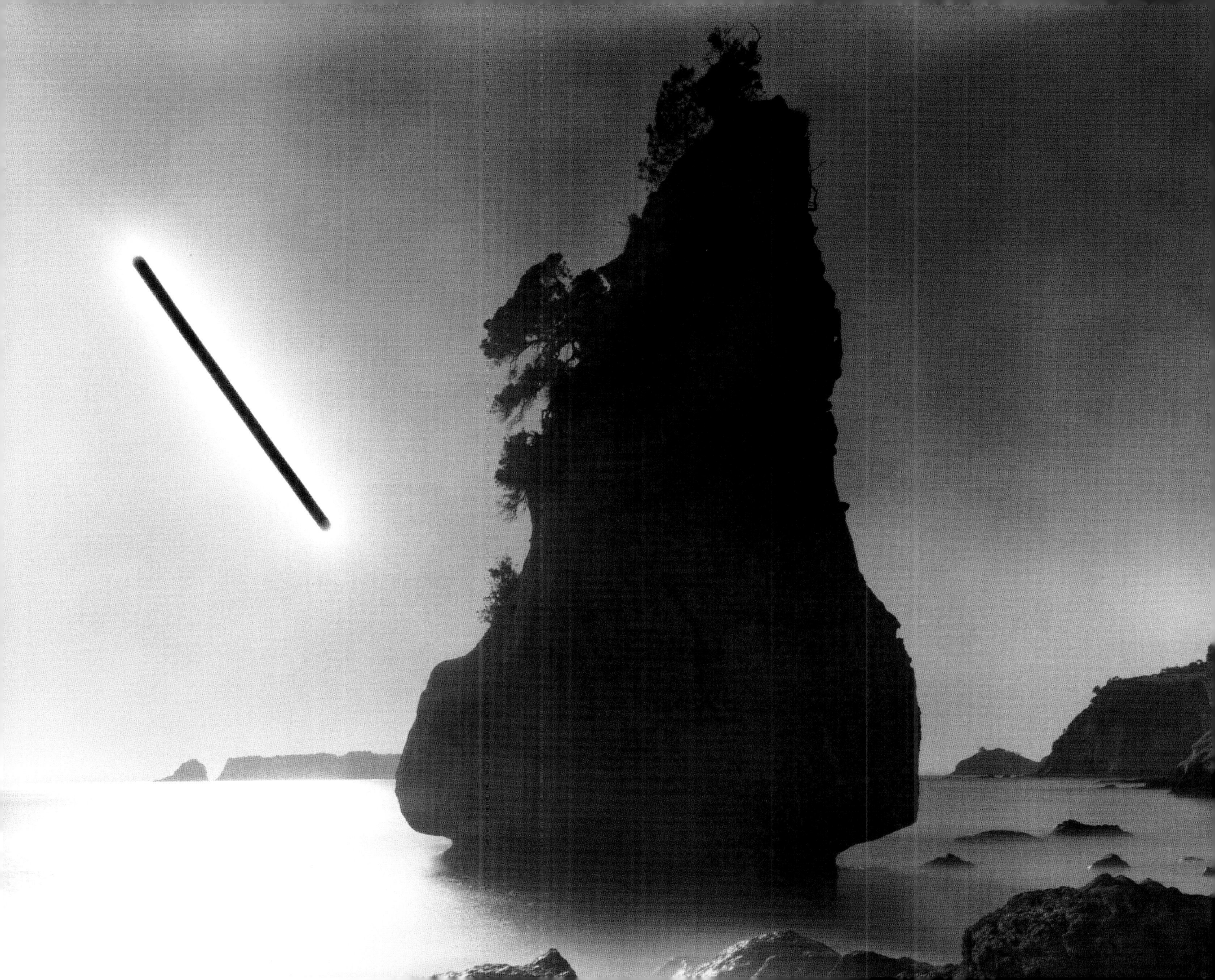

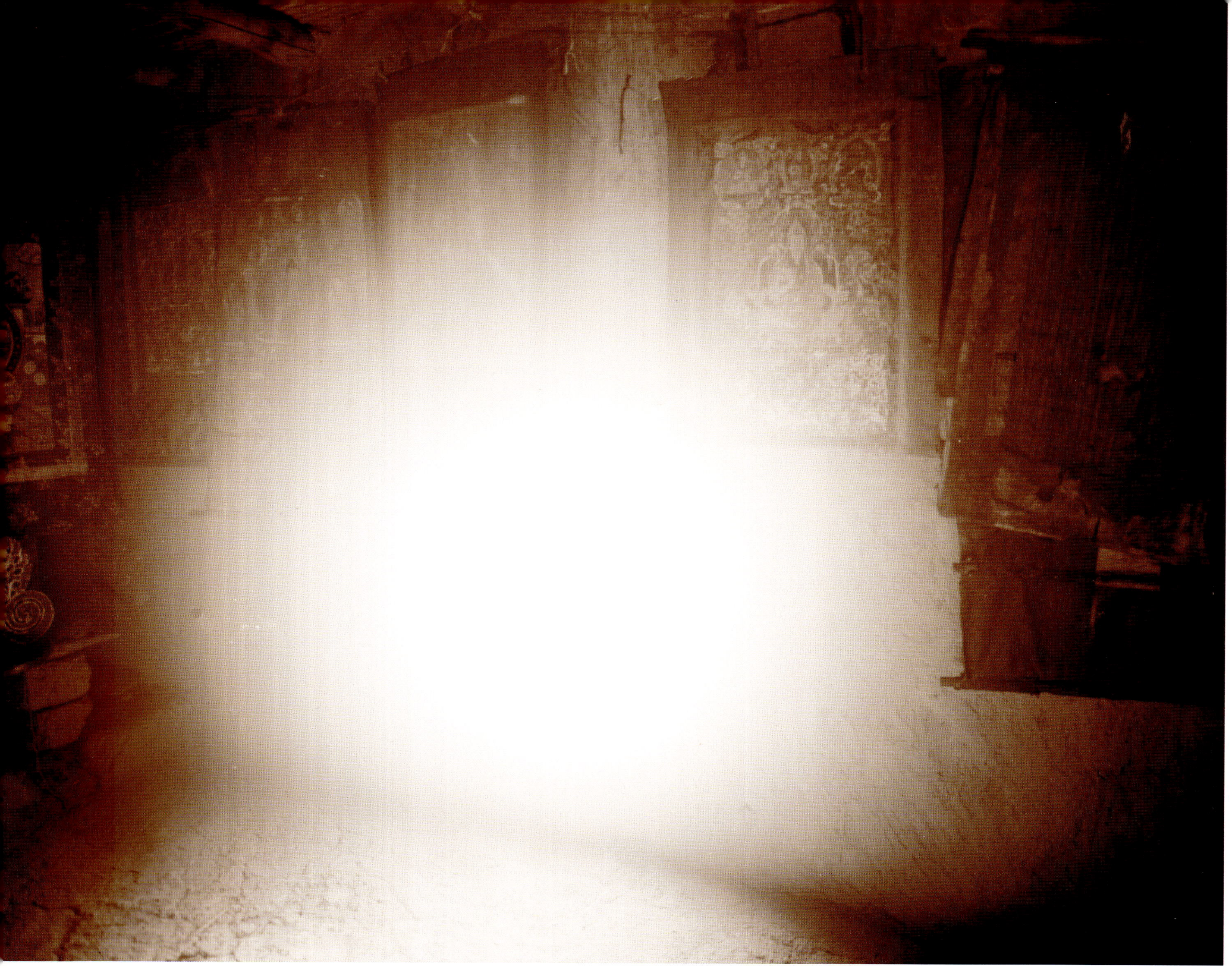

11. Linda CONNOR, *Window with Thangkas, Ladakh, India*, 1994.

102. Jacqueline WOODS, *Black Sun #15*, 2011.

PAGE III

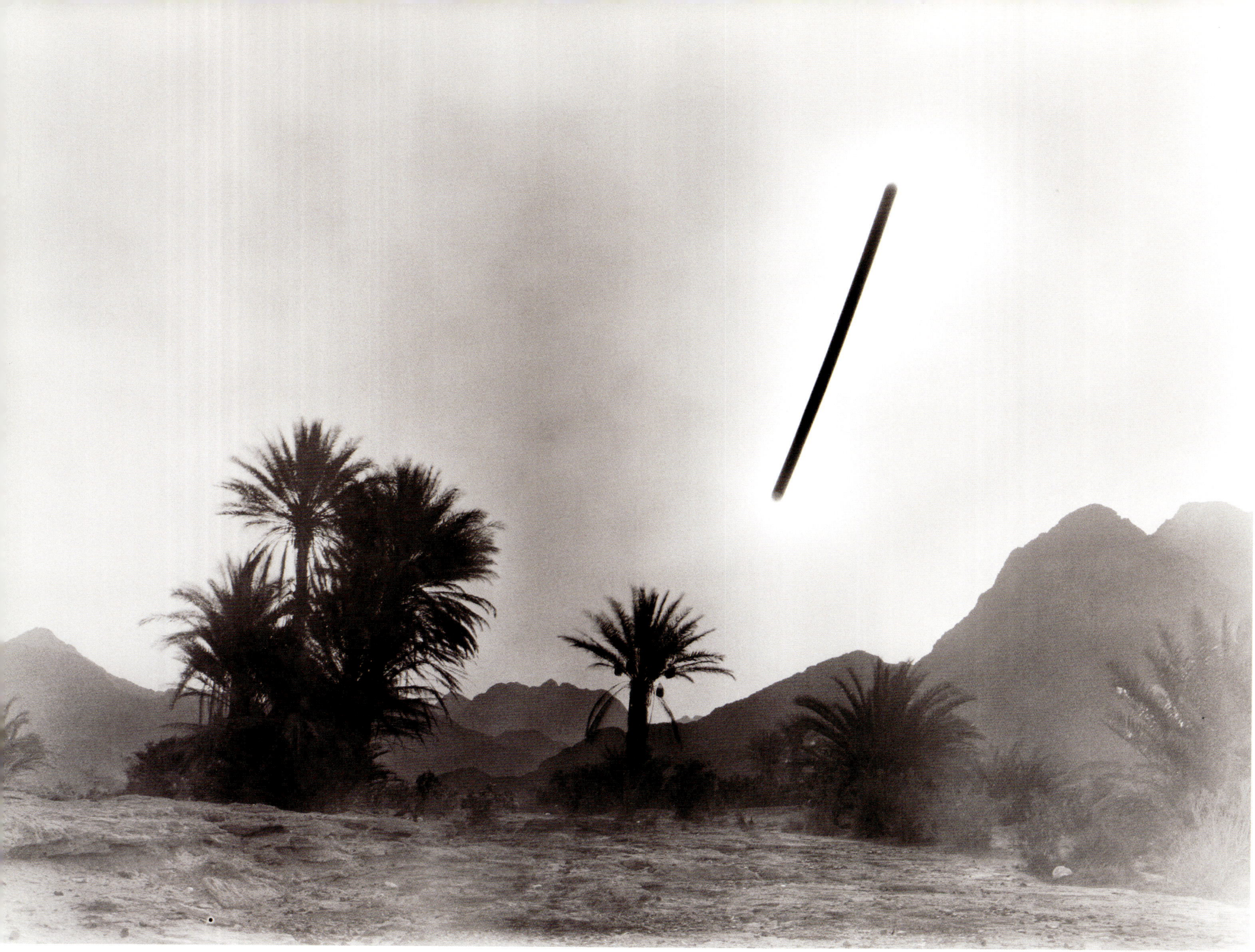

75. Hans-Christian SCHINK, *1hr* series, *5/10/2006, 6:47–7:47 a.m., N 24°47.435' E 055°56.493'*; and

89. *3/28/2010, 6:43–7:43 a.m., S 08°27.131' E 119°52.396'*. Courtesy Wilson Centre for Photography.

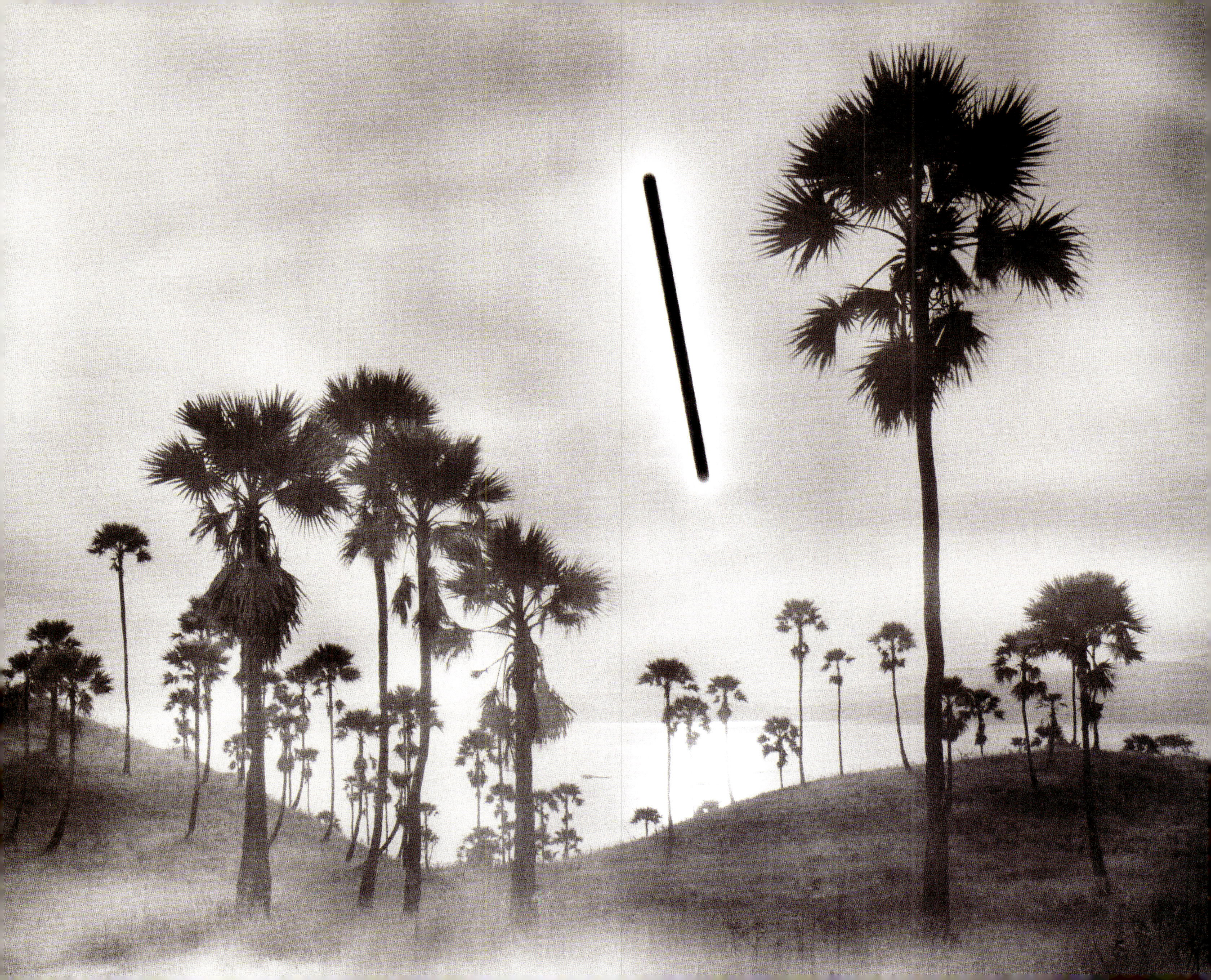

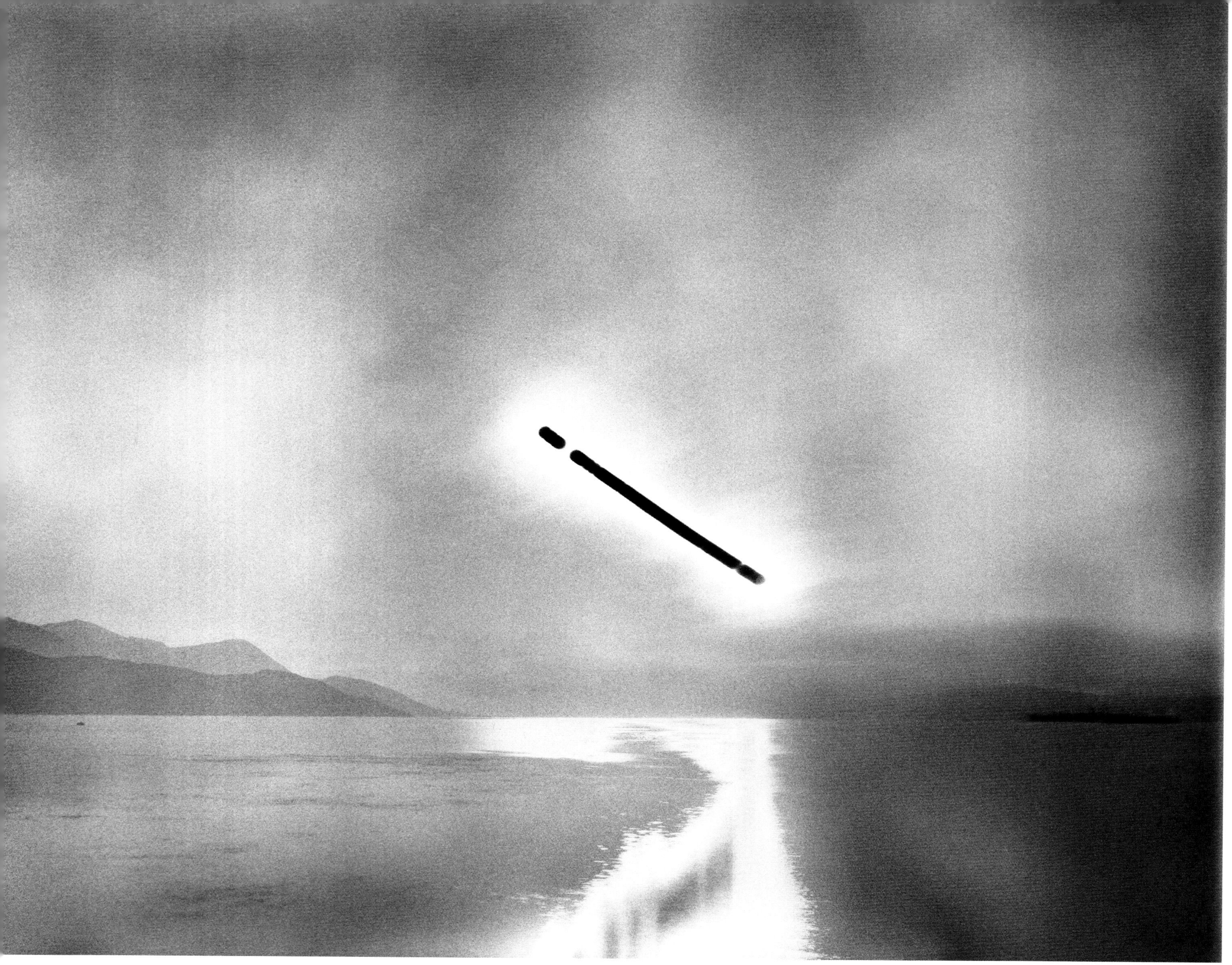

79. Hans-Christian SCHINK, *1hr* series, *1/23/2010, 6:31–7:31 a.m., S 54°48.411' W 068°18.032'*; and

82. *2/20/2010, 6:53–7:53 a.m., S 37°40.831' E 178°32.635'.* Courtesy Wilson Centre for Photography.

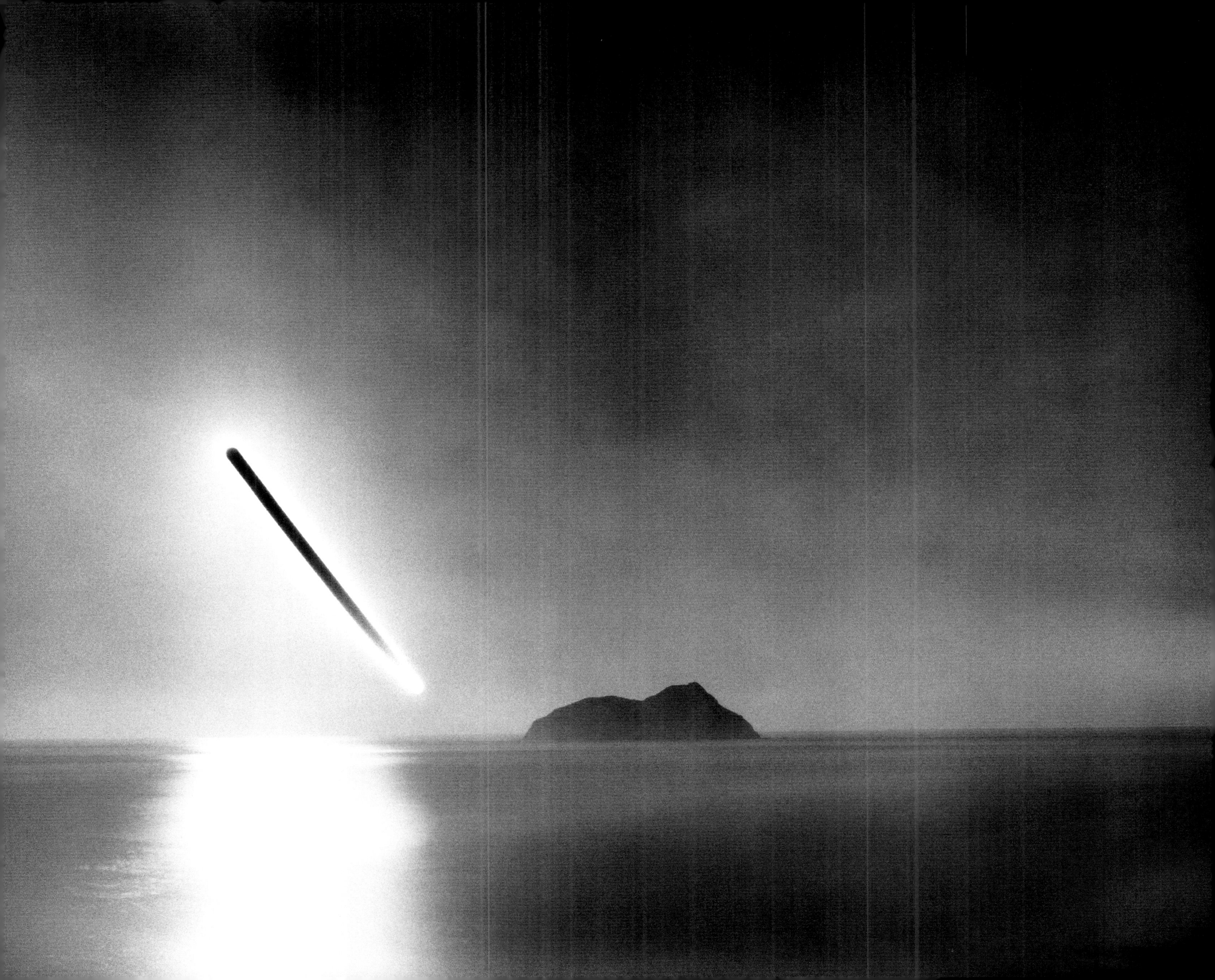

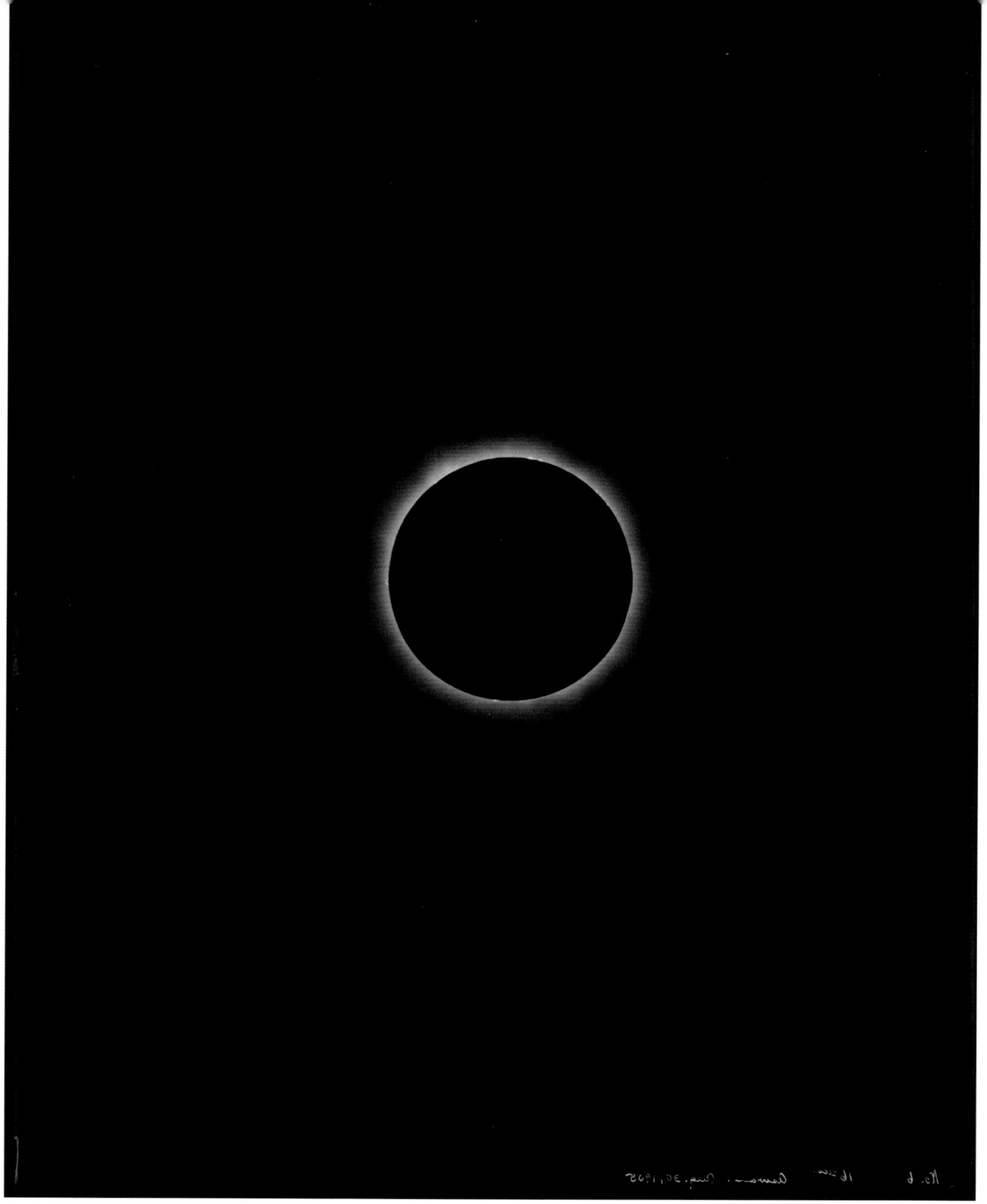

9. Linda CONNOR, *Solar Eclipse, Egypt, August 30, 1905,* 1998. Printed from Lick Observatory Archive negative.
Original photo by William Hussey. © UC Regents/Lick Observatory.

6. Ken Crawford, *IC 5067*.

Star Birth

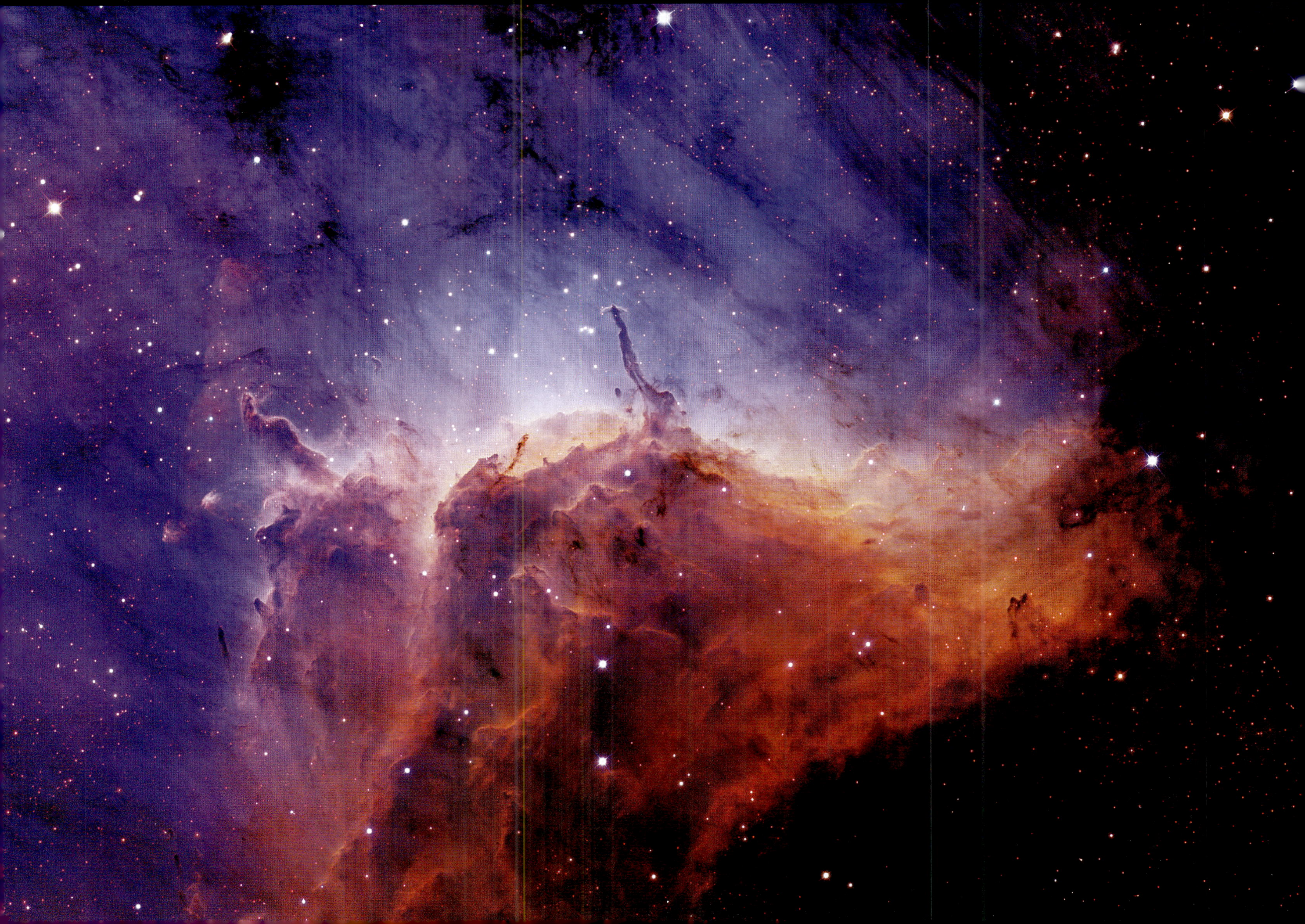

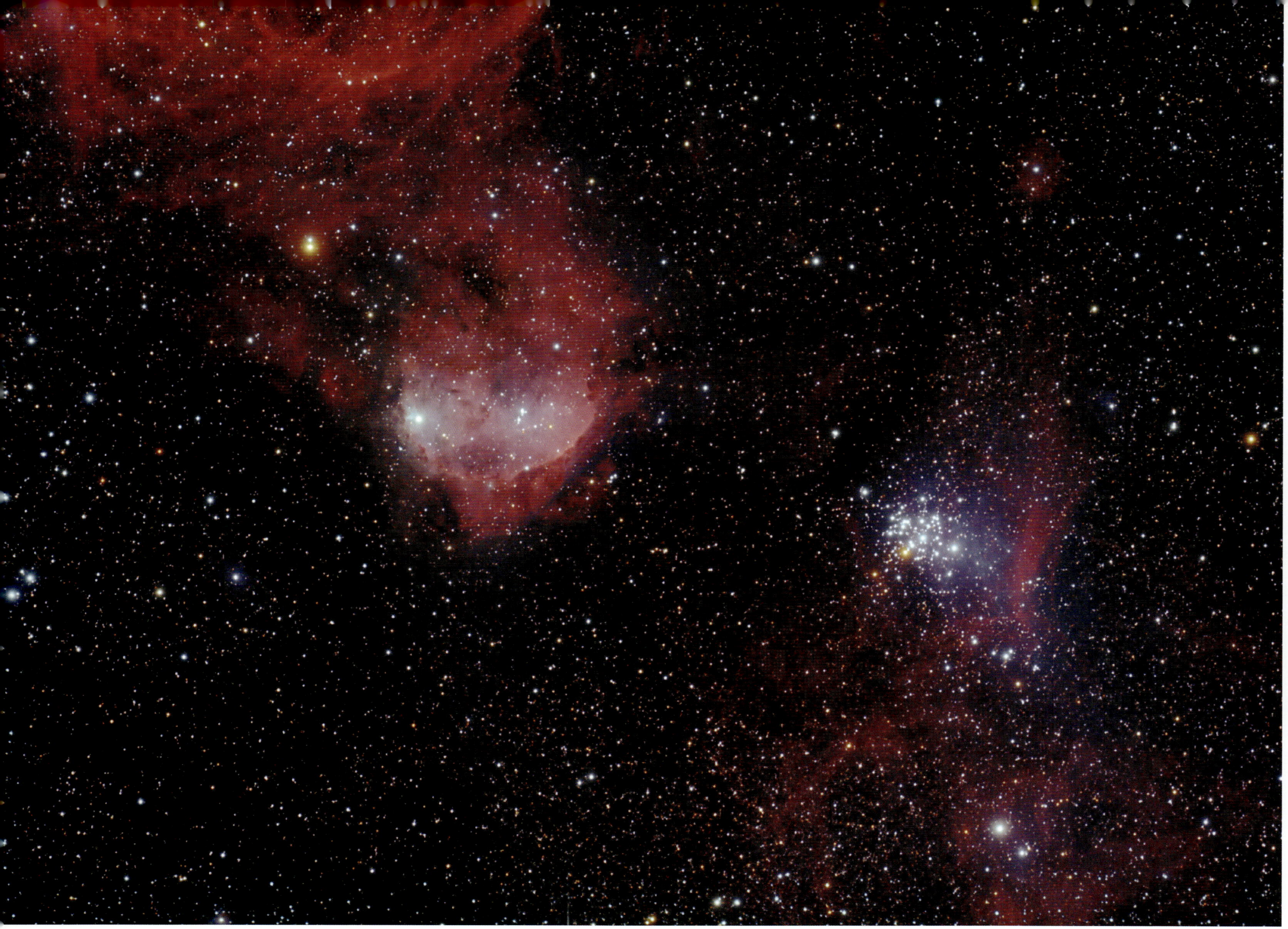

20. Ignacio DIAZ-BOBILLO, *Light Drops #1-A*; and

21. *Light Drops #1-B*.

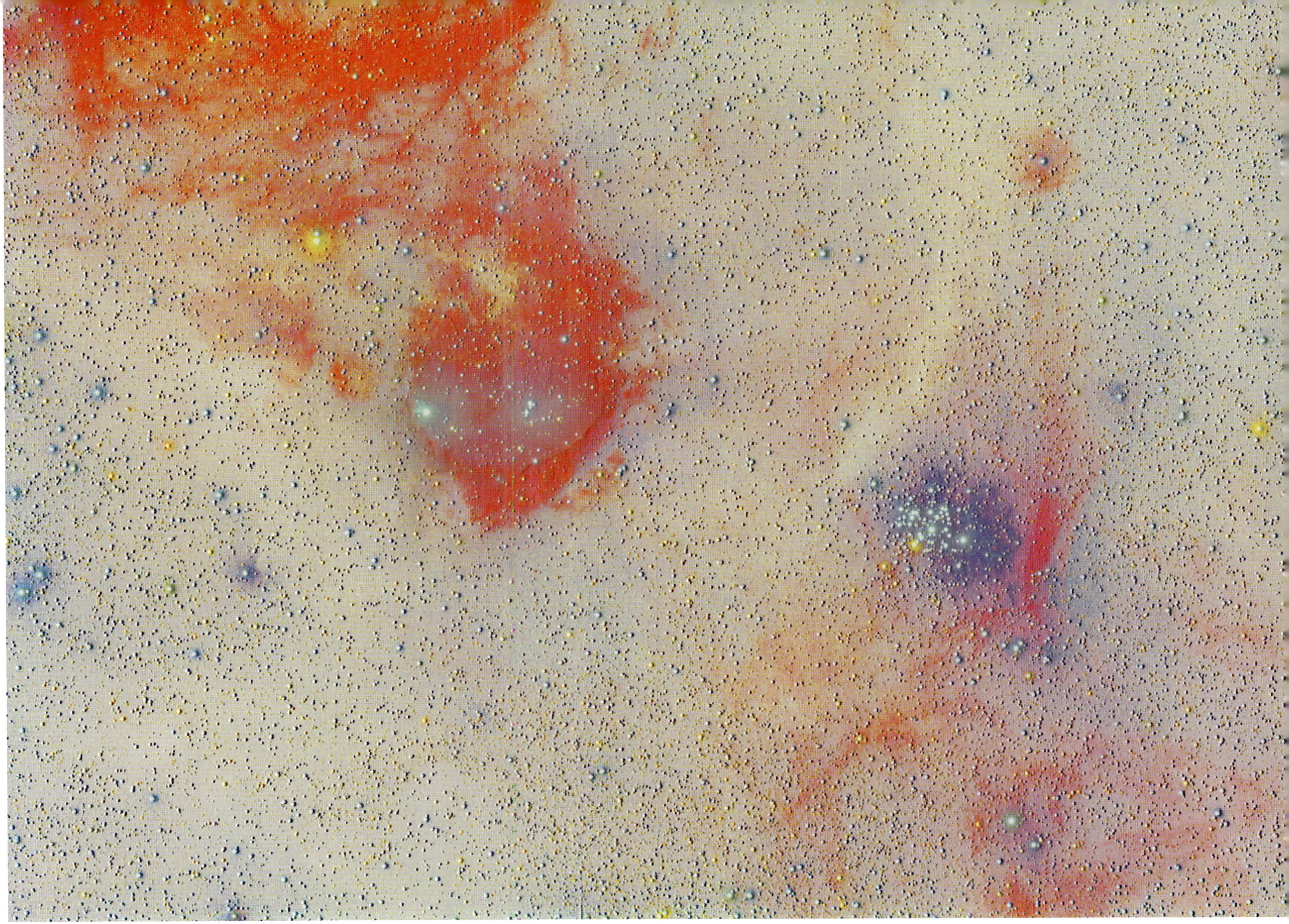

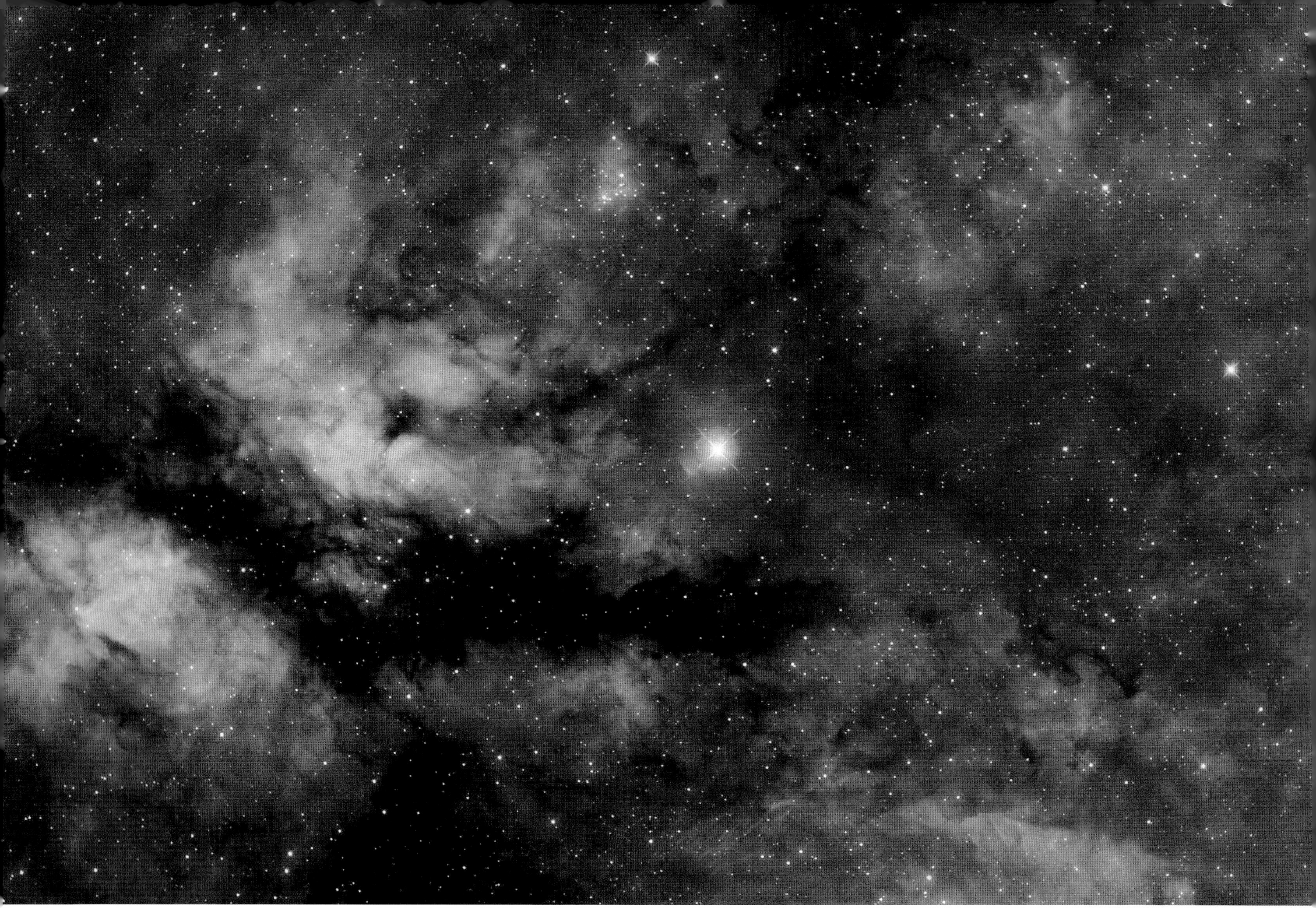

44. Walter KOPROLIN, *Cygni Nebulosity.*

60. Kirk ROGERS, *IC 2177, Seagull in Monoceros in Hydrogen-alpha.*

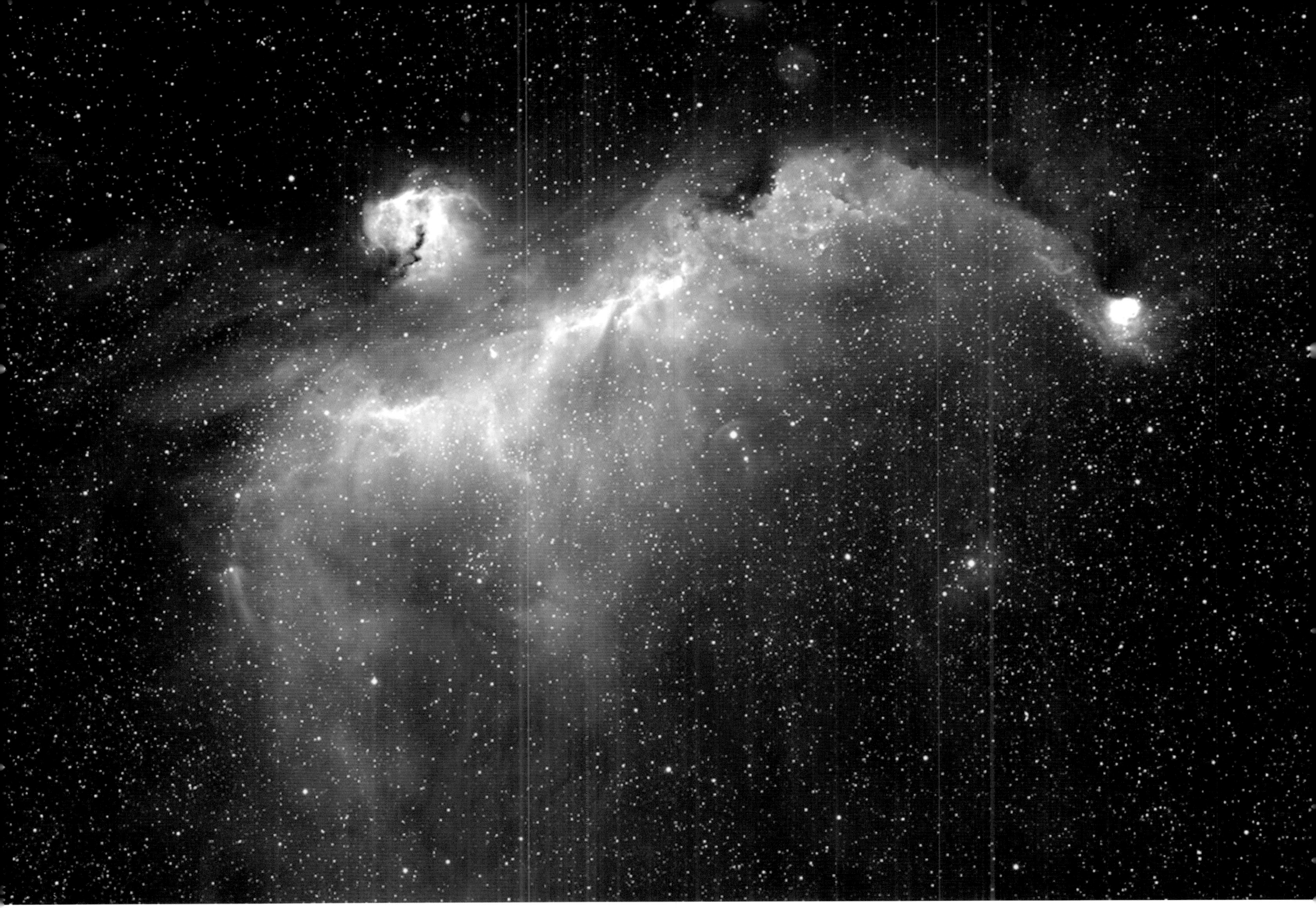

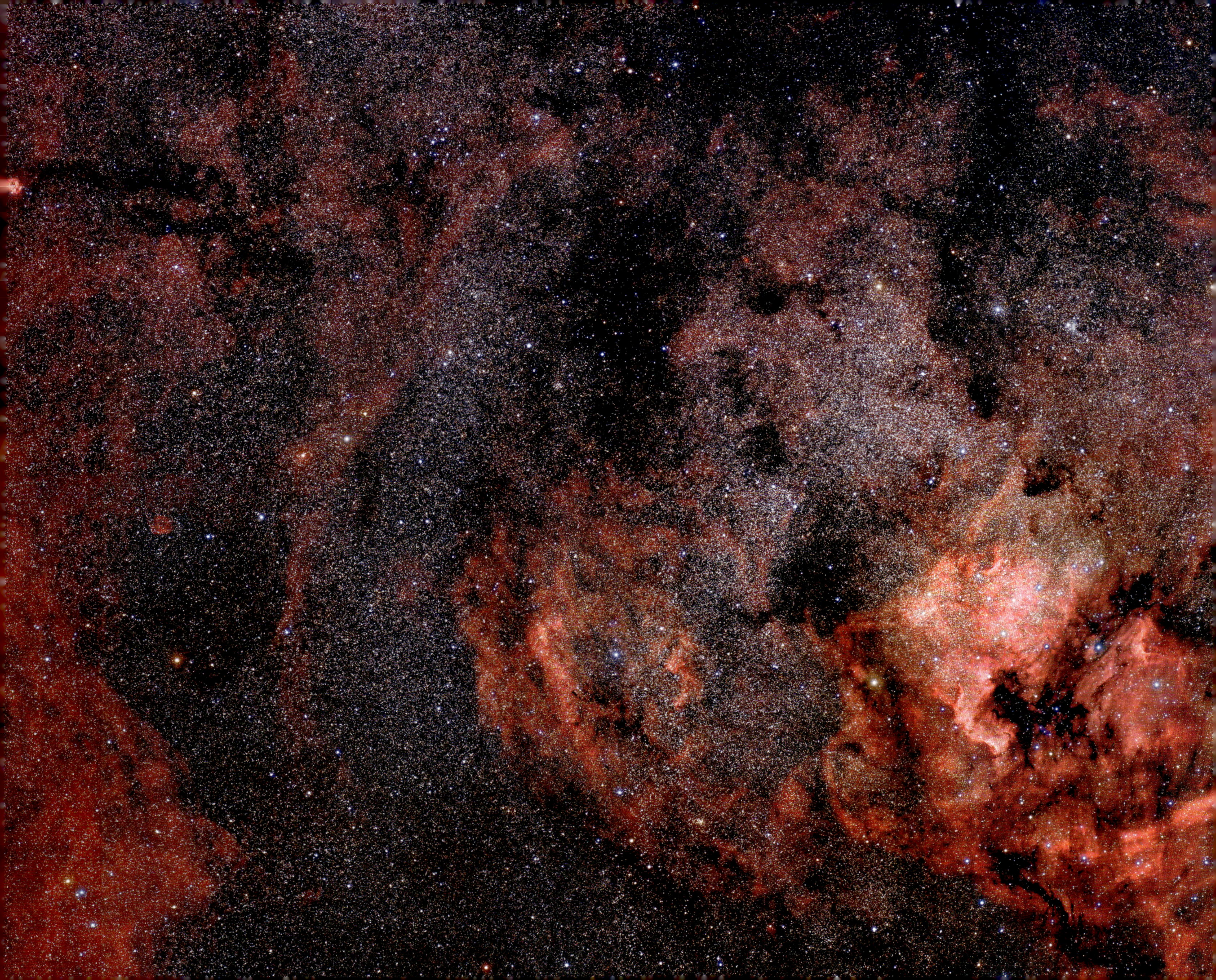

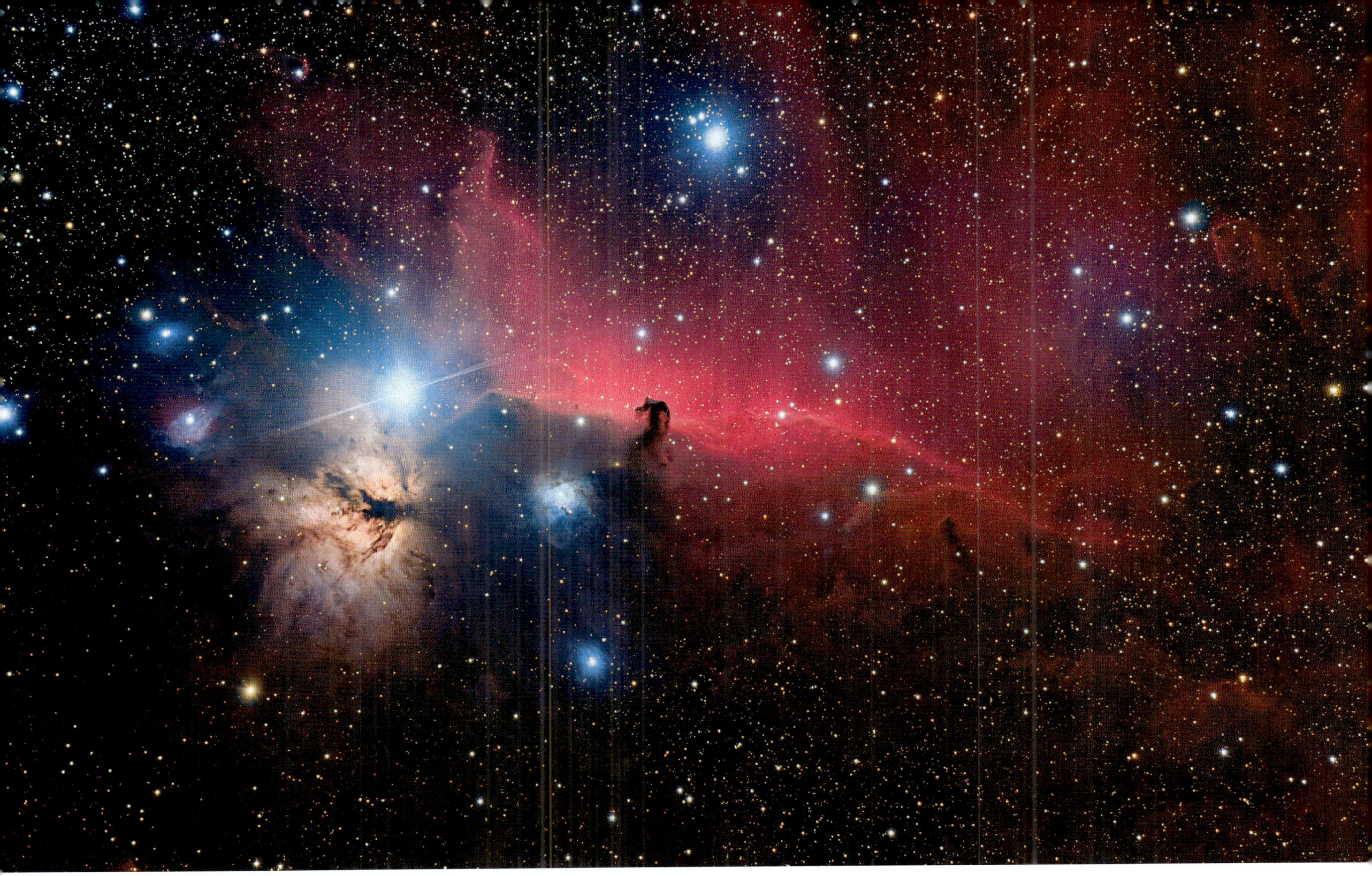

42. Warren KELLER, *Alnitak and Horsehead Region of Orion*, 2011.

18. John A. DAVIS, *Cocoon over North America, June-July*, 2011.

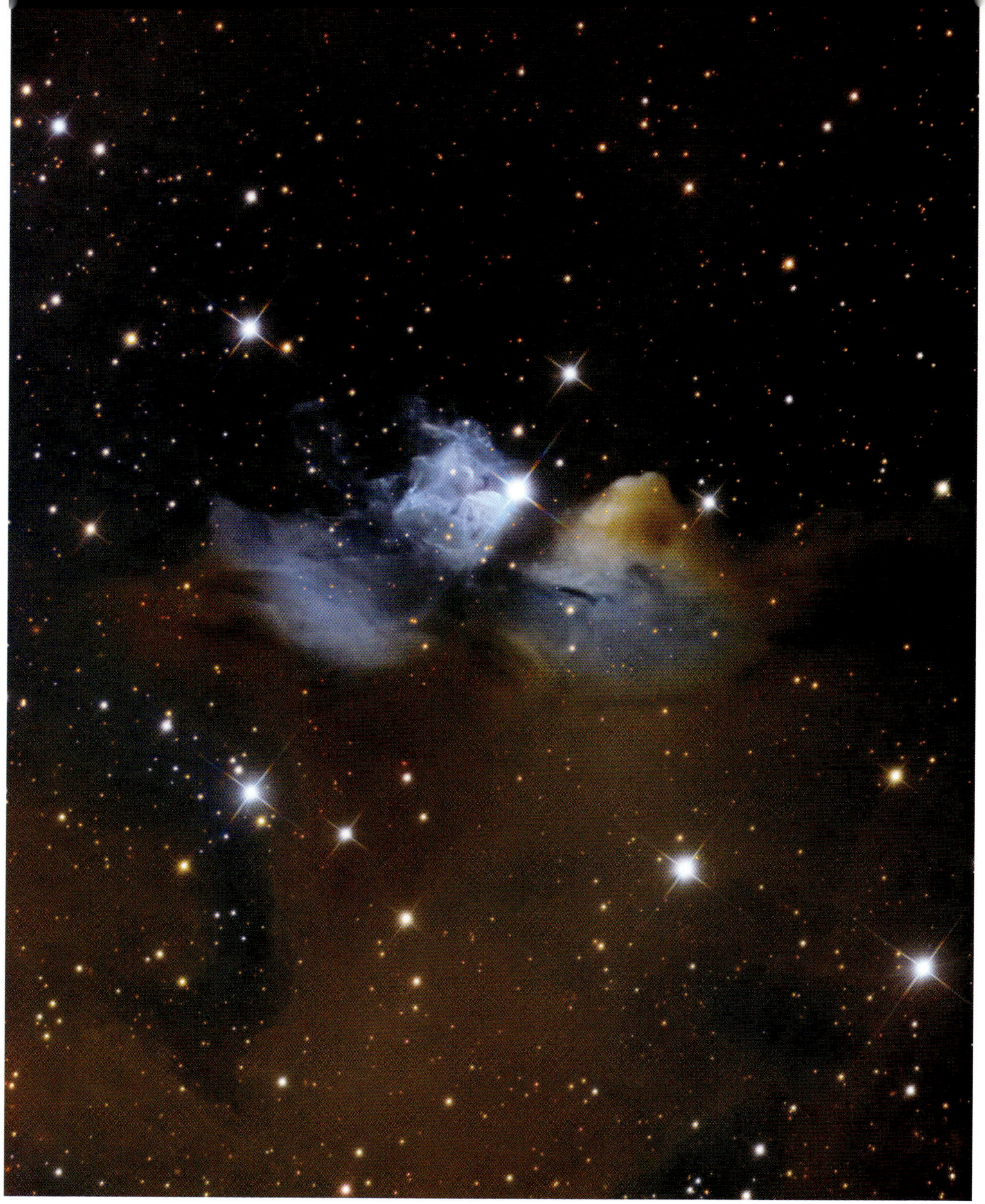

5. Adam D. BLOCK, *Gift (vdB 24)*.

34. Don GOLDMAN, *Cosmic Dust*.

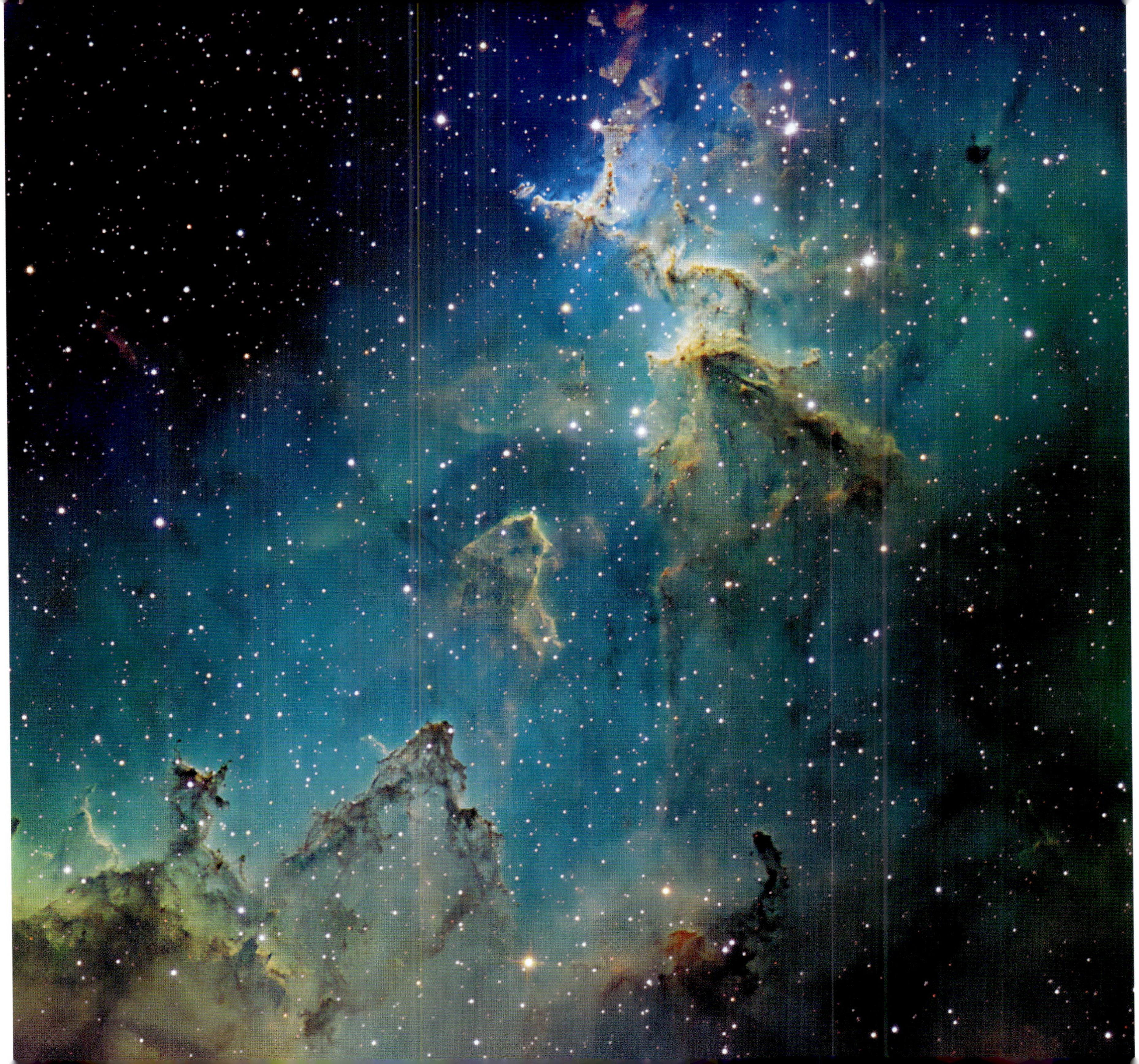

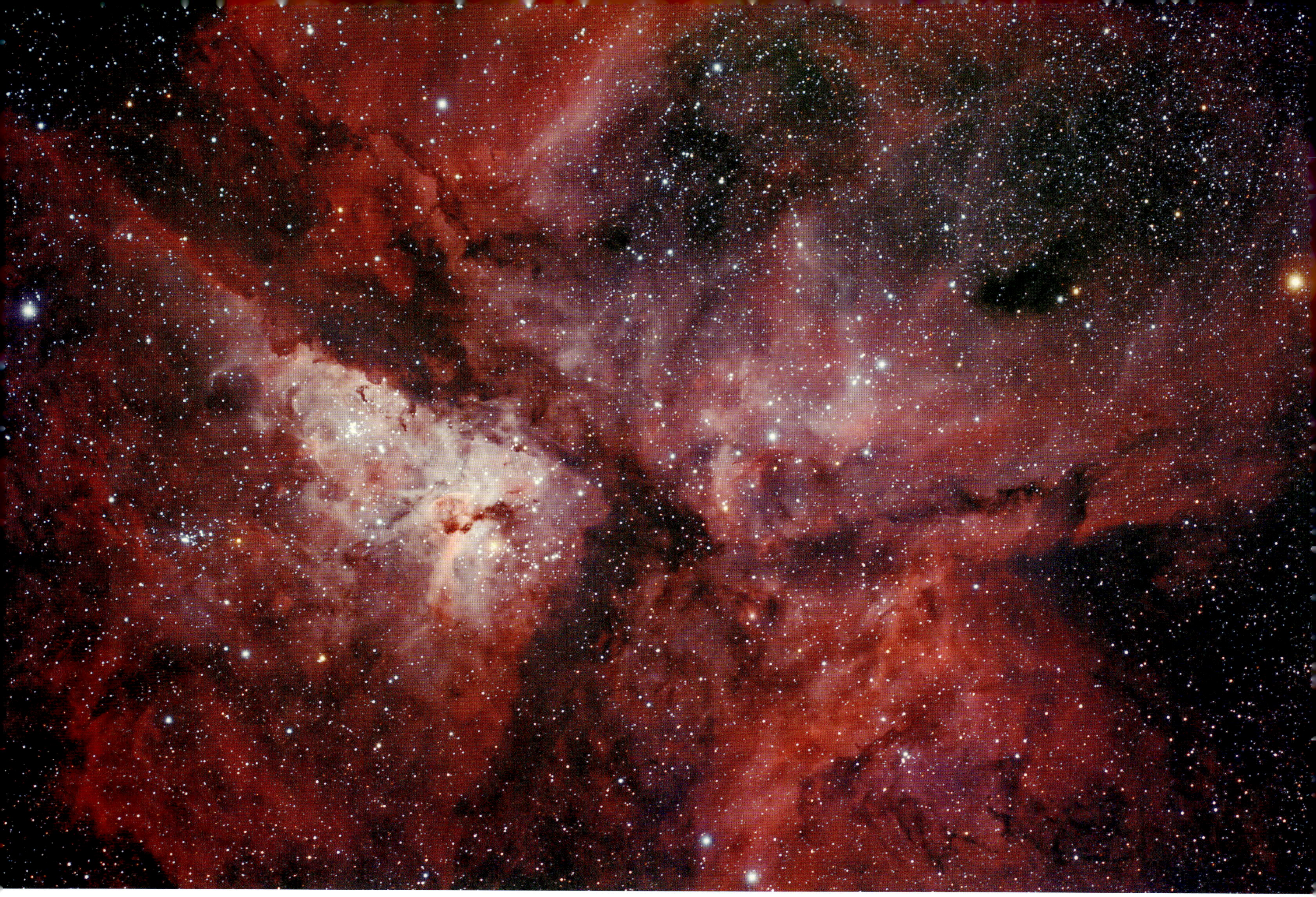

22. Ignacio DIAZ-BOBILLO, *Light Drops #2-A*; and

23. *Light Drops #2-B*.

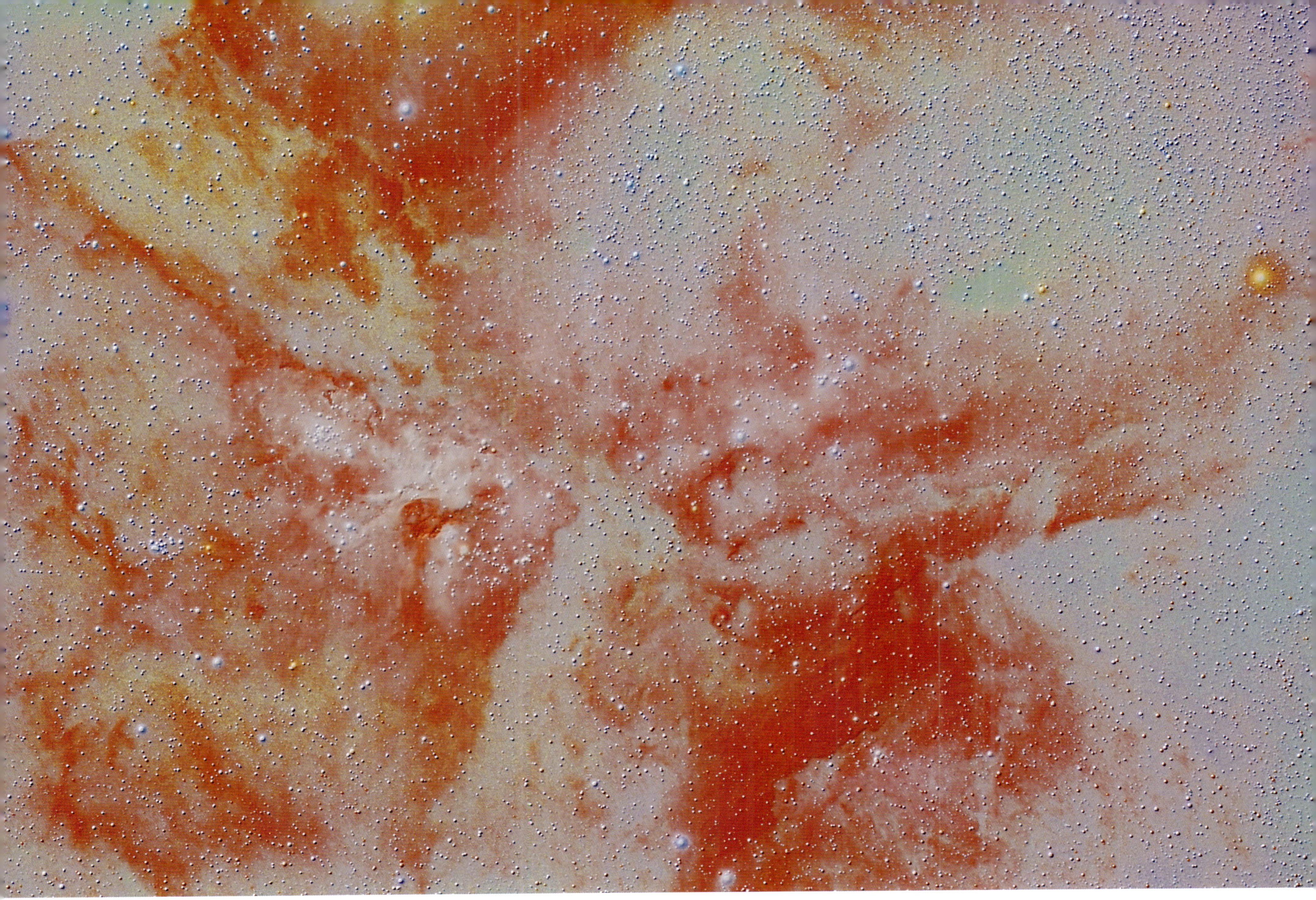

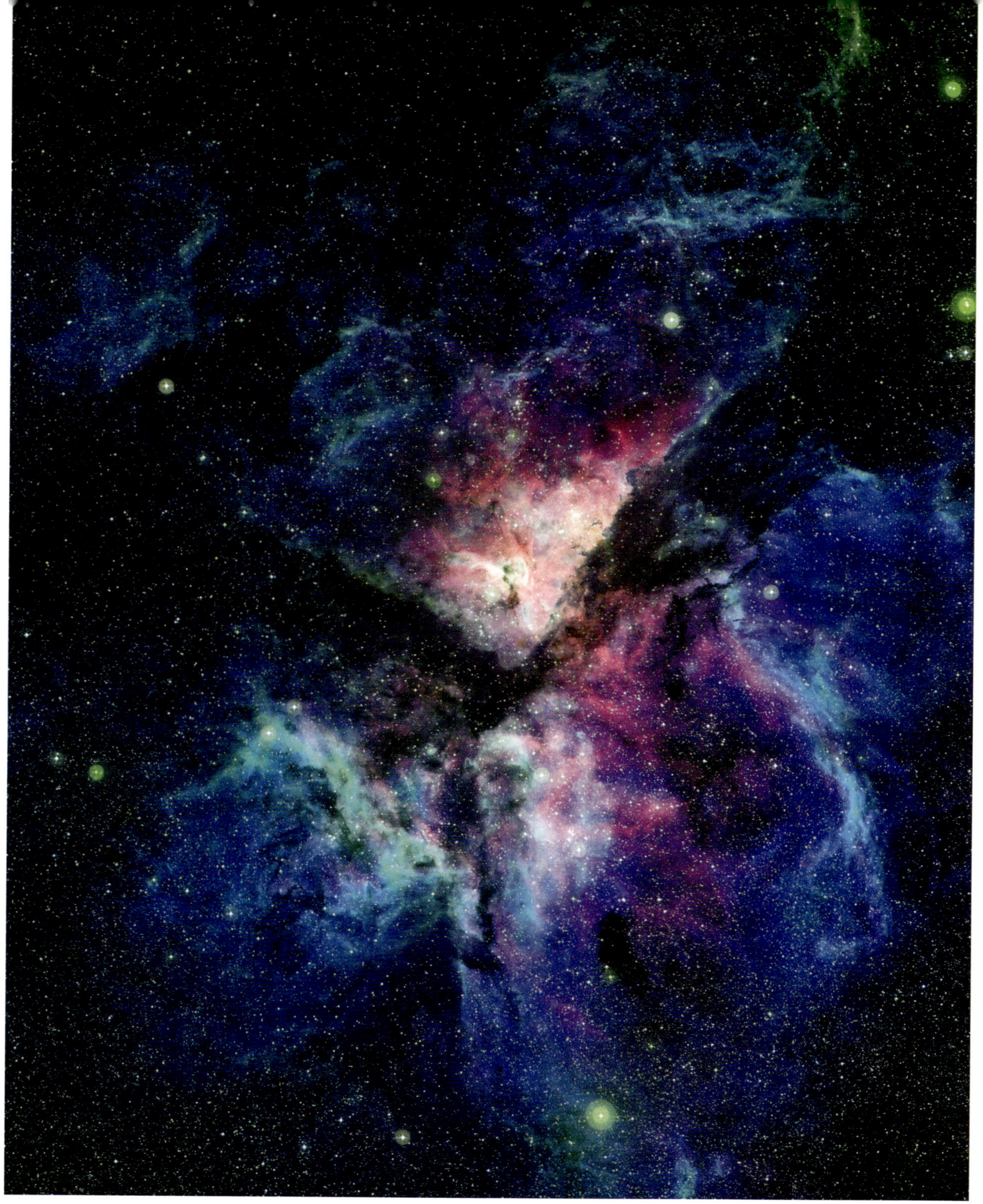

46. David MALIN, *The Elemental Structure of the Carina Nebula*. © Australian Astronomical Observatory.

35. Don GOLDMAN, *Statue of Liberty in Space*.

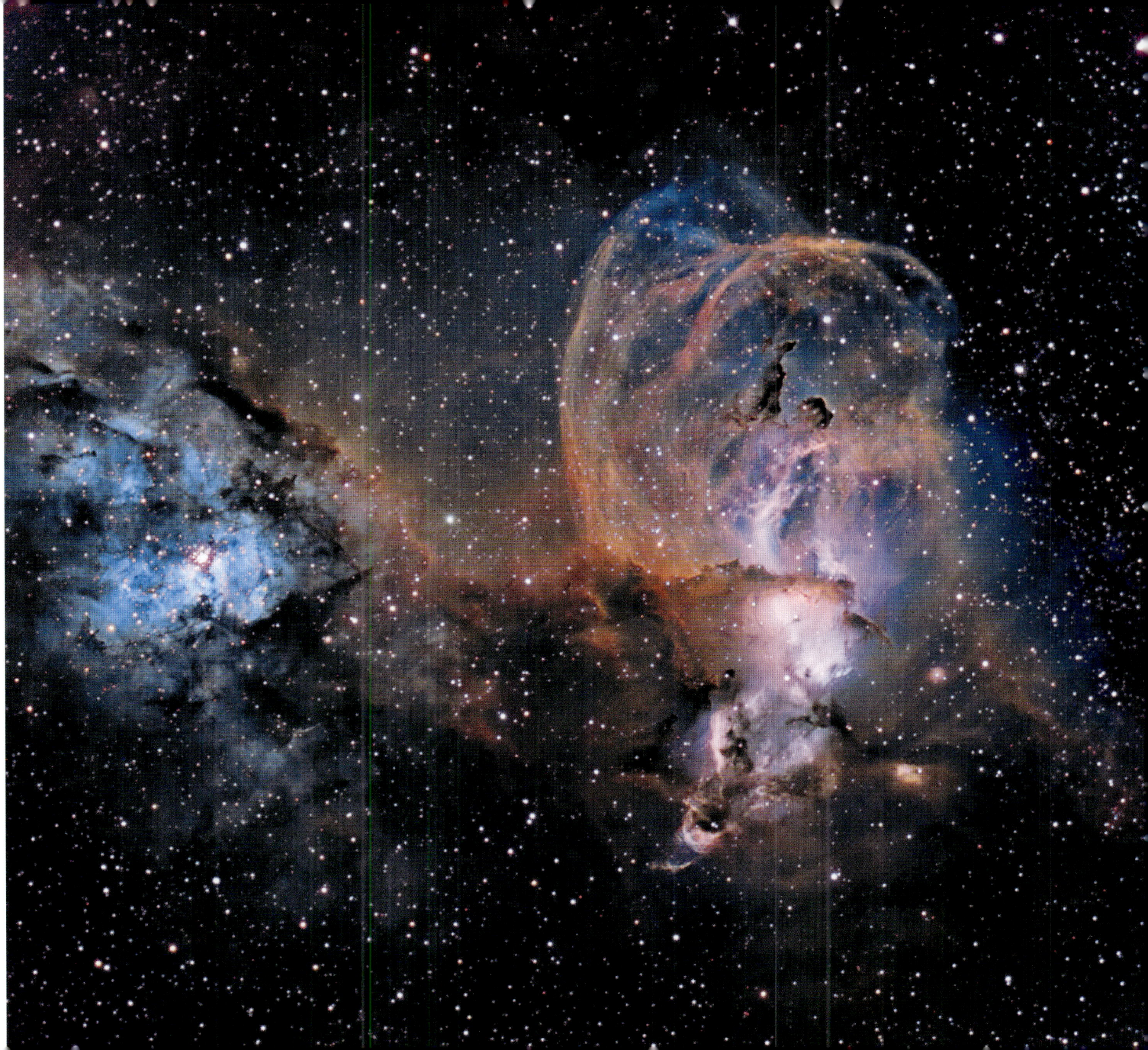

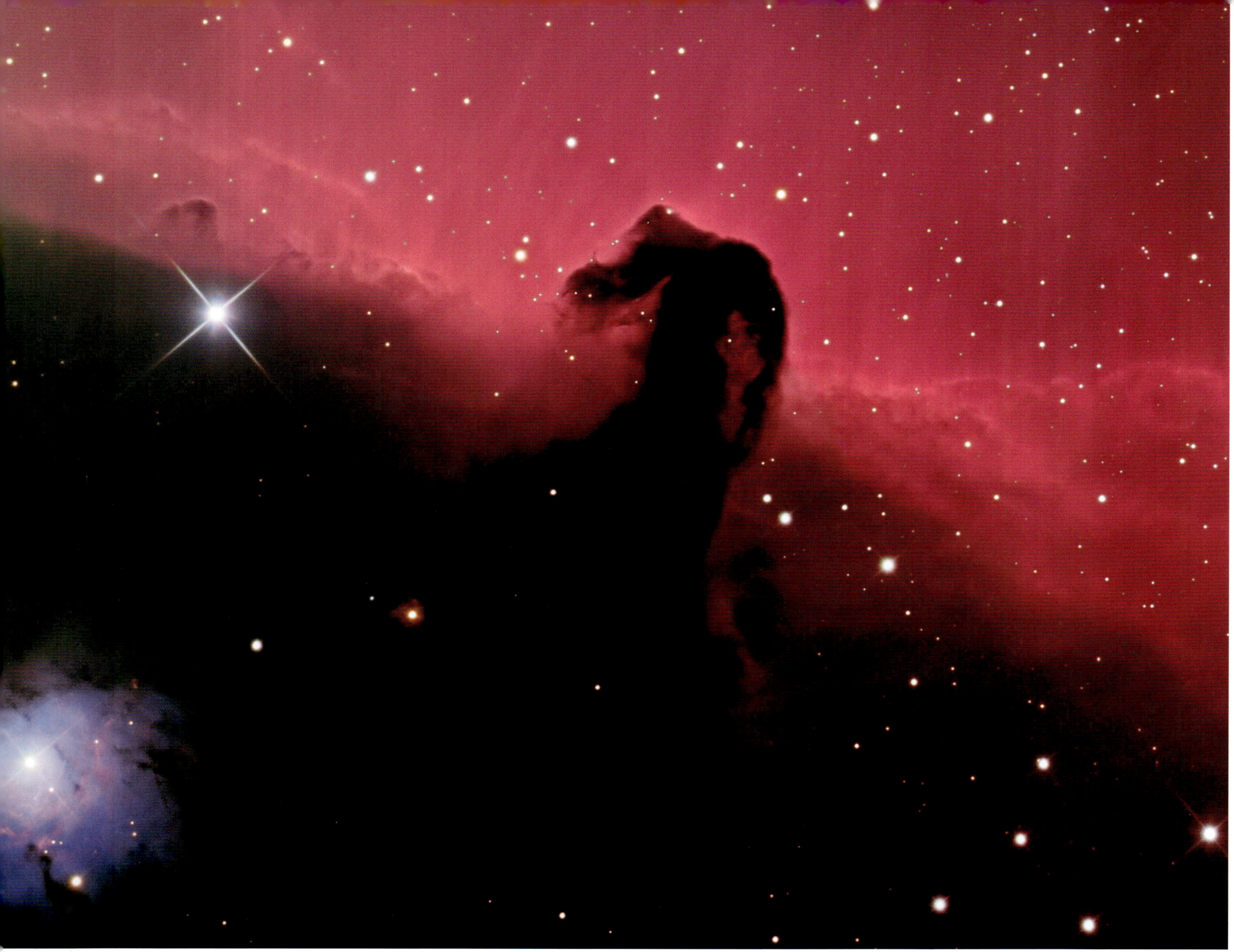

57. Martin PUGH, *IC 434 in Orion (The Horsehead Nebula)*.

29. Robert GENDLER, *M42, The Great Nebula in Orion*, 2005.

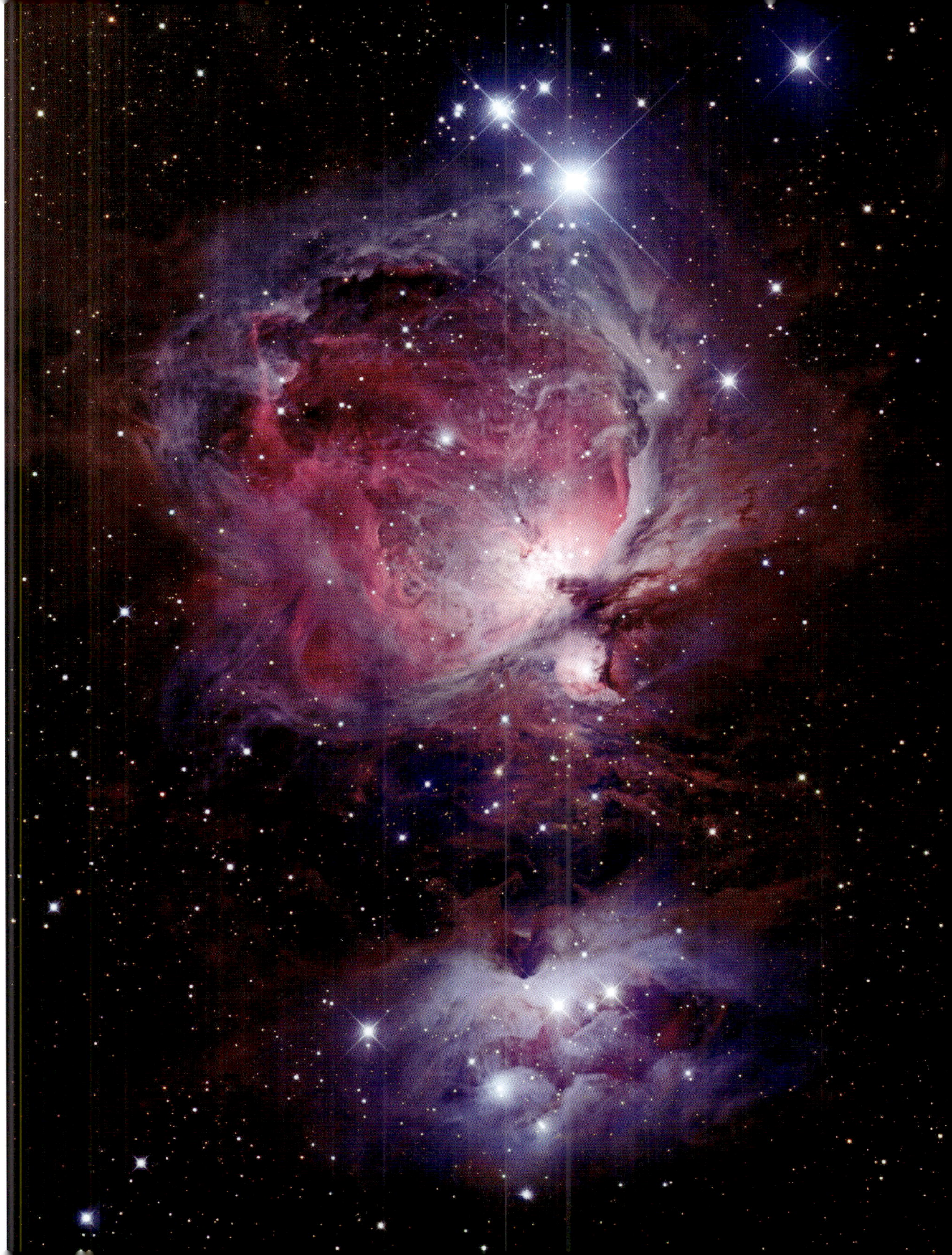

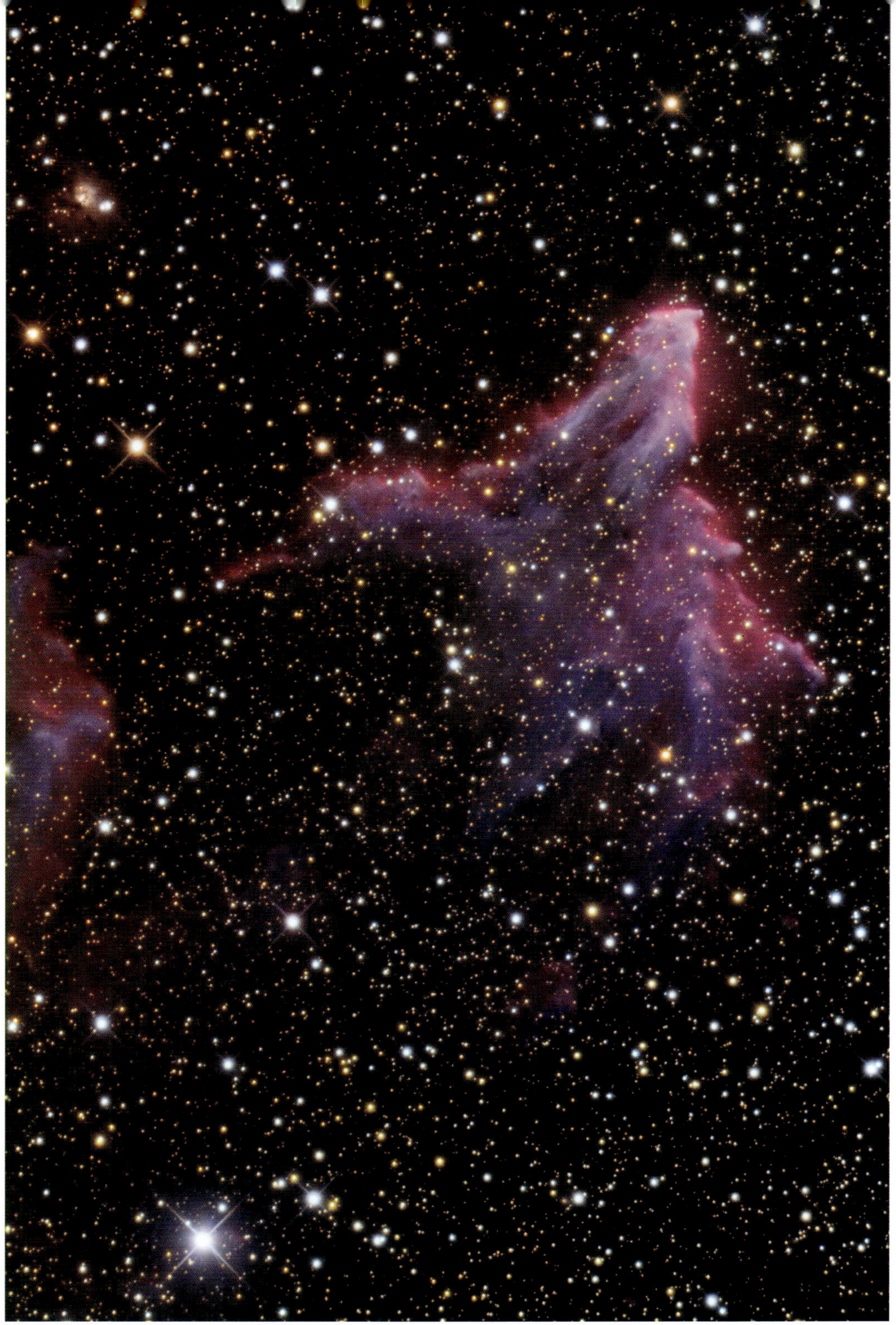

6. Adam D. BLOCK, *Galactic Glide (IC 63)*.

41. Warren KELLER, *Seven Sisters*, 2011.

Stars

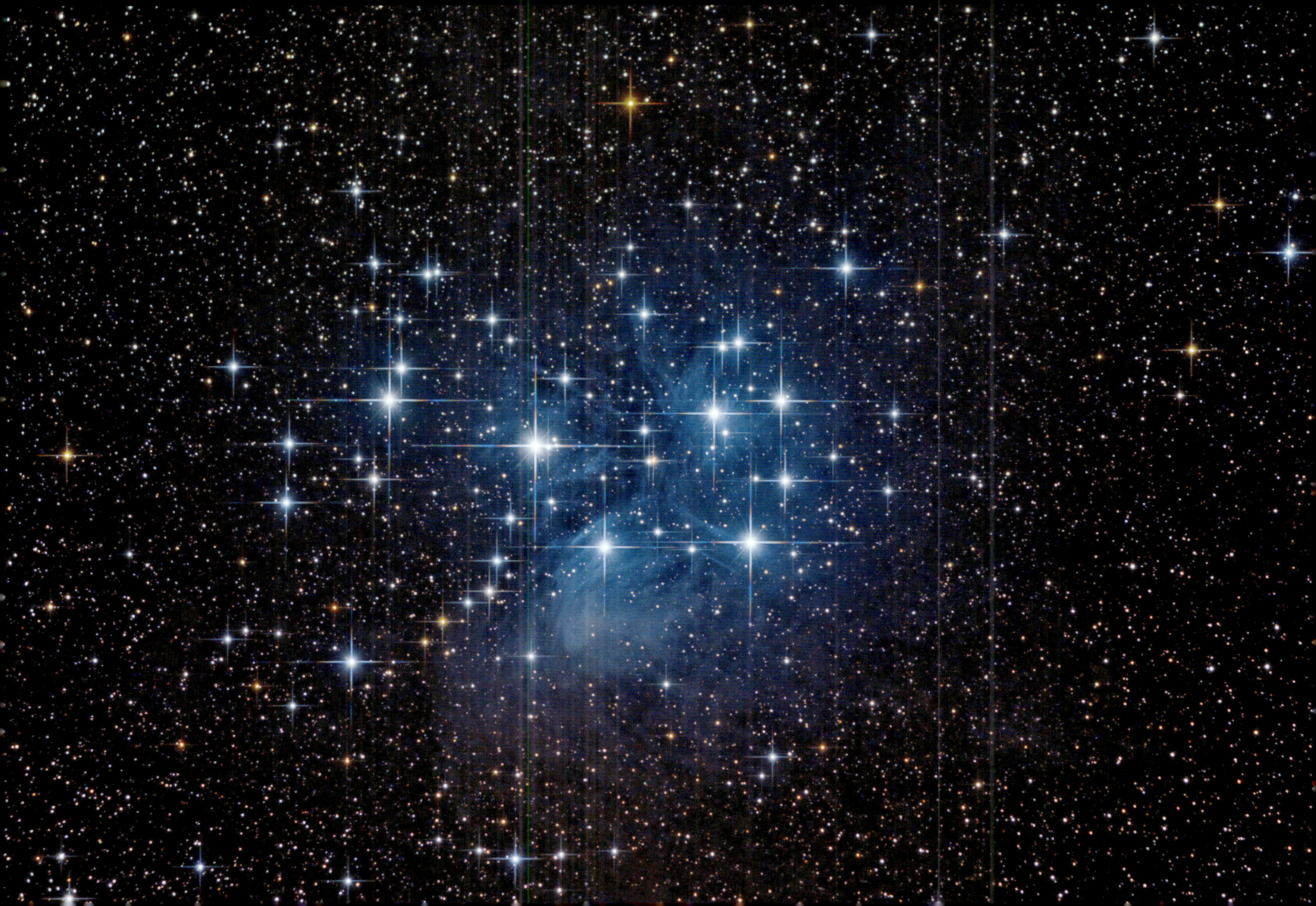

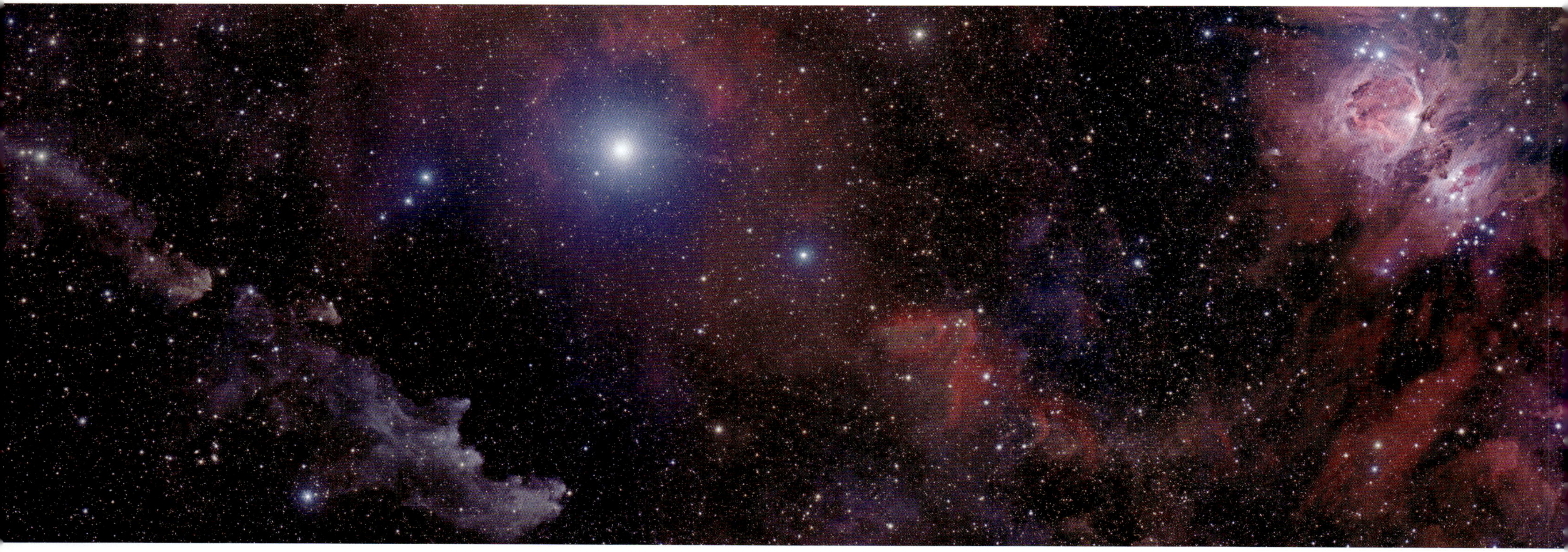

19. John A. DAVIS, *Sword to Witch*, 2010.

47. David MALIN, *The Globular Cluster Omega Centauri, NGC 5139*. © Australian Astronomical Observatory.

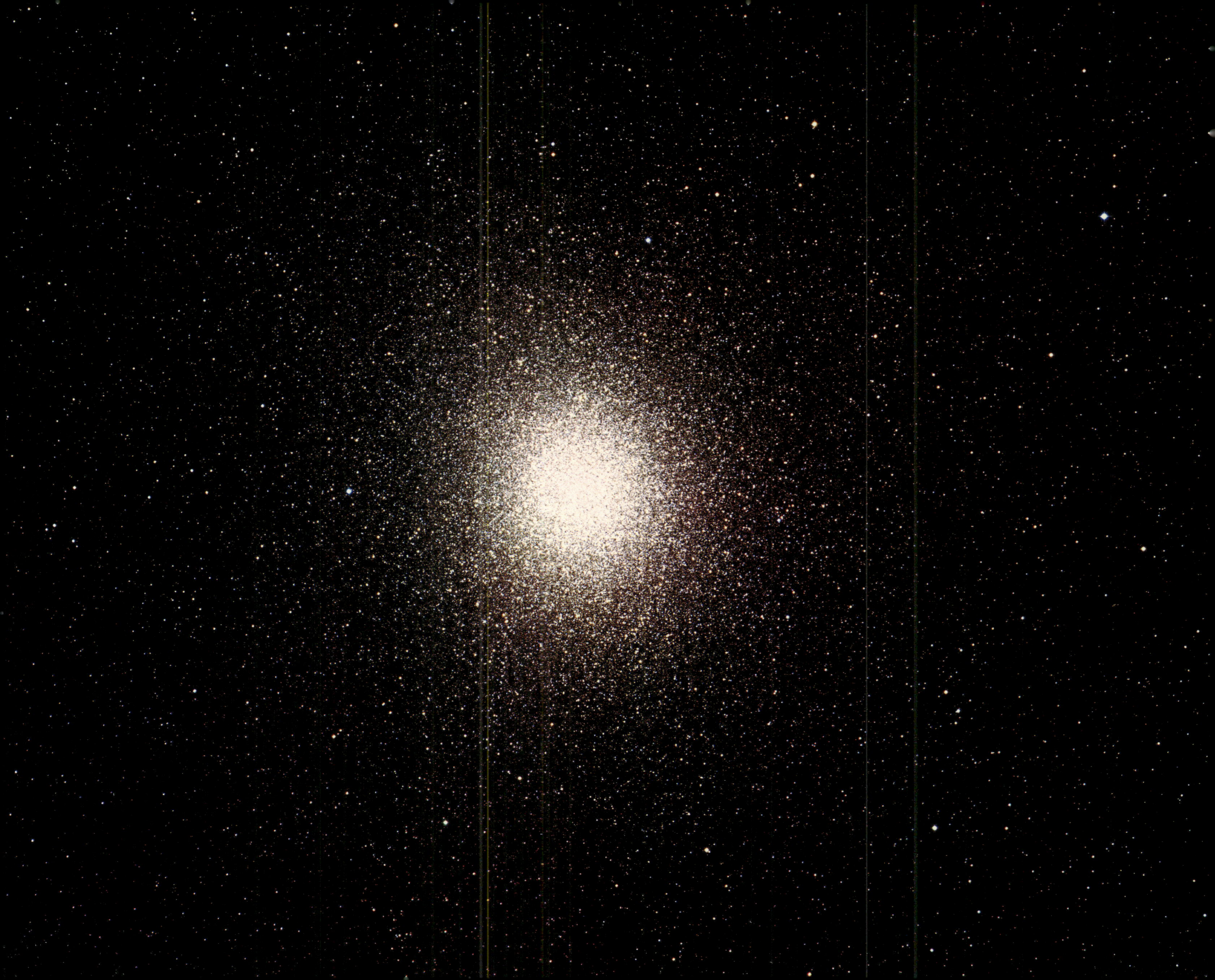

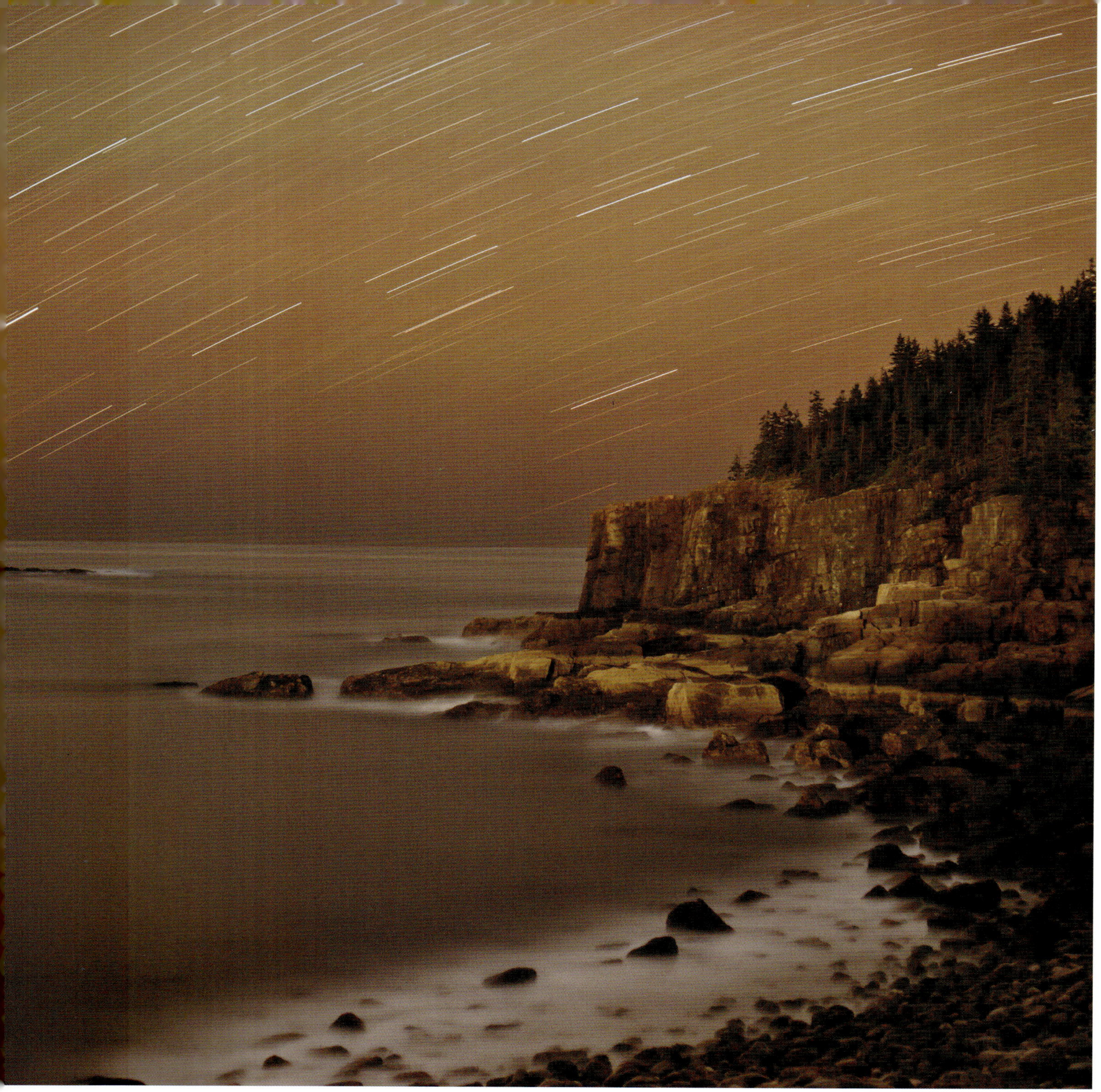

51. Jim NICKELSON, *Nightfall, Otter Cliffs*, 2011.

10. Linda CONNOR, *Open Dome with Star Trails, Caravansary, Turkey*, 2004.

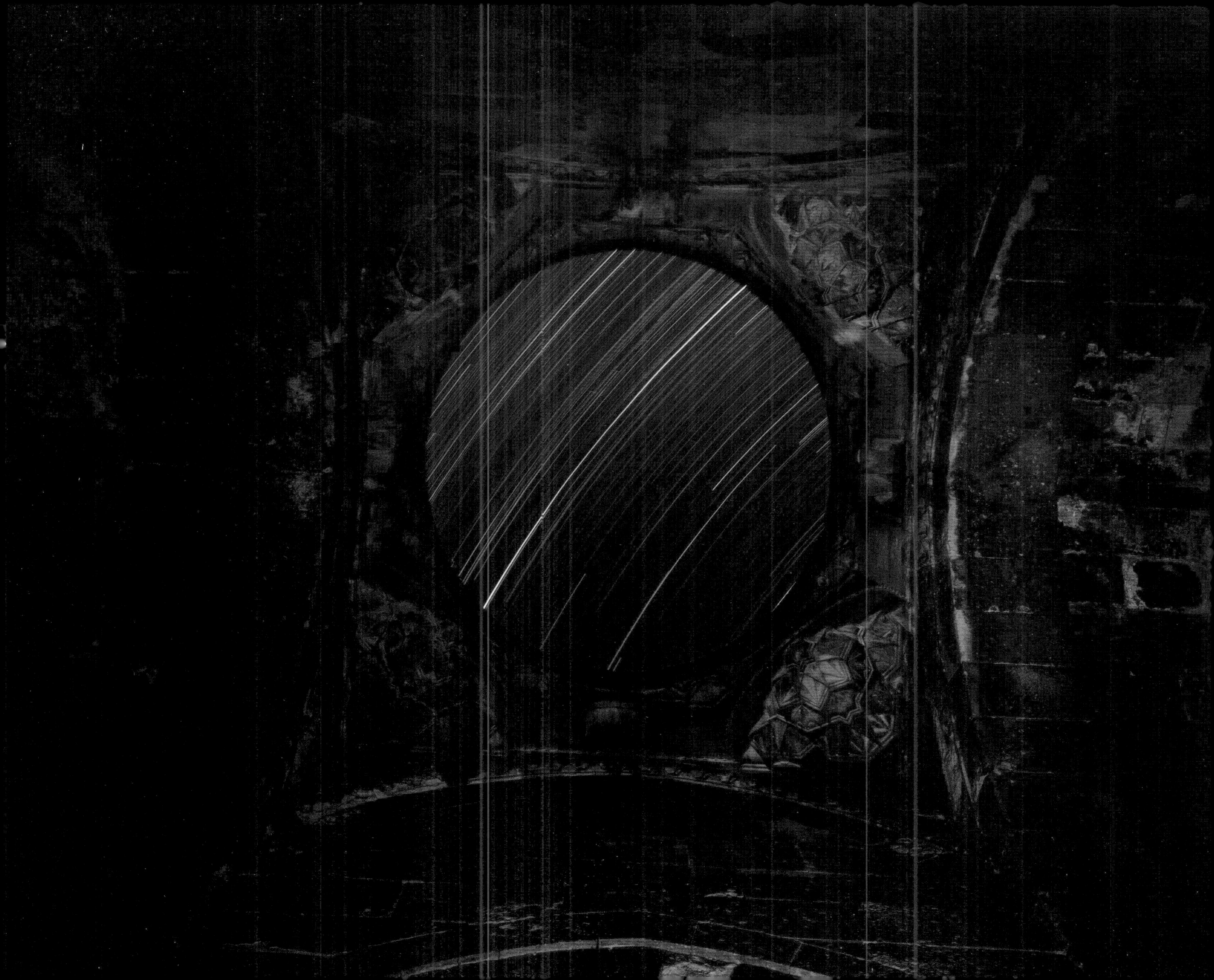

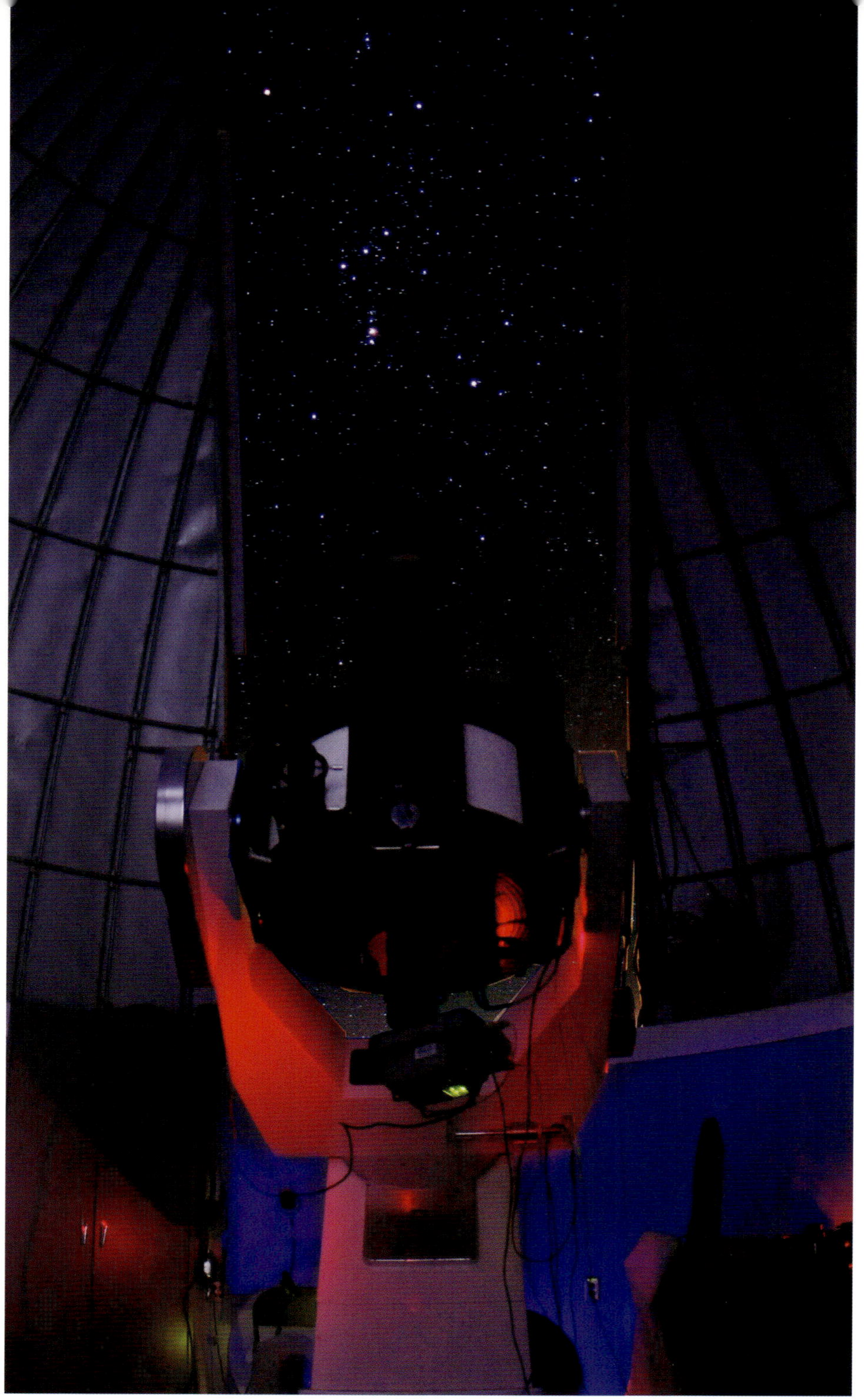

42. Warren KELLER, *Alnitak and Horsehead Region of Orion*, 2011.

Star Death

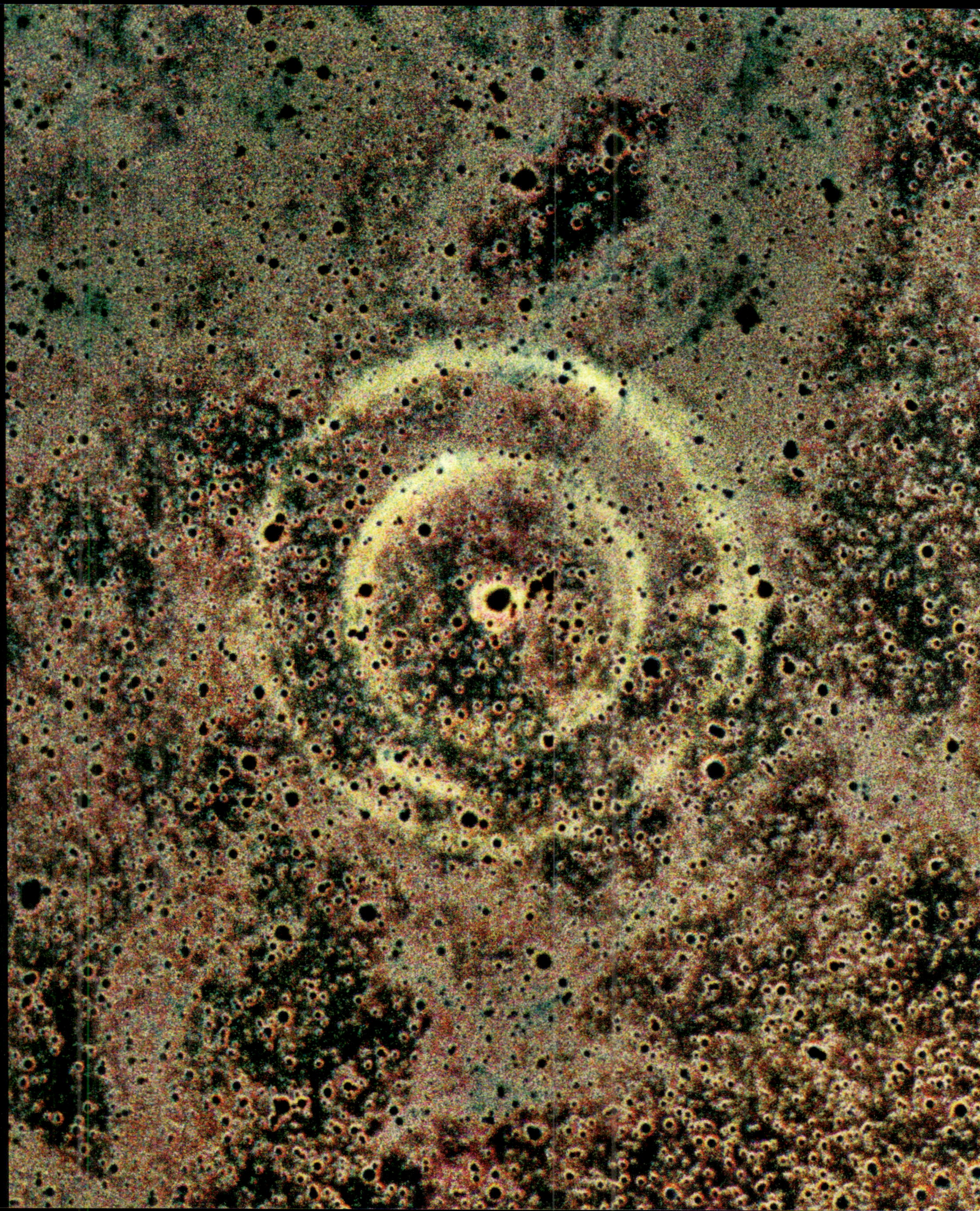

45. David MALIN, *The Light Echo of Supernova 1987A*, © Australian Astronomical Observatory.

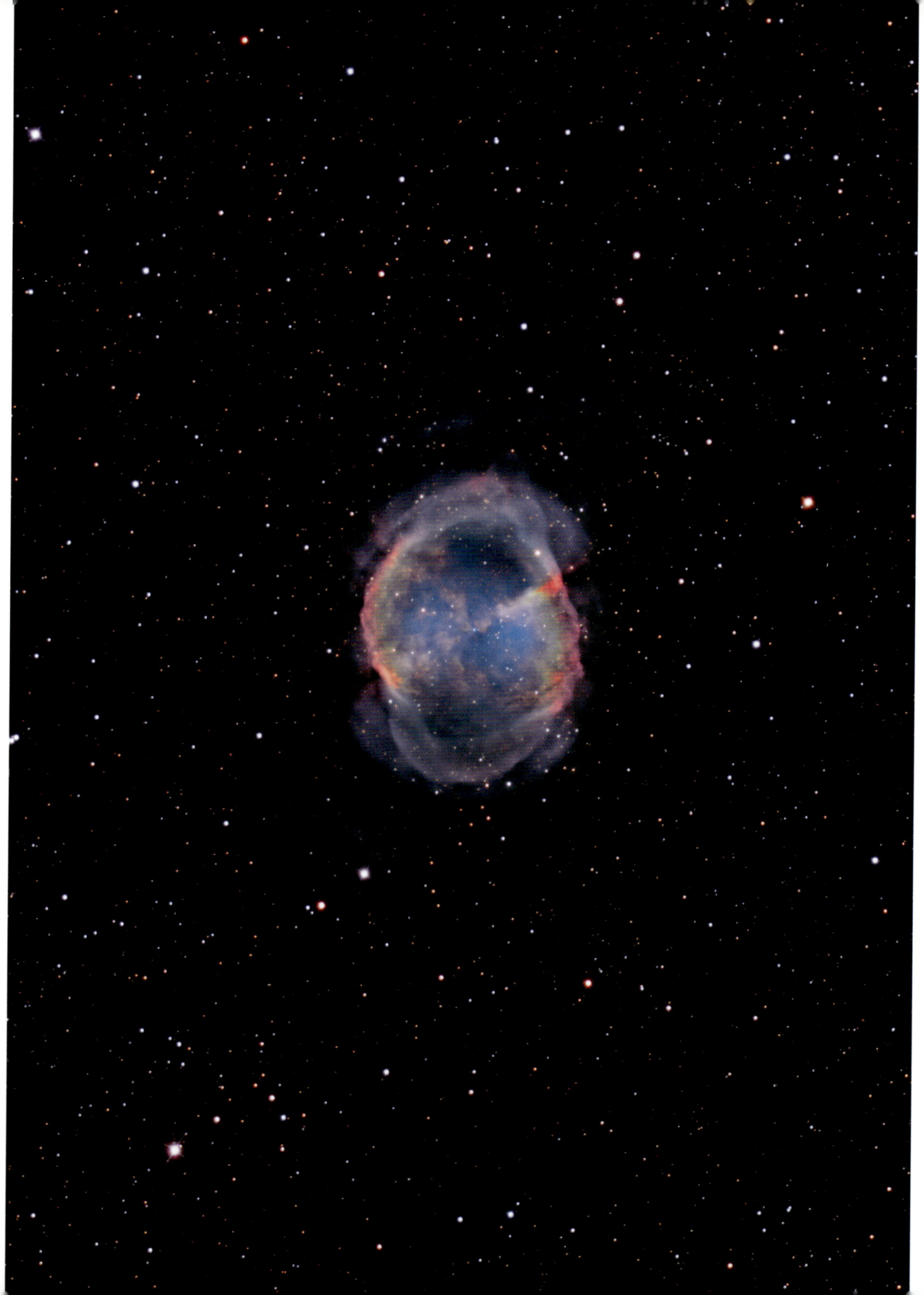

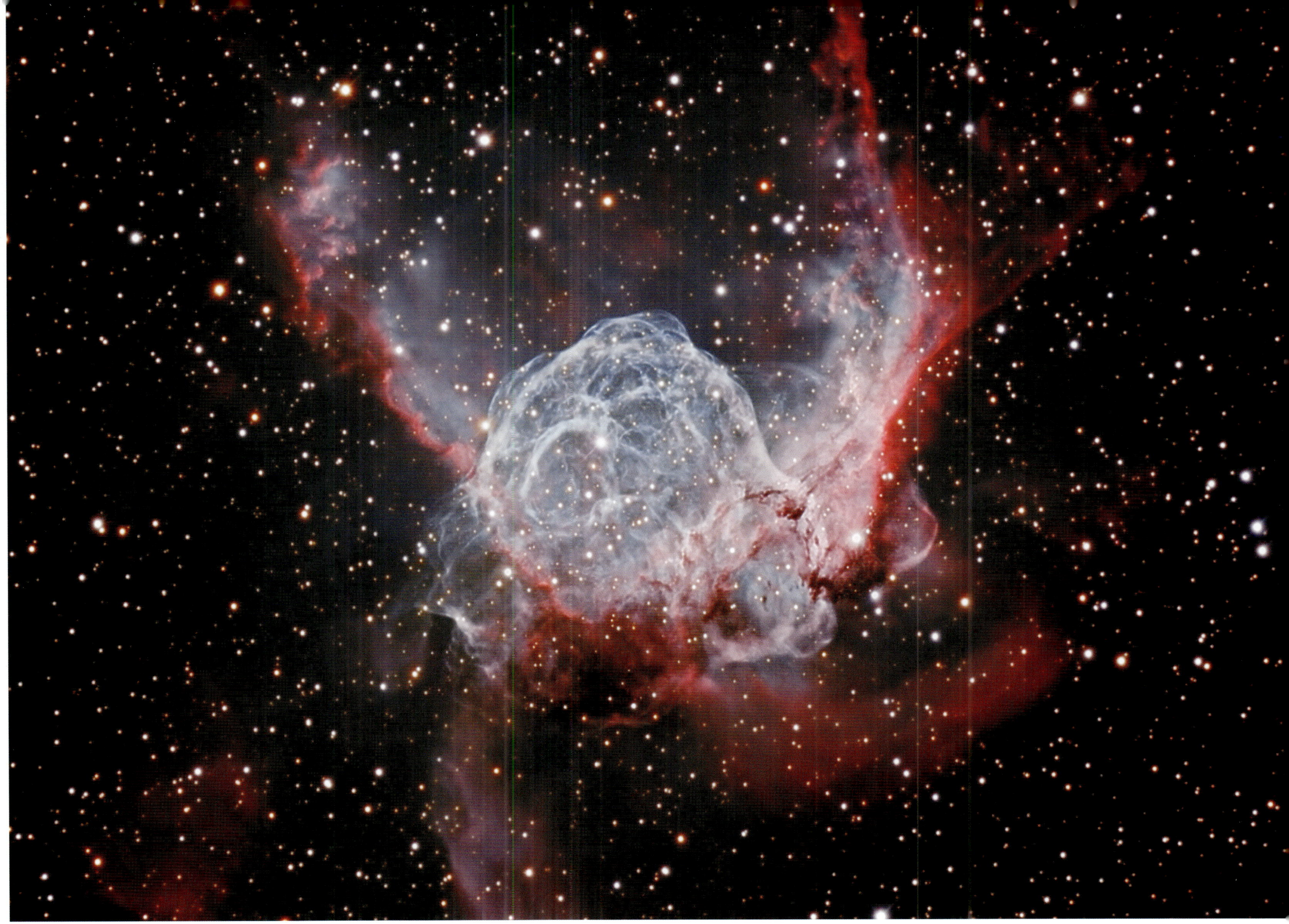

48. Steven MAZLIN & Jack HARVEY, *Thor's Helmet in Canis Major (aka NGC 2359).*

[Within Thor's Helmet is a Wolf-Rayet star believed to be in a pre-supernova stage of evolution.—Editor]

 Martin Pugh, *M27 in Vulpecula (The Dumbbell Nebula).*

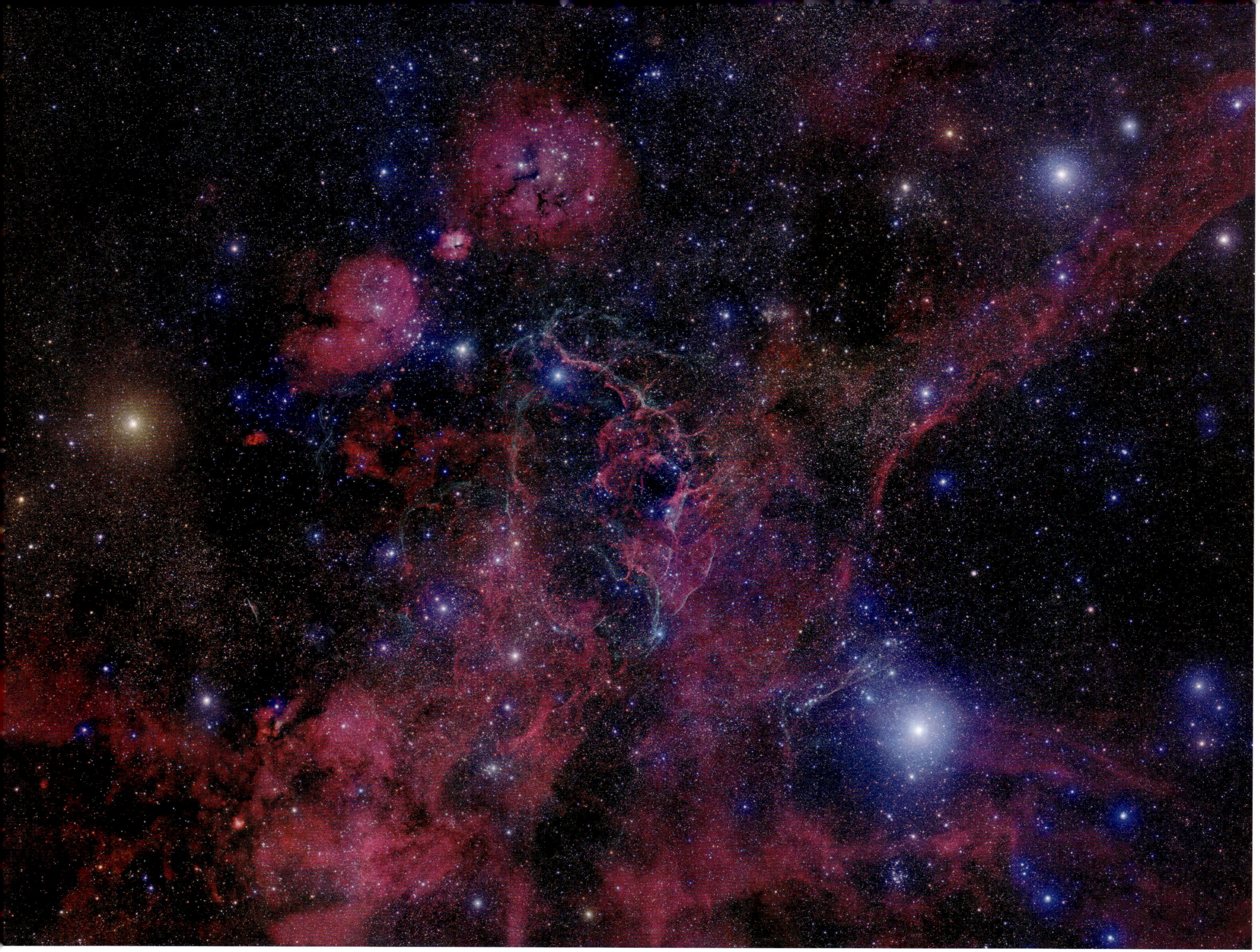

32. Robert GENDLER, *Vela Supernova Remnant*, 2008.

36. Don GOLDMAN, *Witch's Broom.*

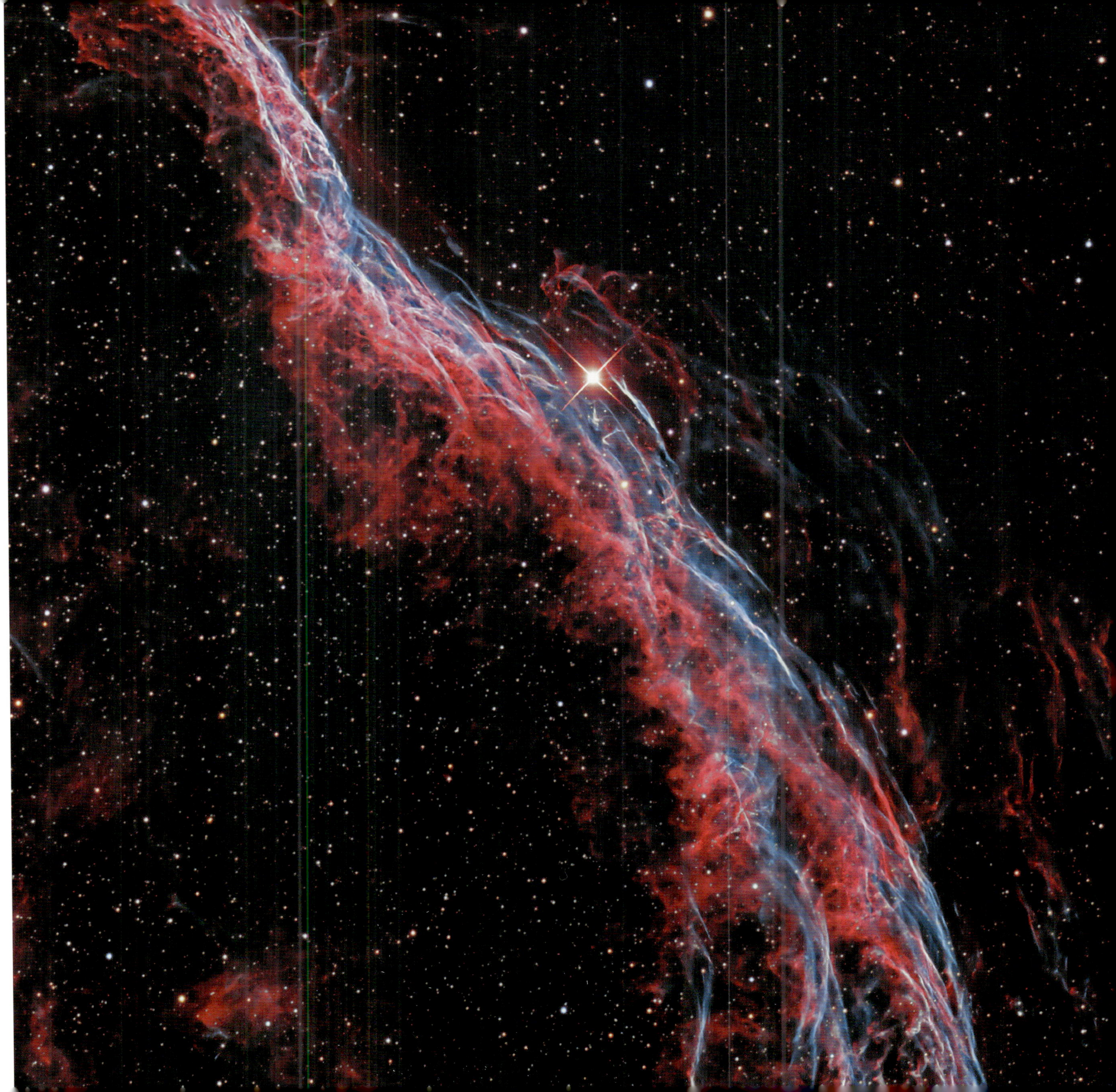

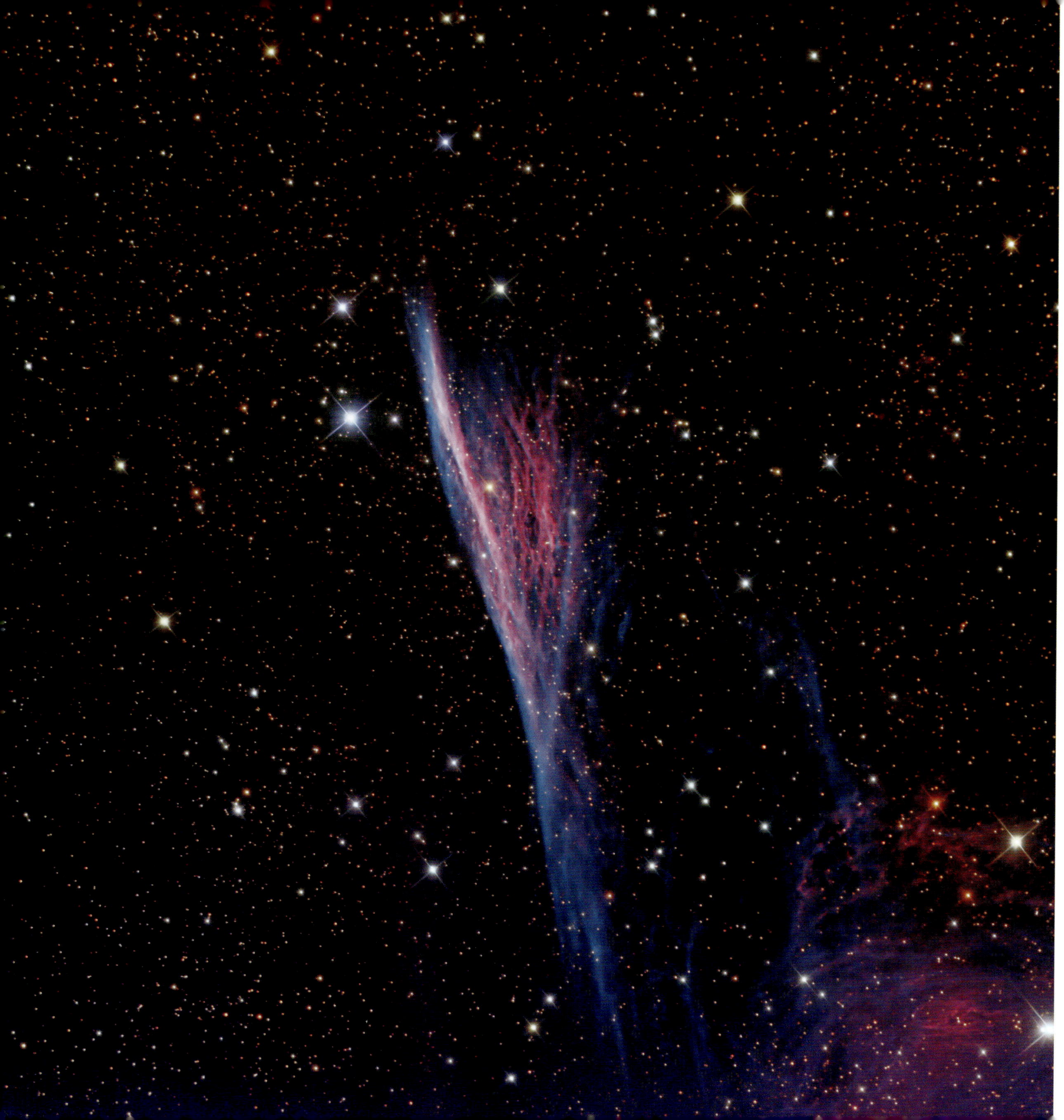

17. Ken CRAWFORD, *NGC 2736*.

Galaxies

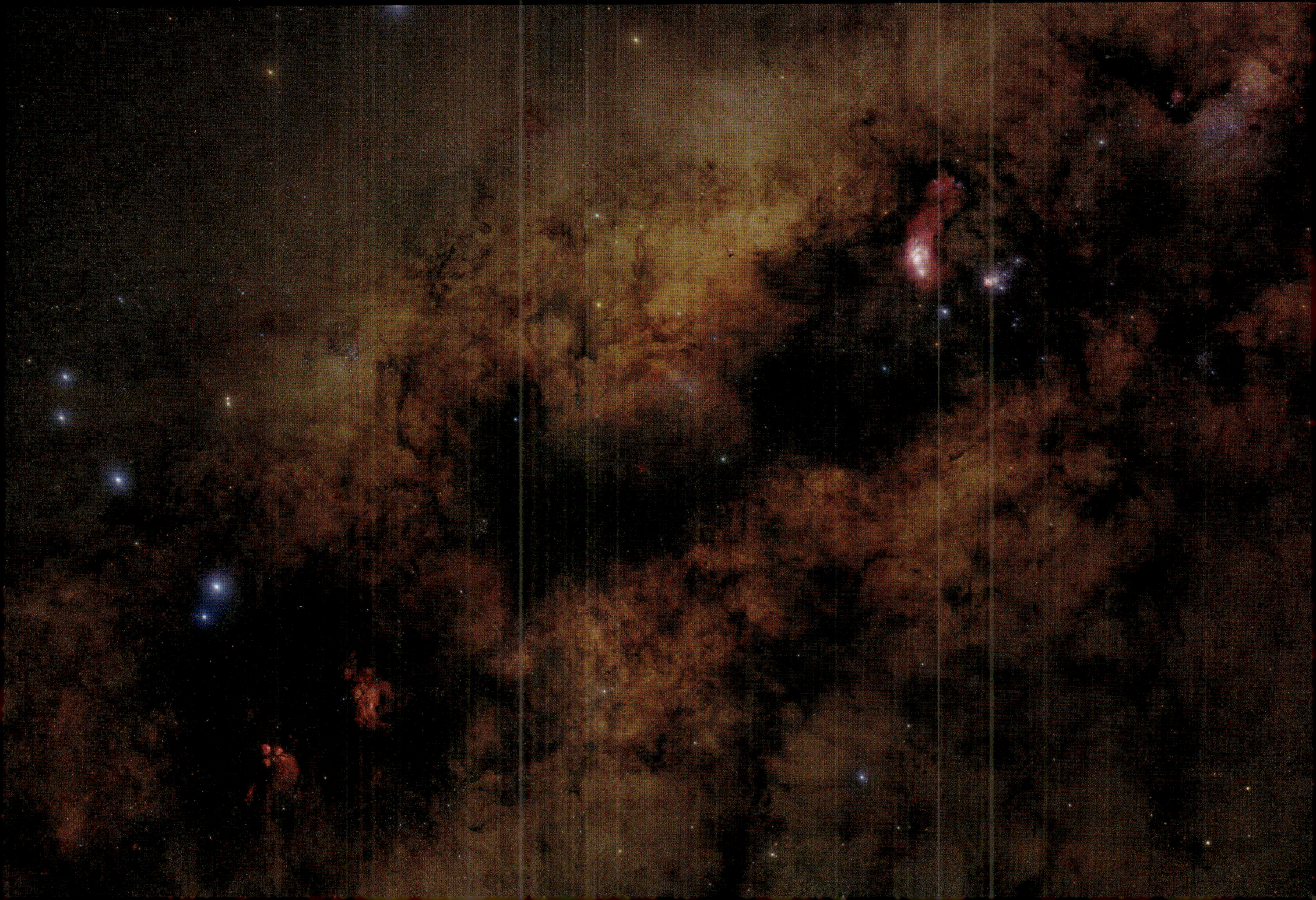

30. Robert GENDLER, *The Galactic Center*, 2009.

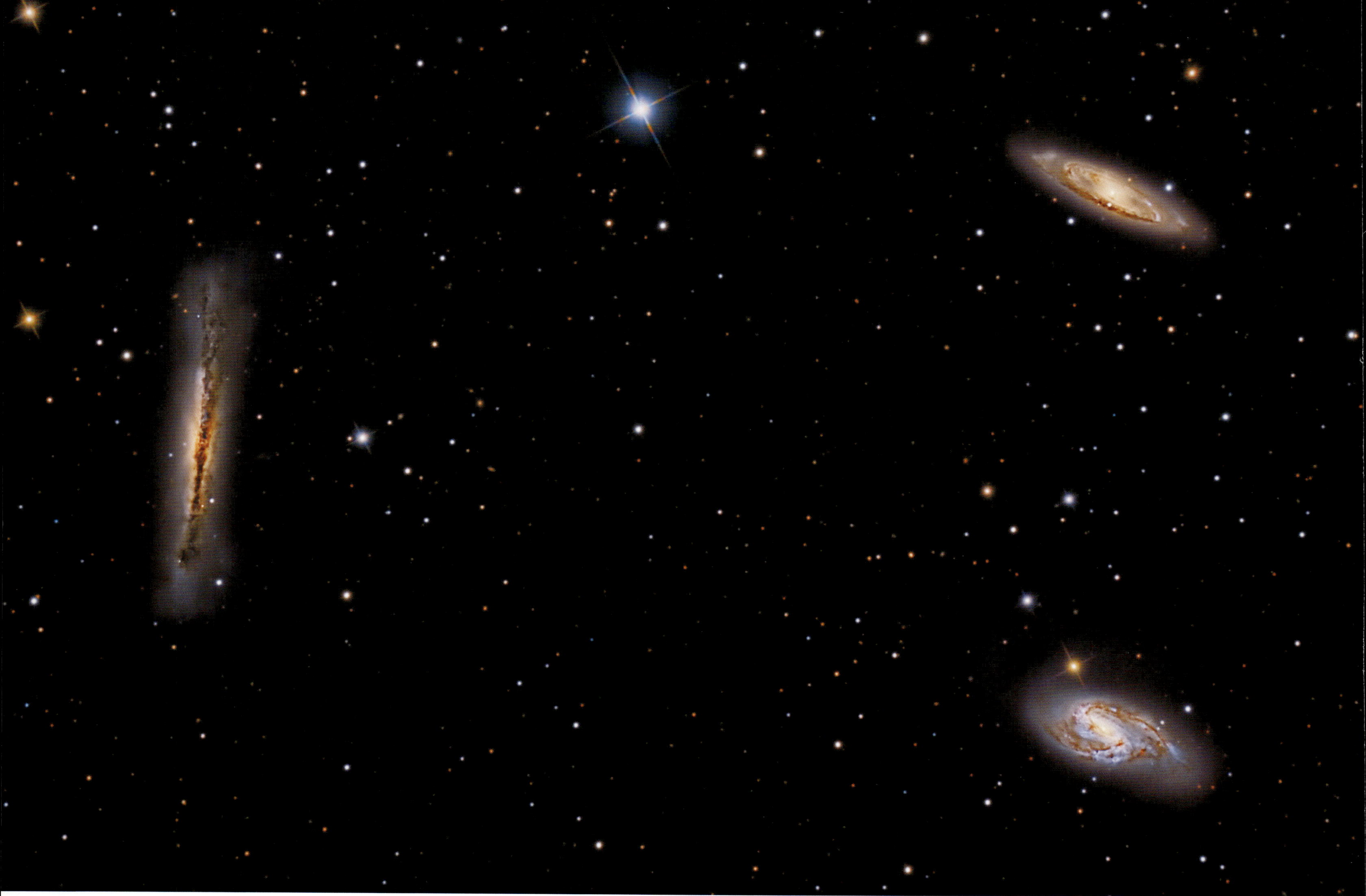

43. Warren KELLER, *Leo Triplet*, 2011.

31. Robert GENDLER, *The Andromeda Galaxy*, 2005.

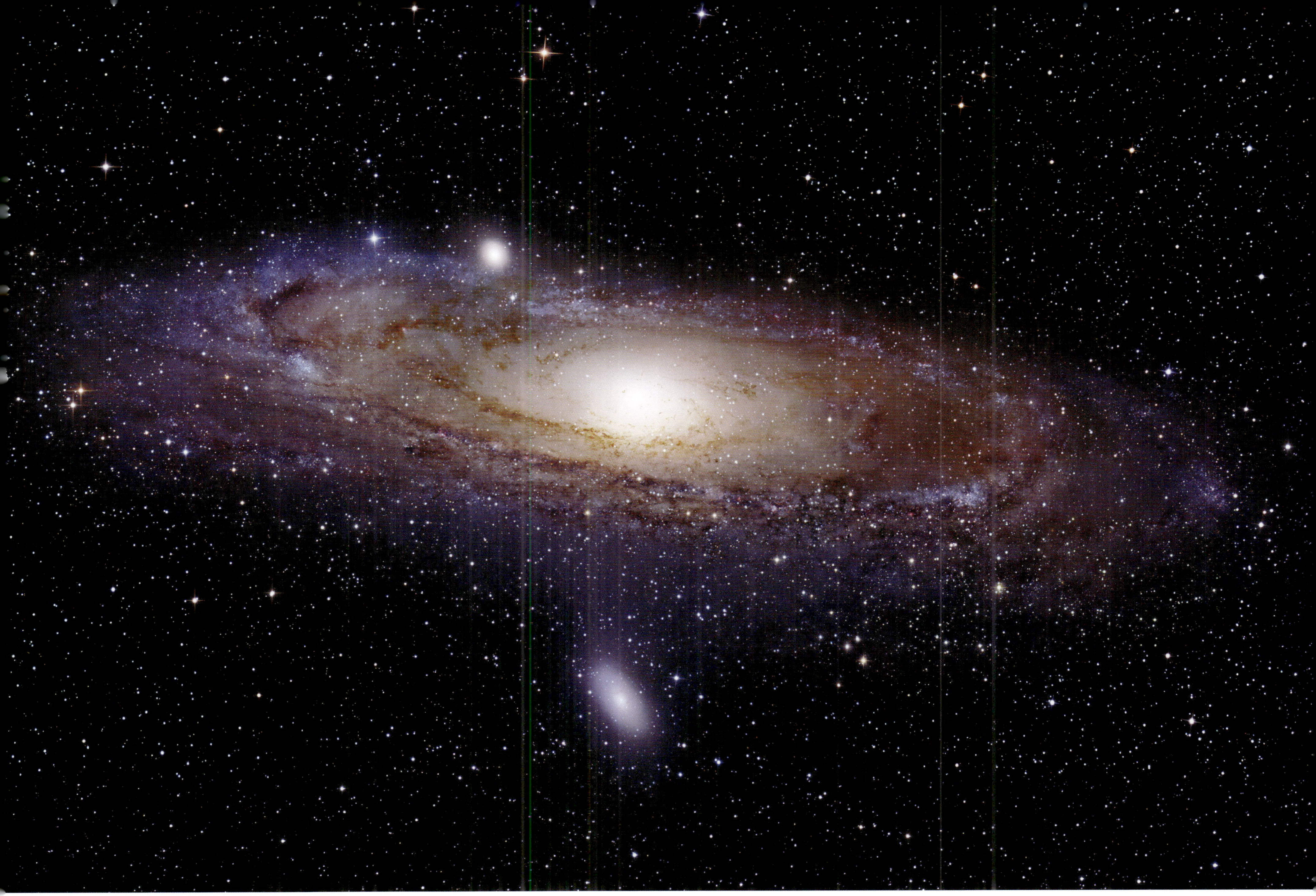

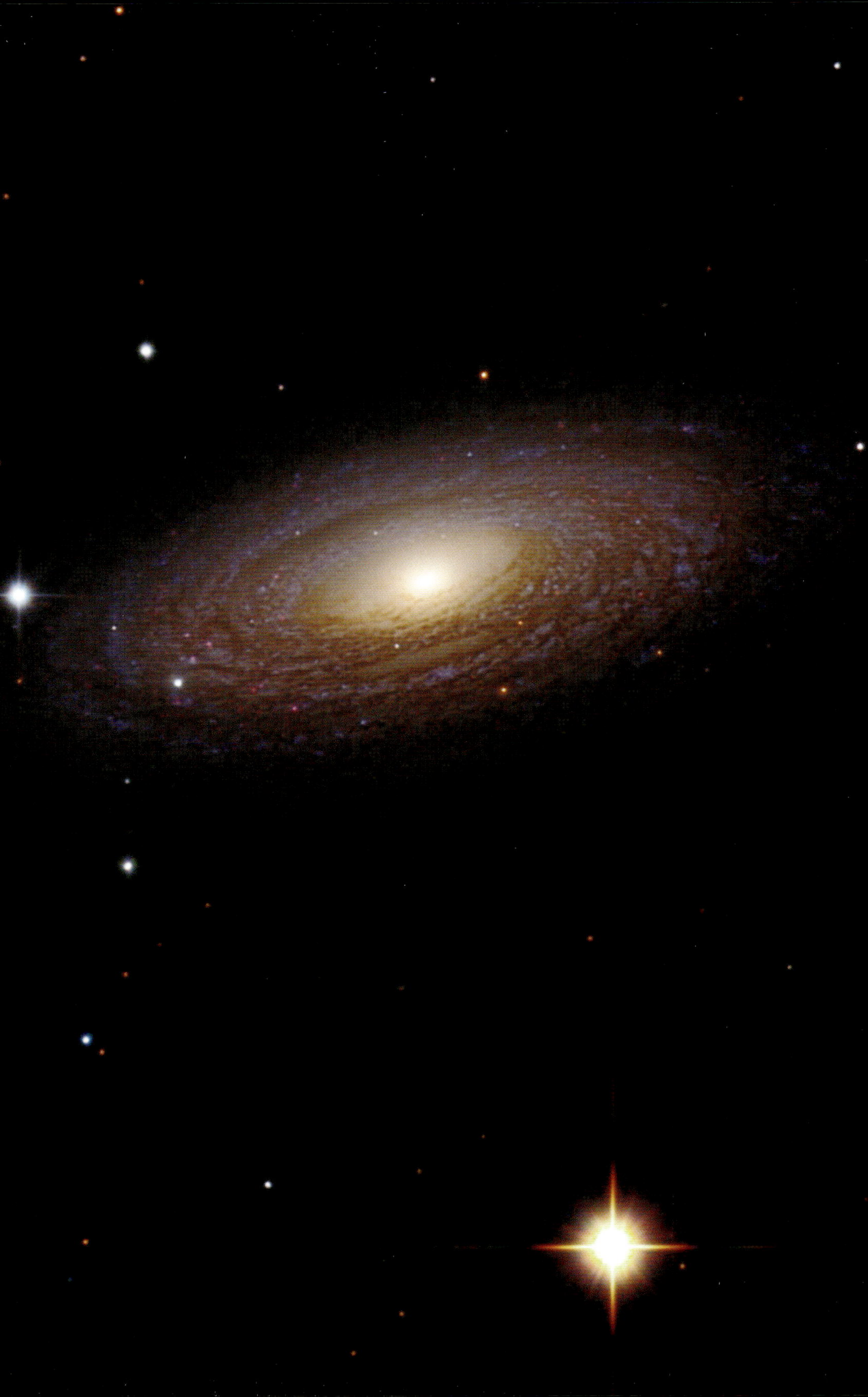

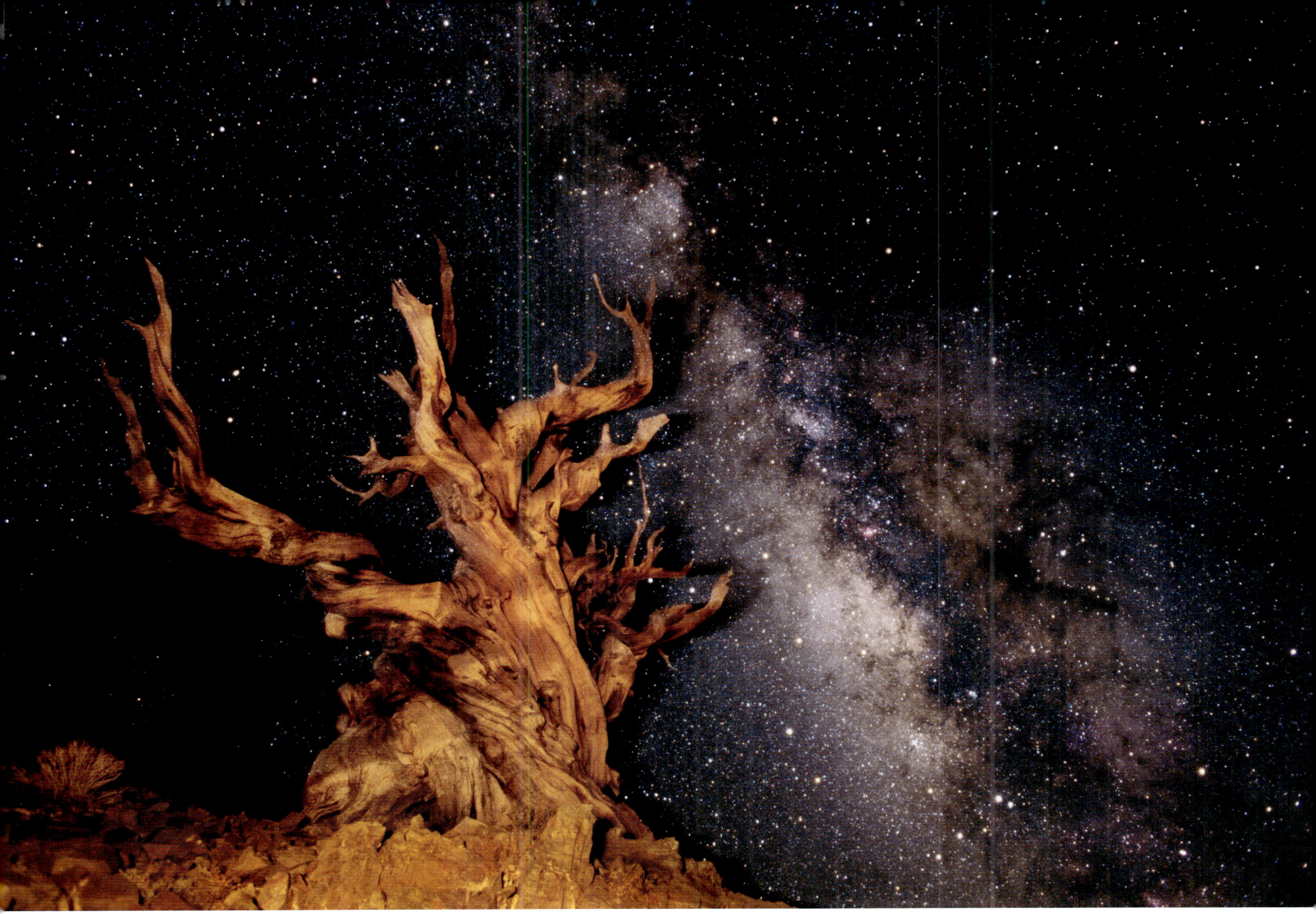

62. Tony ROWELL, *Bristlecone Pine and Milky Way, White Mountains, California*, 2008.

4. Adam D. BLOCK, *Delicate Perfection (NGC 2841)*.

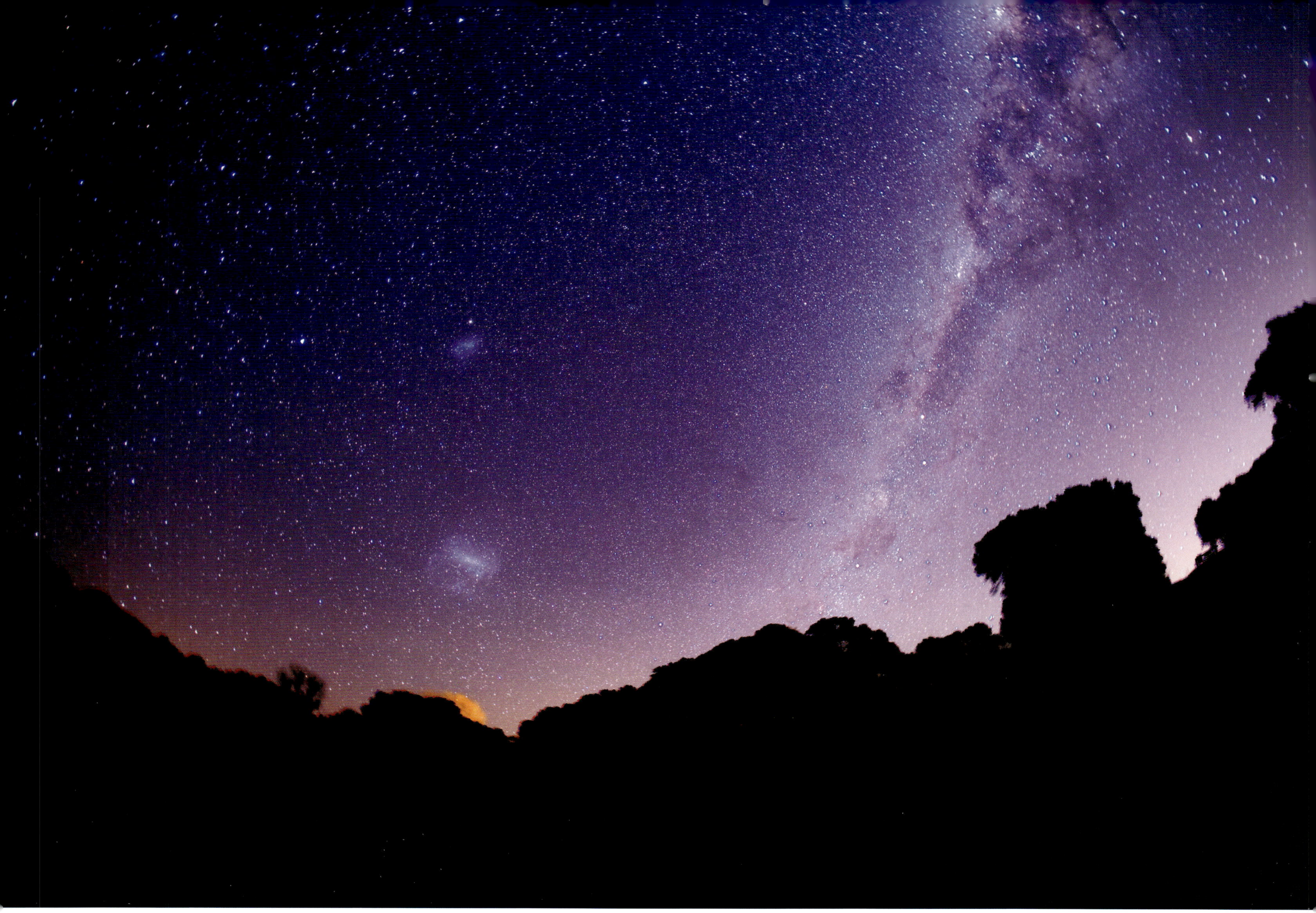

33. Jacob GERRITSEN, *A Southern View of the Southern Skies and the Magellanic Clouds*, 2011.

49. Steven MAZLIN & Jack HARVEY, *NGC 4945 in Centaurus.*

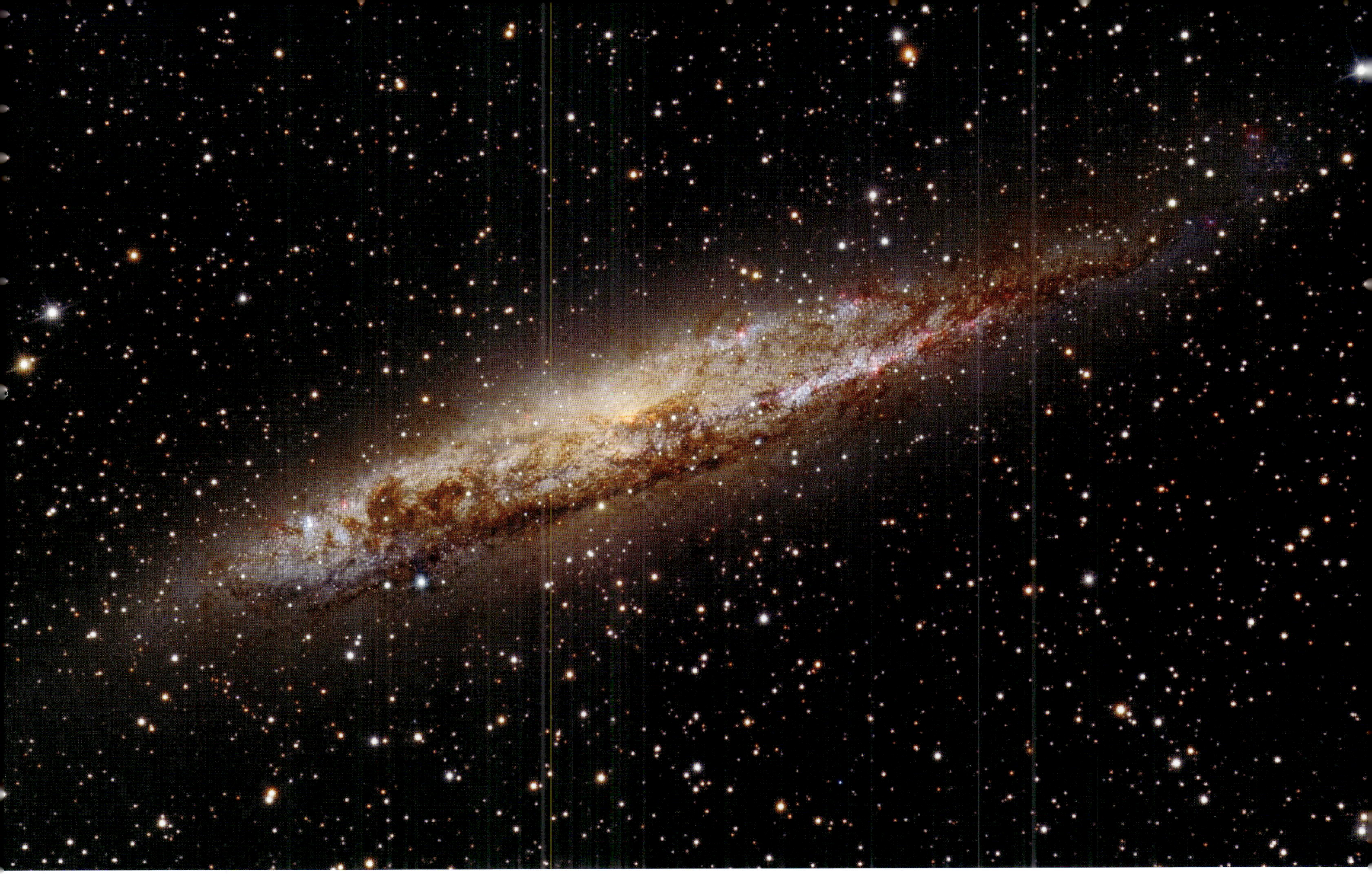

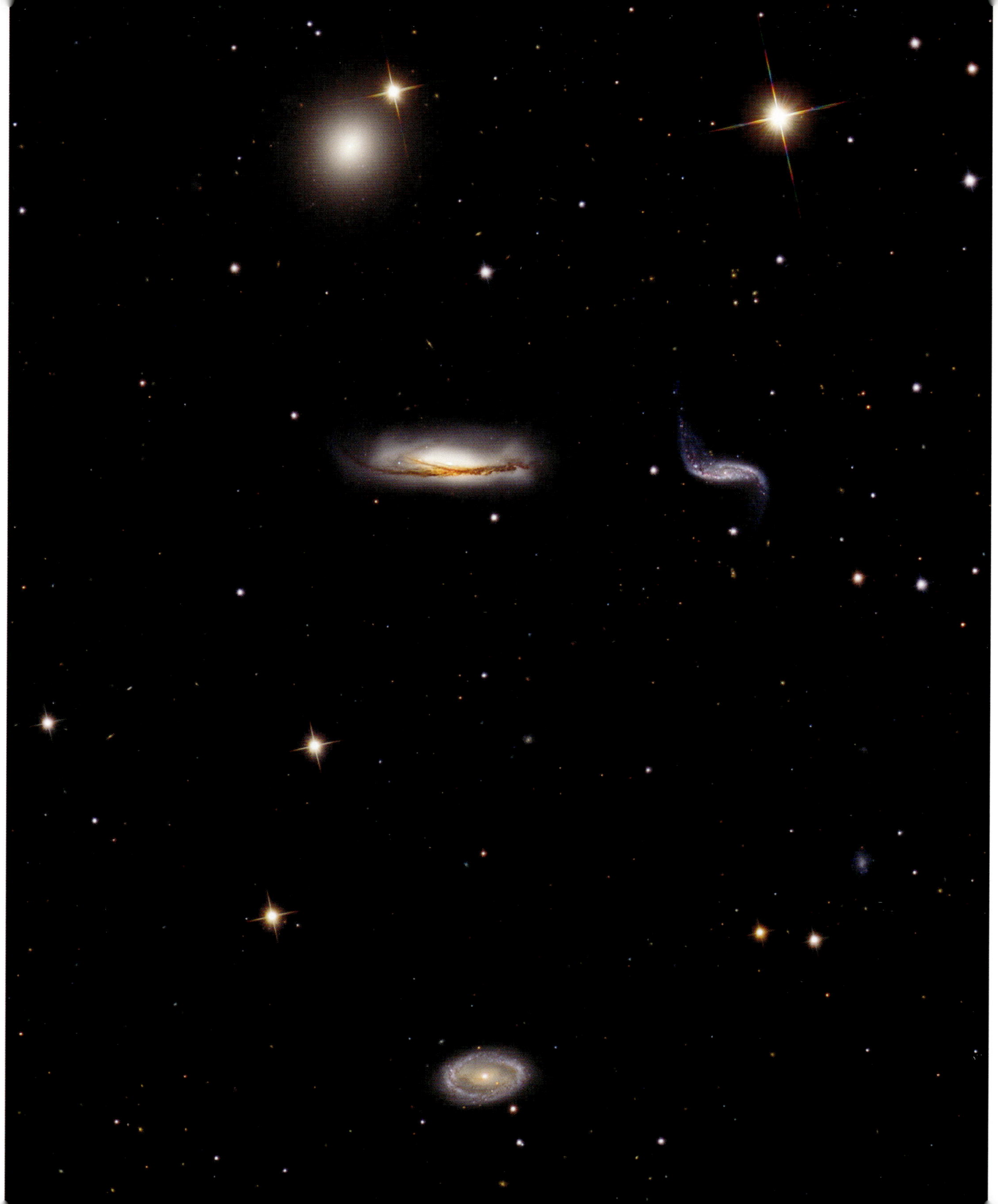

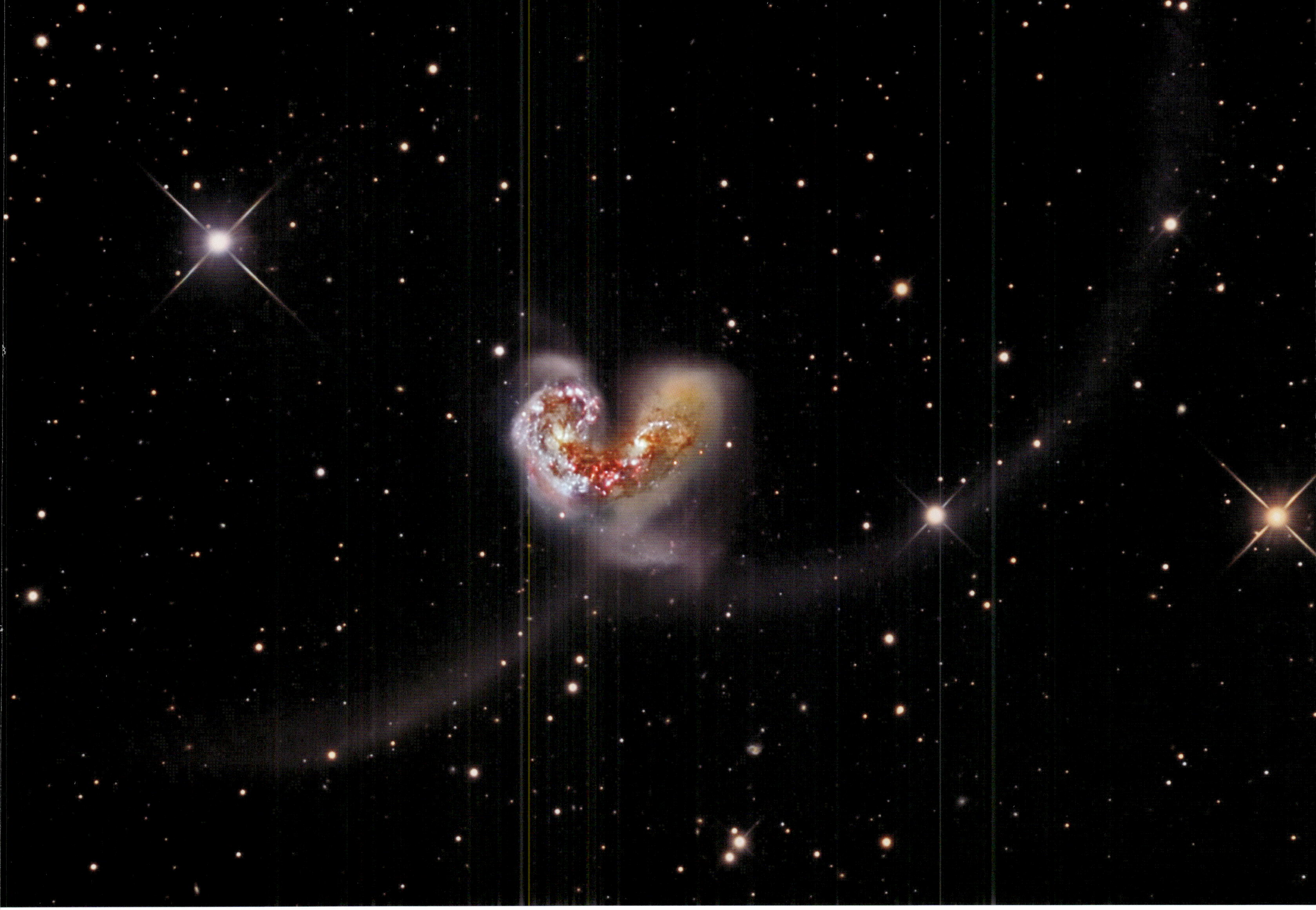

50. Steven MAZLIN & Jack HARVEY, *The Antennae Galaxies in Corvus (aka NGC 4038 & 4039).*

3. Adam D. BLOCK, *Many and Beyond (NGC 3190).*

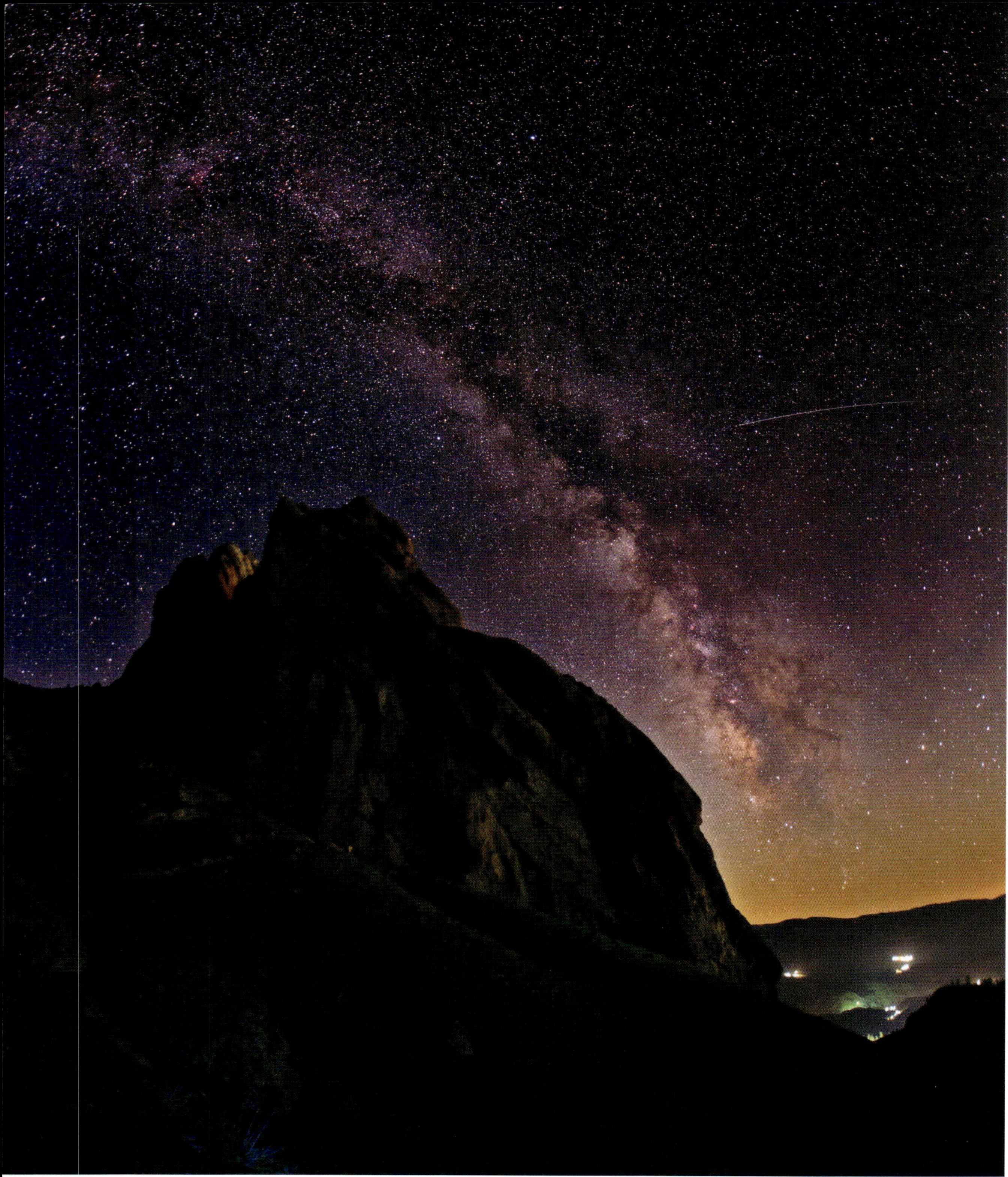

95. Babak TAFRESHI, *Alamut Starry Night.*

Scale & Distance

The Earth is roughly 7,926 miles in diameter. If the width of this page (11½ inches) represented that distance—illustrated on page 162—and a series of pages were laid end-to-end, our closest astronomical neighbor, the Moon, would be about thirty pages away, and about 3 ⅛ inches across (see page 183).

The Earth

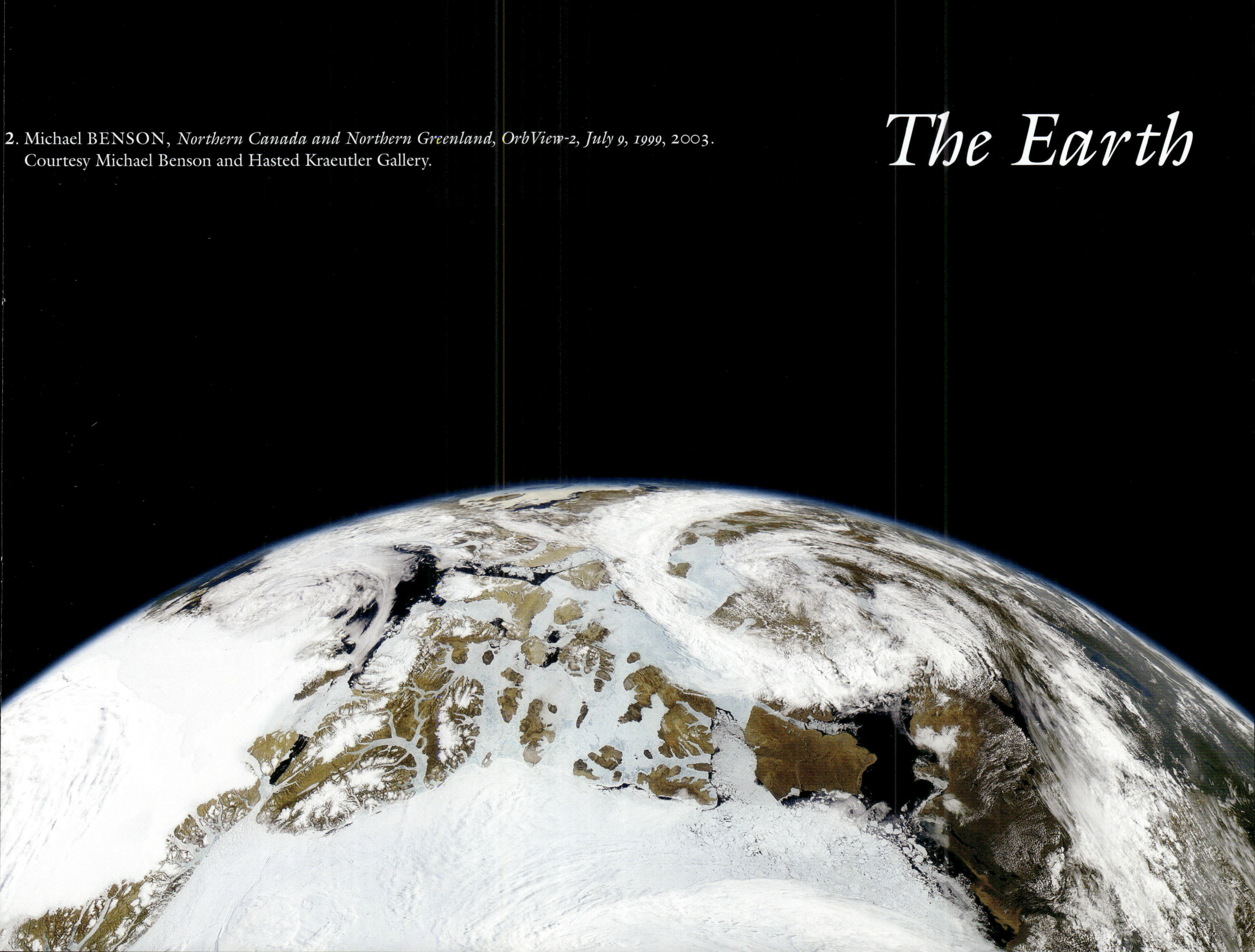

2. Michael BENSON, *Northern Canada and Northern Greenland, OrbView-2, July 9, 1999*, 2003.
Courtesy Michael Benson and Hasted Kraeutler Gallery.

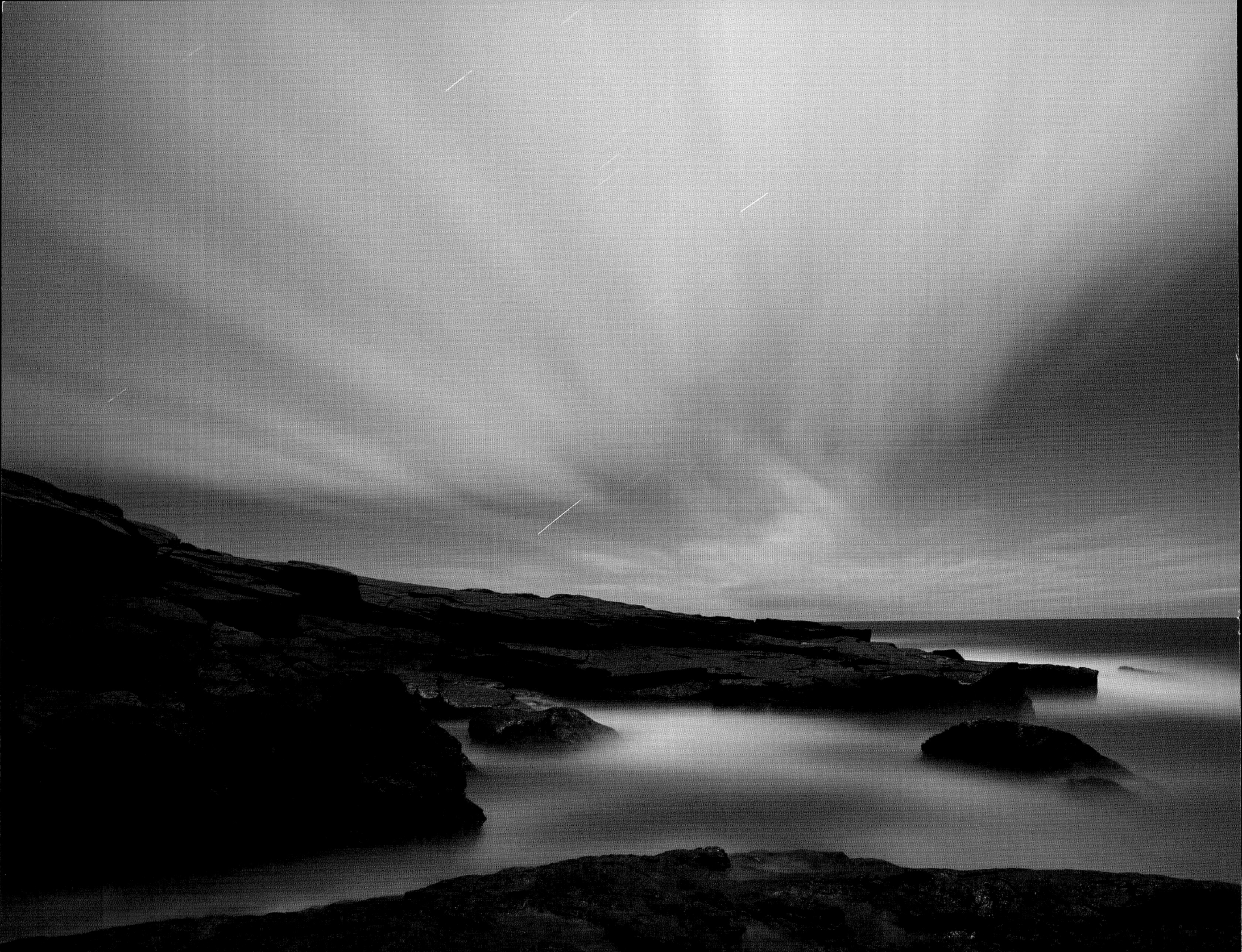

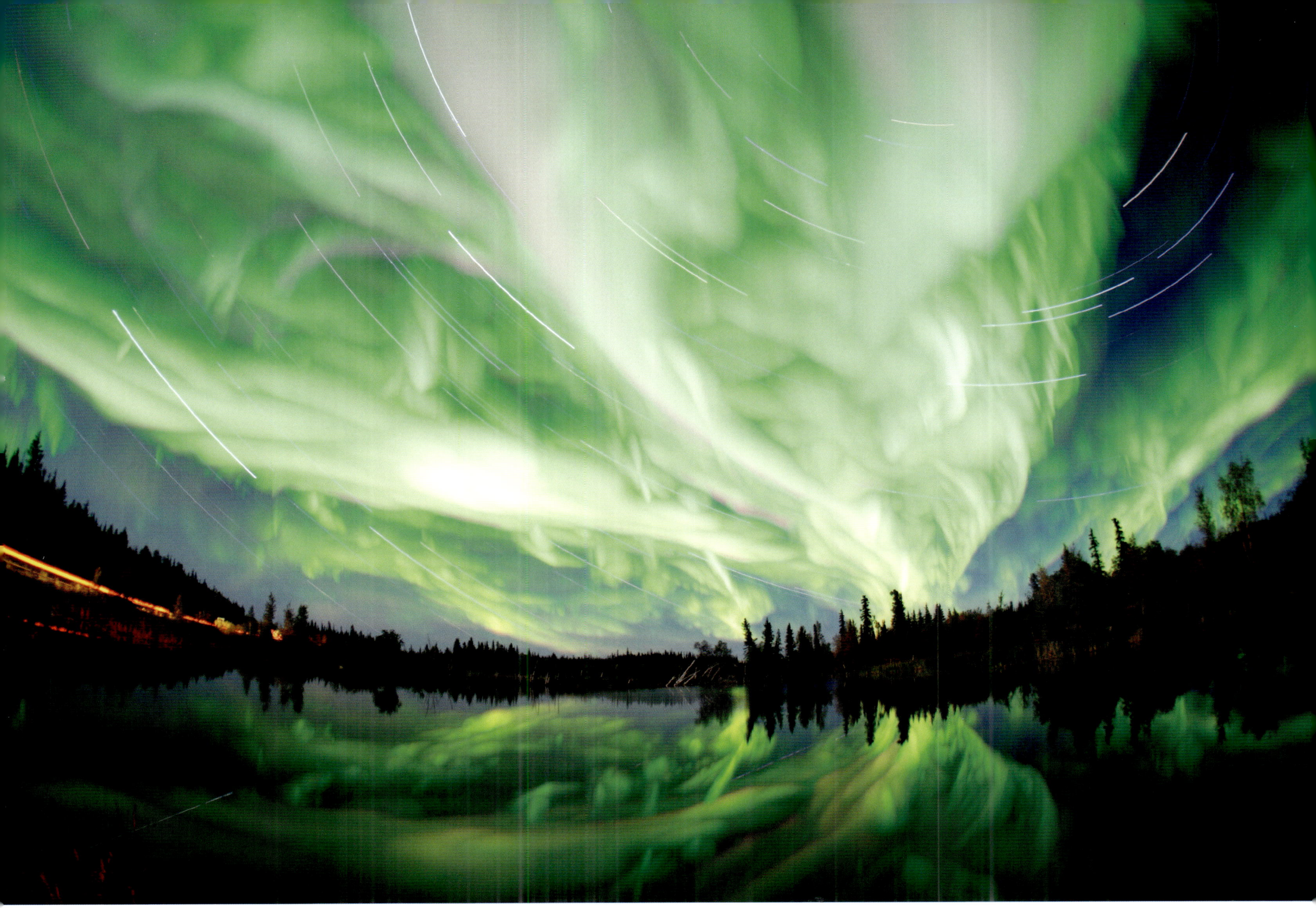

98. Yuichi TAKASAKA, *Startrails, September 12*, 2011.

14. James CORMIER, *Departing Clouds*, 2010.

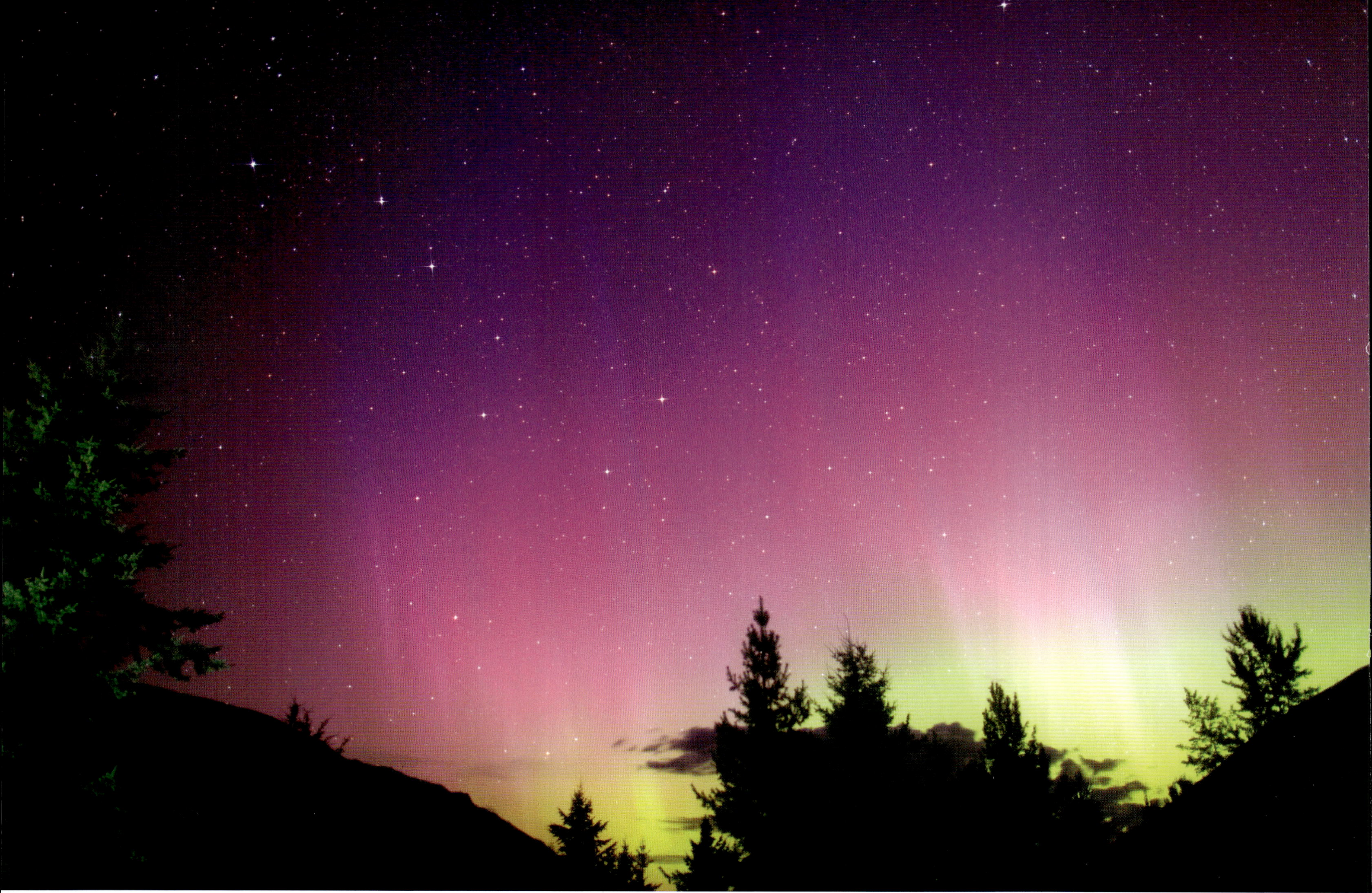

97. Yuichi TAKASAKA, *Big Dipper, August 5*, 2011.

63. Tony ROWELL, *Moonbow and Stars at Yosemite Falls, Yosemite National Park*, 2010.

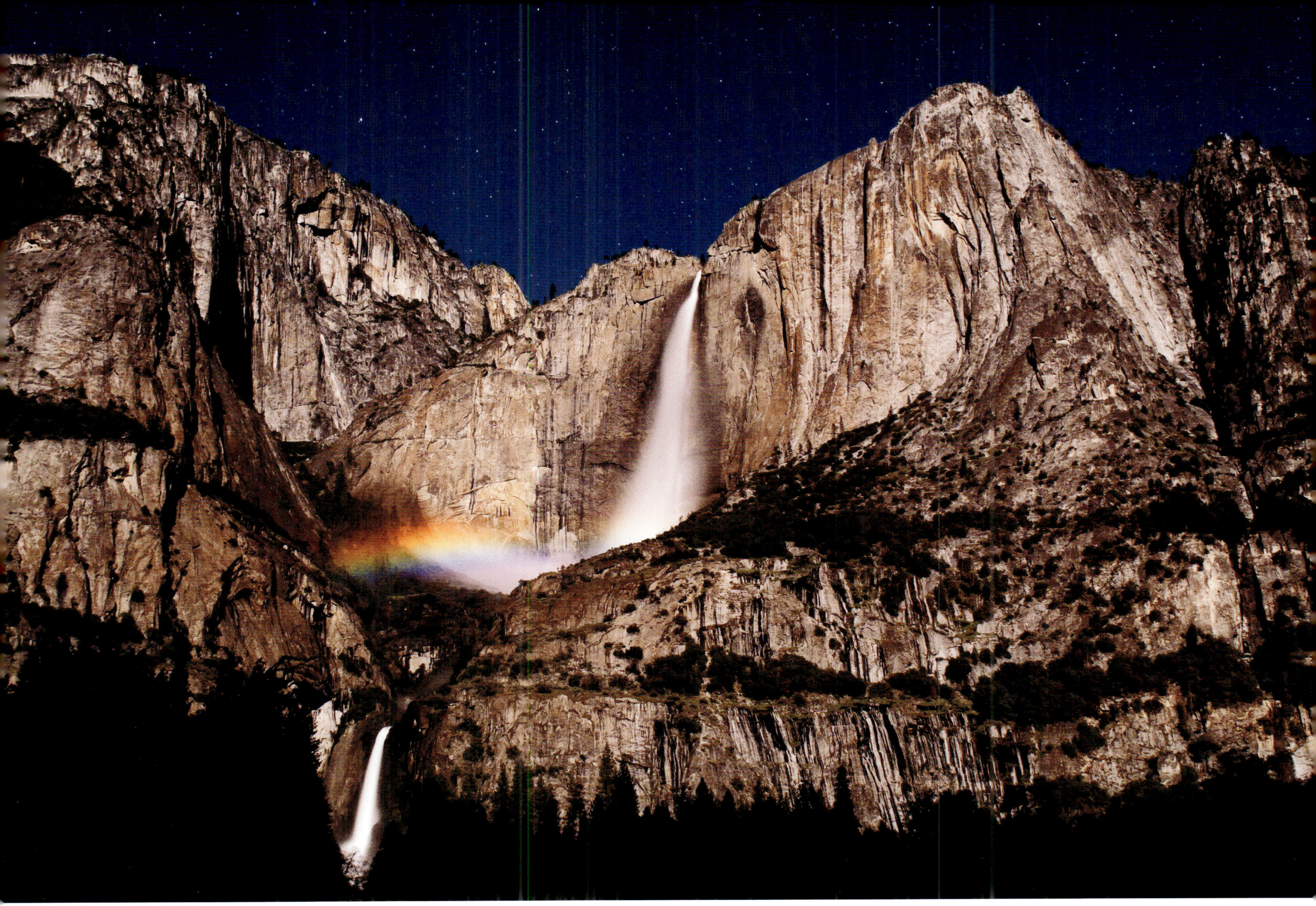

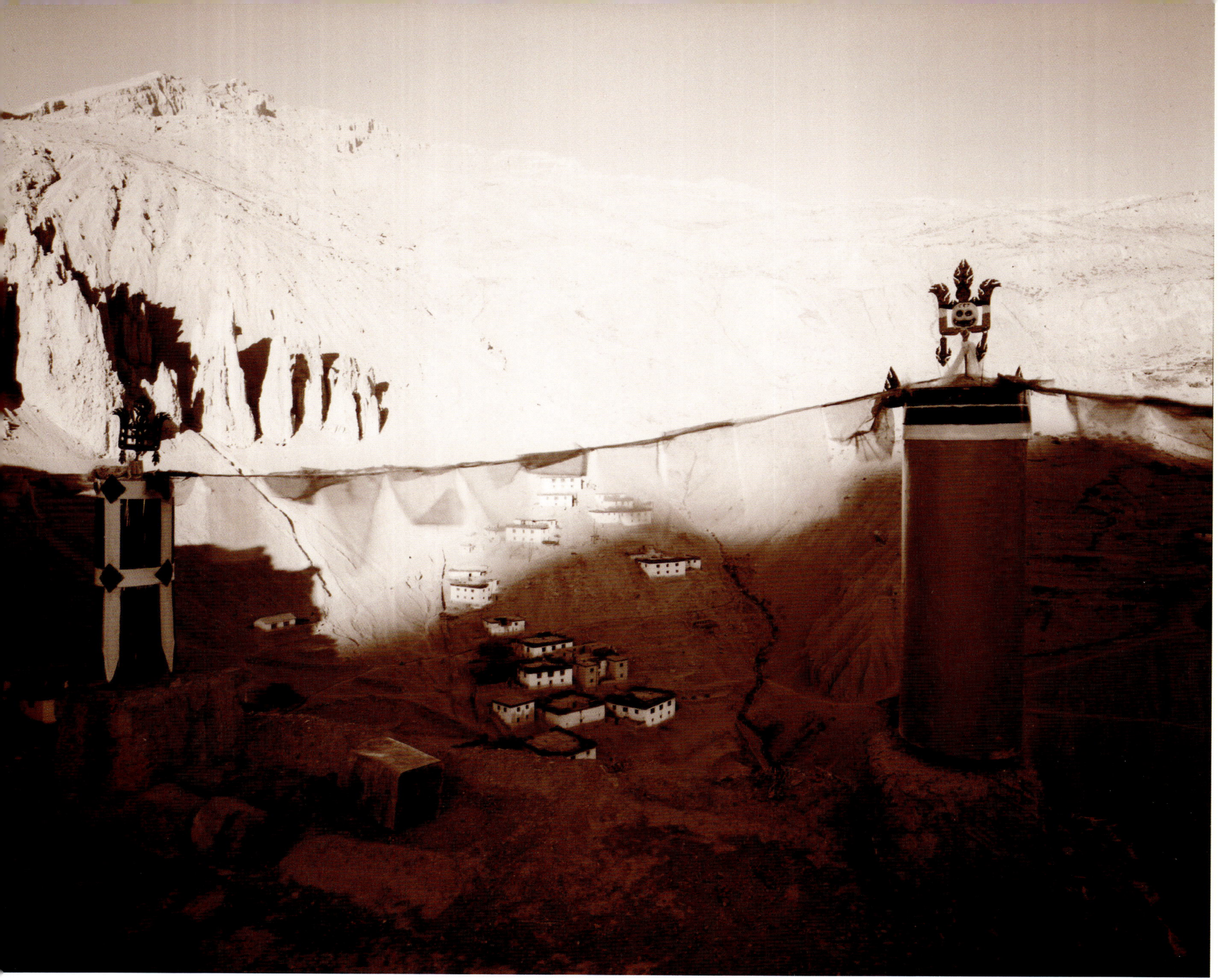

13. Linda CONNOR, *From the Monastery Roof, Spiti, Ladakh, India*, 1994.

15. James CORMIER, *Incoming Tide*, 2010.

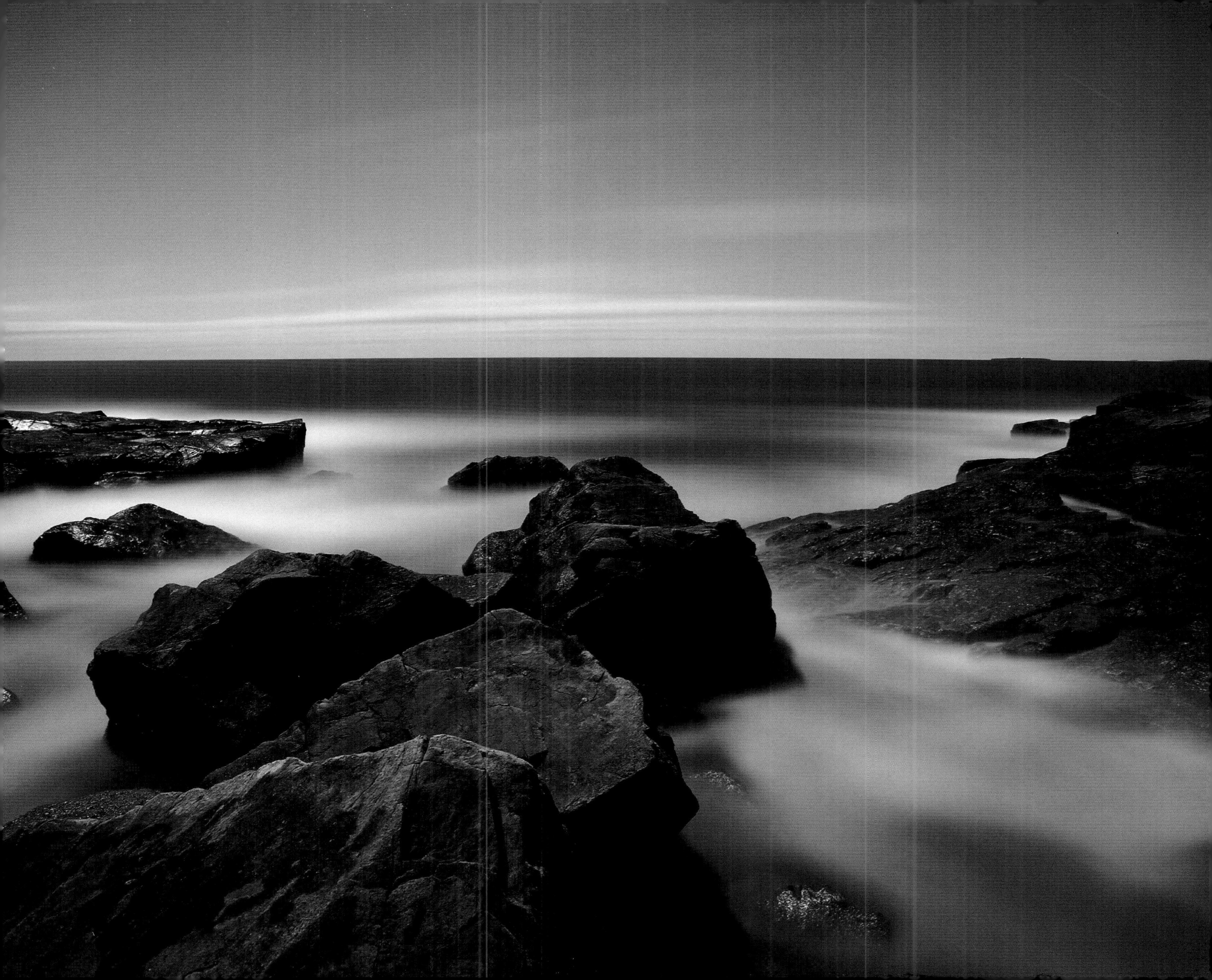

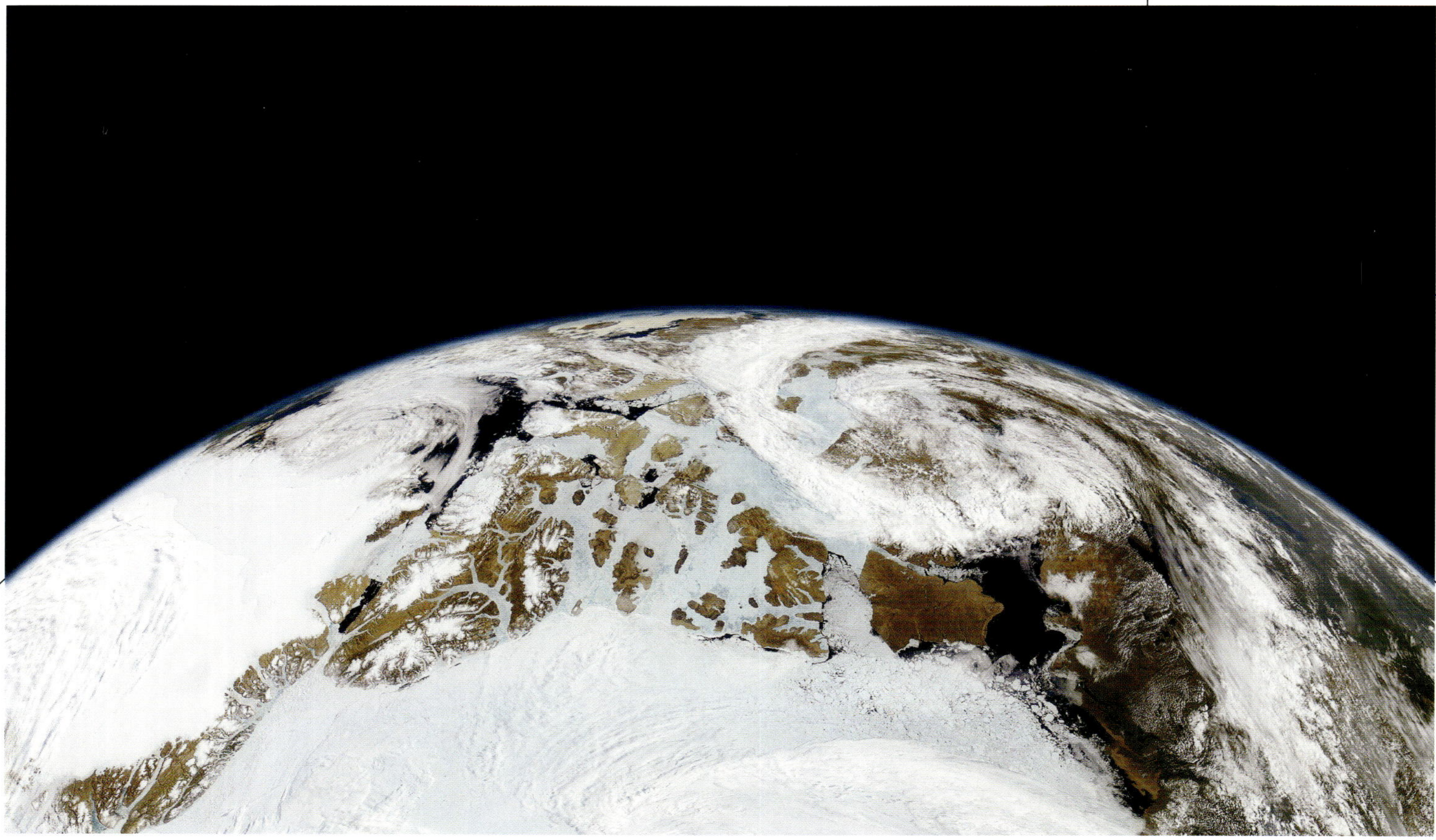

Scale & Distance (cont.)

As described on page 154, one can imagine this page's width representing the Earth's diameter, as illustrated here. To view the Moon at the same scale, see page 183 (scale also delineated at upper right).

2. Michael BENSON, *Northern Canada and Northern Greenland, OrbView-2, July 9, 1999*, 2003. Courtesy Michael Benson and Hasted Kraeutler Gallery.

The Moon

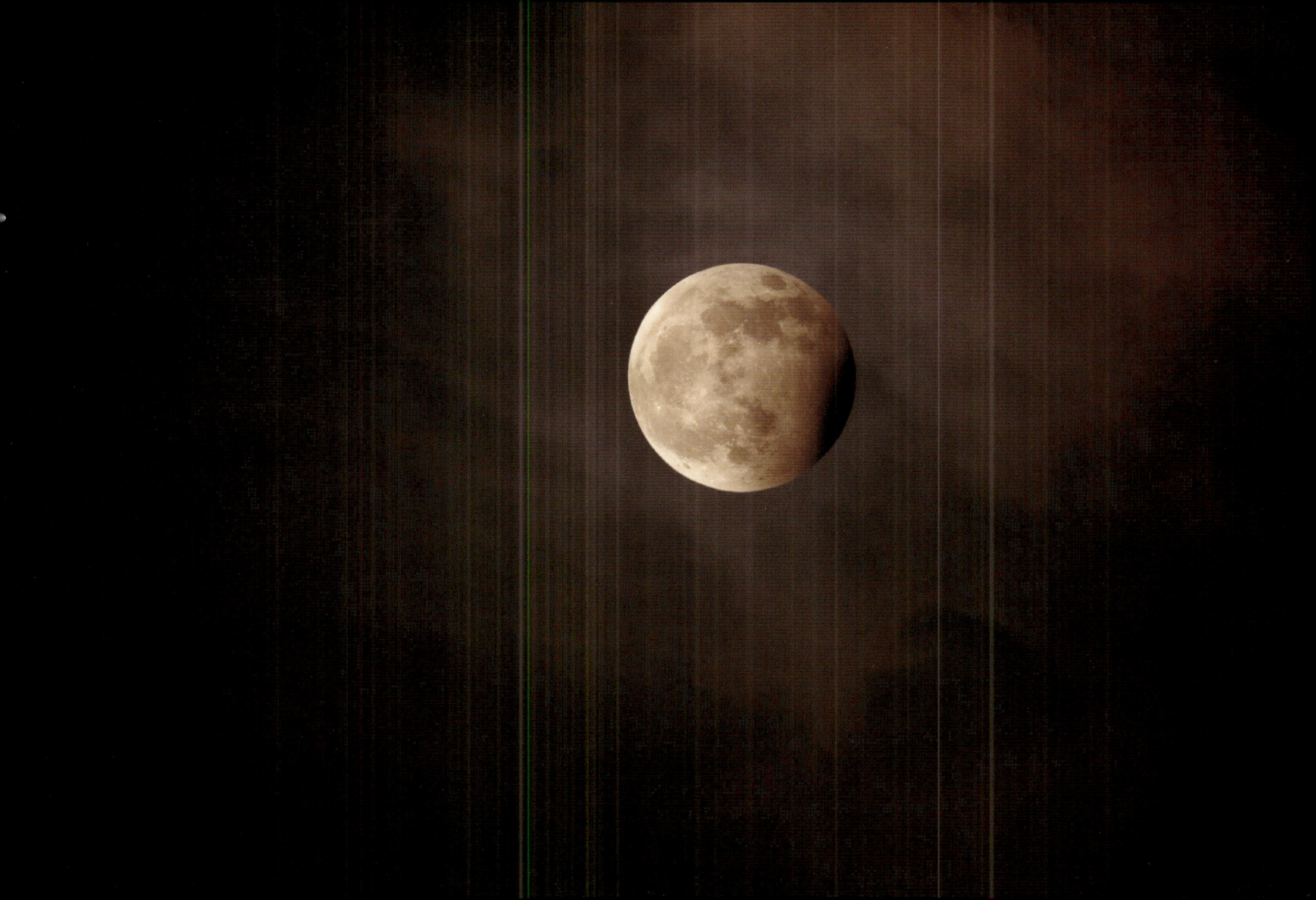

1. Jean-Paul ROUX, *Blue Moon Eclipse*, 2010.

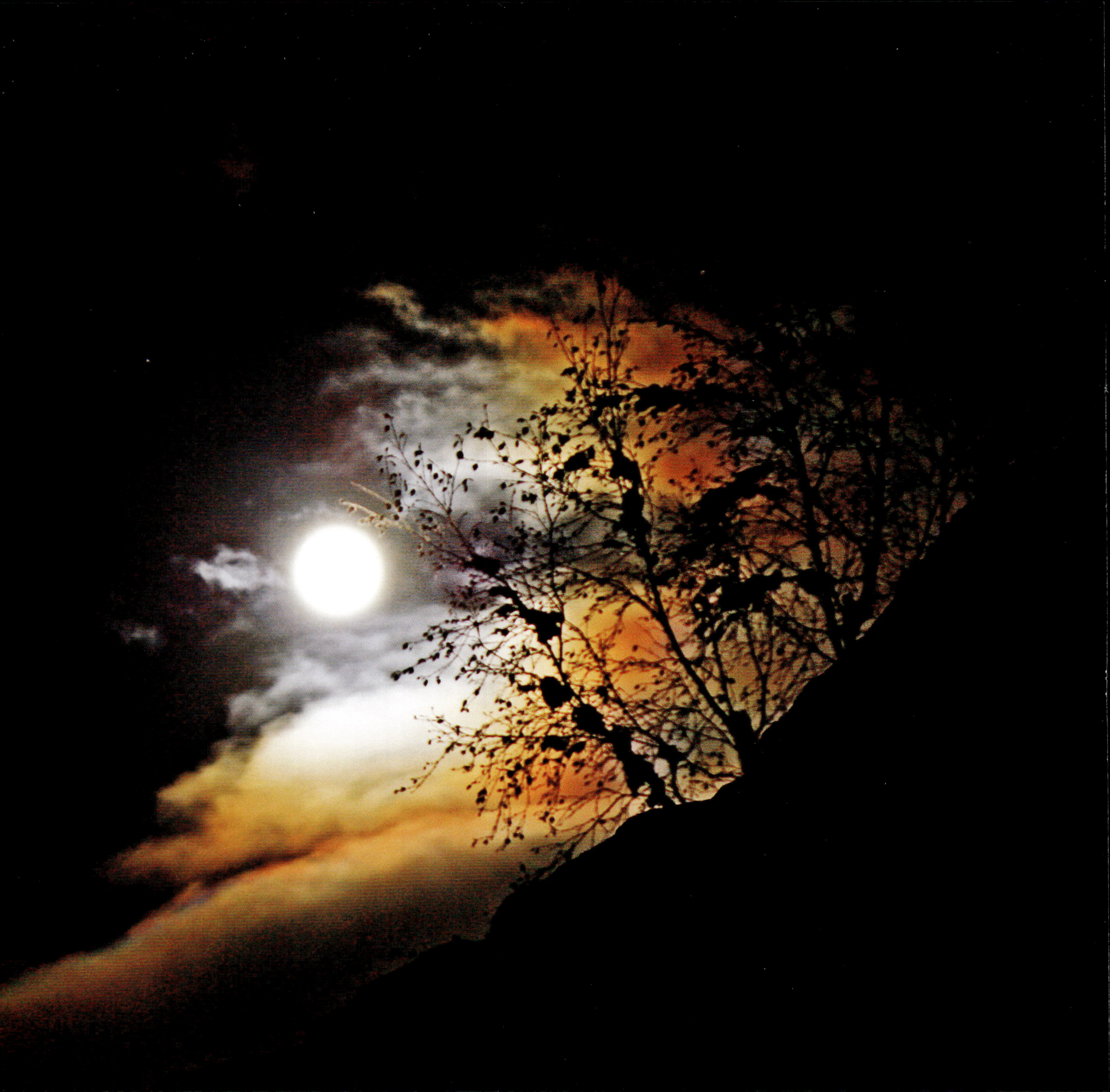

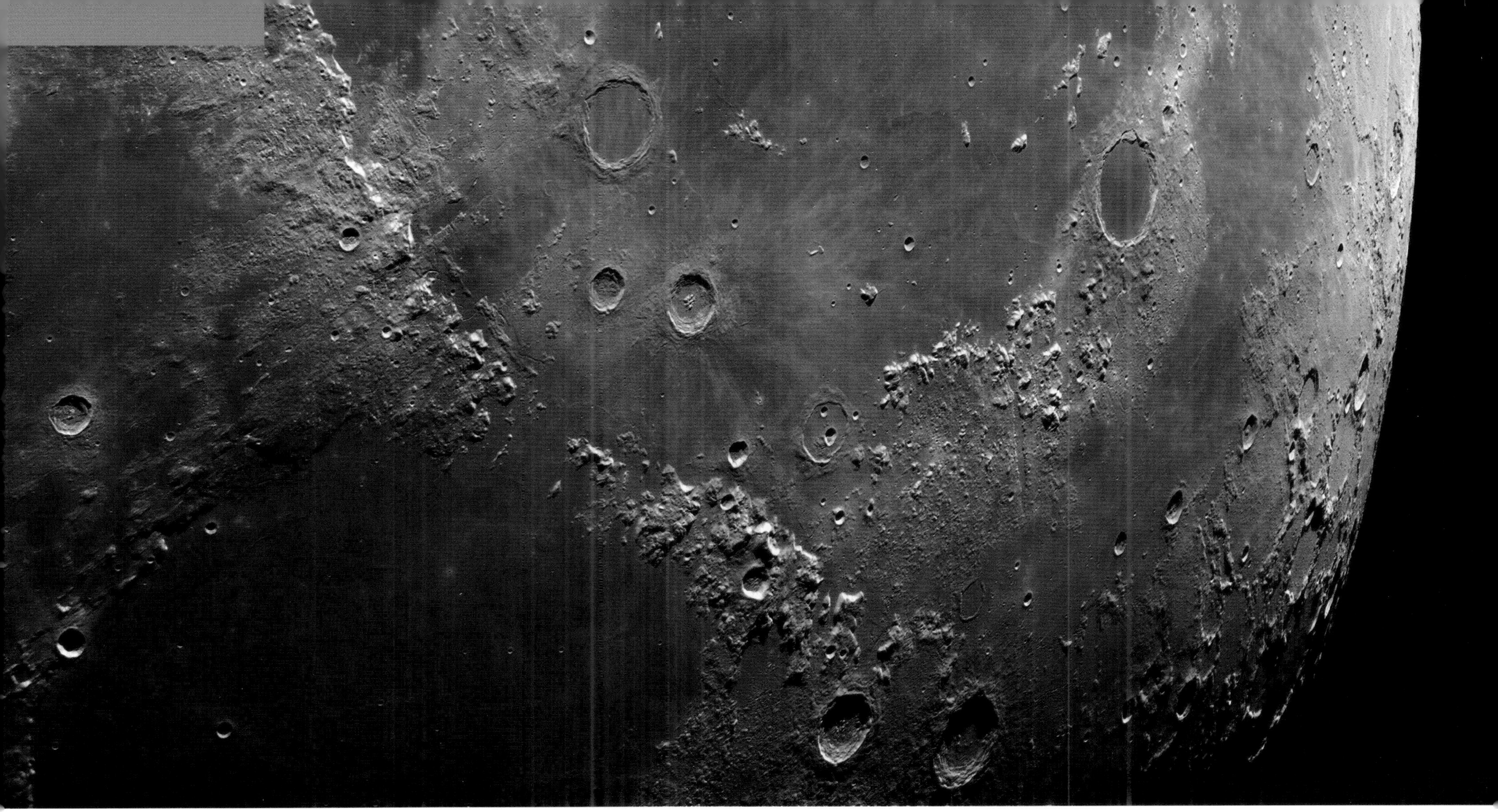

This photograph's title refers to the Moon's Ina Caldera.—Editor]

28. Alan FRIEDMAN, *Kleine Ina Nachtmusik.*

53. Geir ØYE, *Moon Iridescence over Snow Covered Tree Branches.*

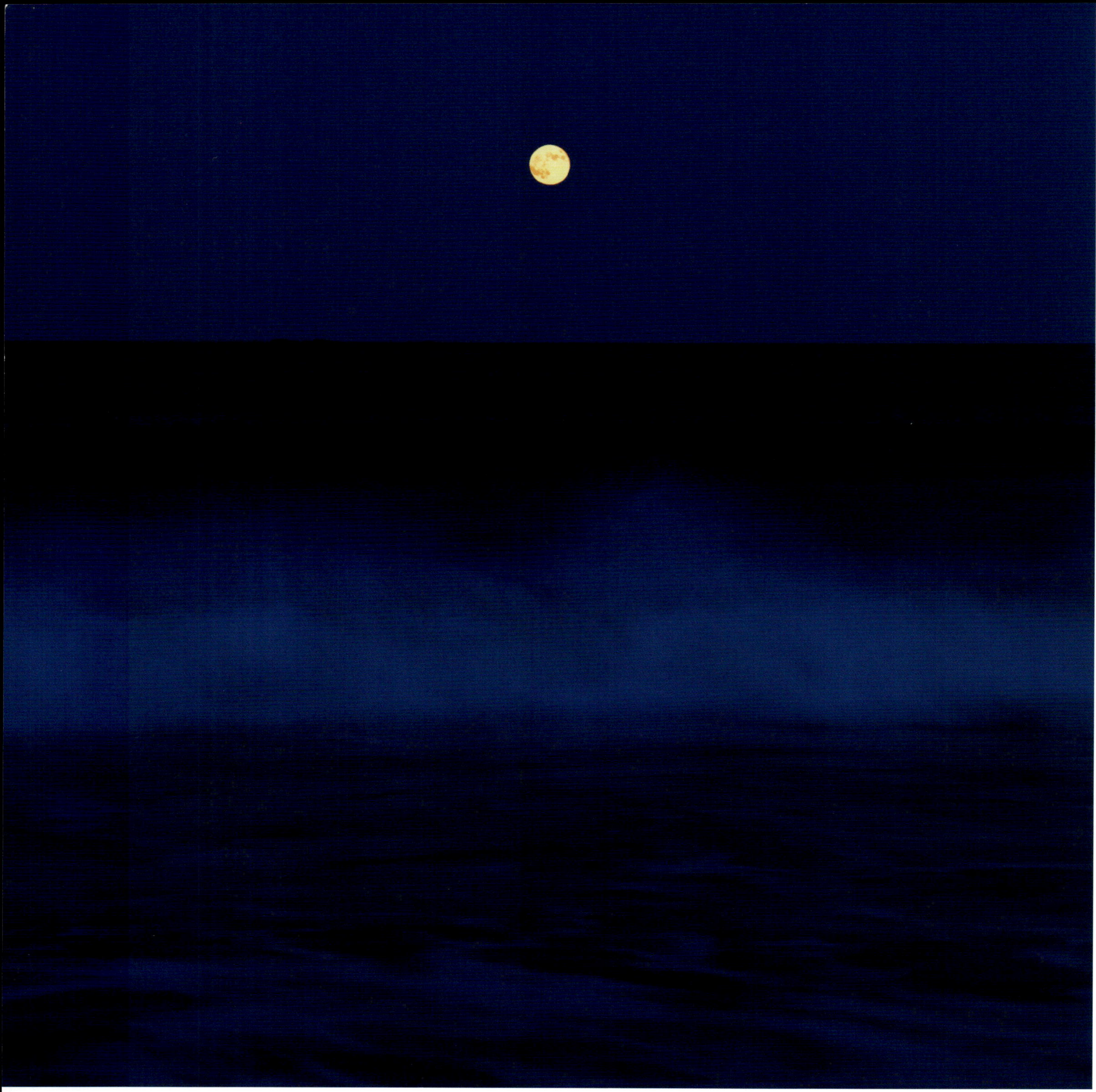

52. Jim NICKELSON, *Thunder Moon II*, 2011.

39. Sharon HARPER, *Night Sky Over Banff, Alberta, September 12–October 10, 2007, 25 September*, 2007.

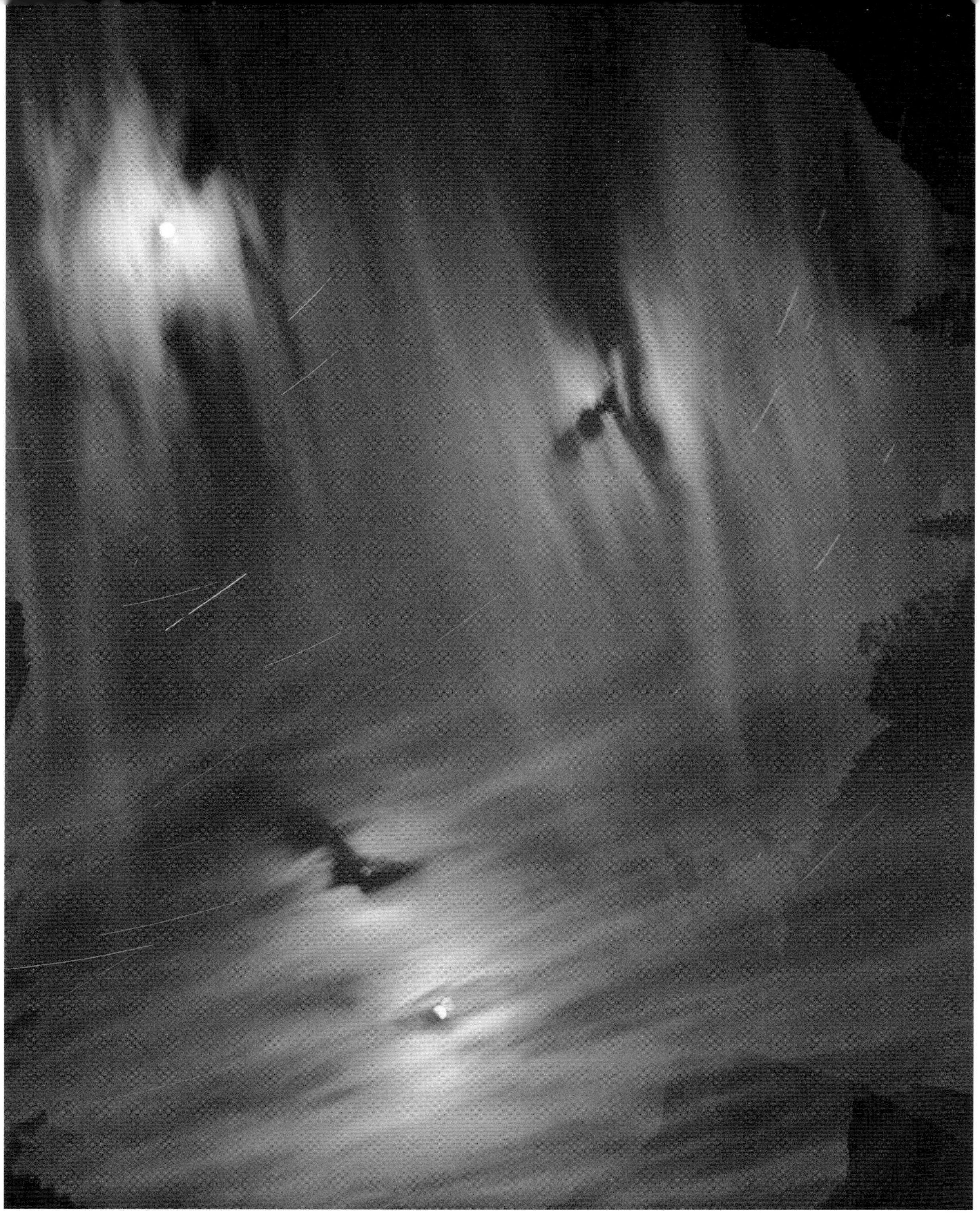

40. Sharon HARPER, *Night Sky Over Banff, Alberta, September 12–October 10, 2007, 27–29 September*, 2007.

99. Sean WALKER, *Rupes Recta*.

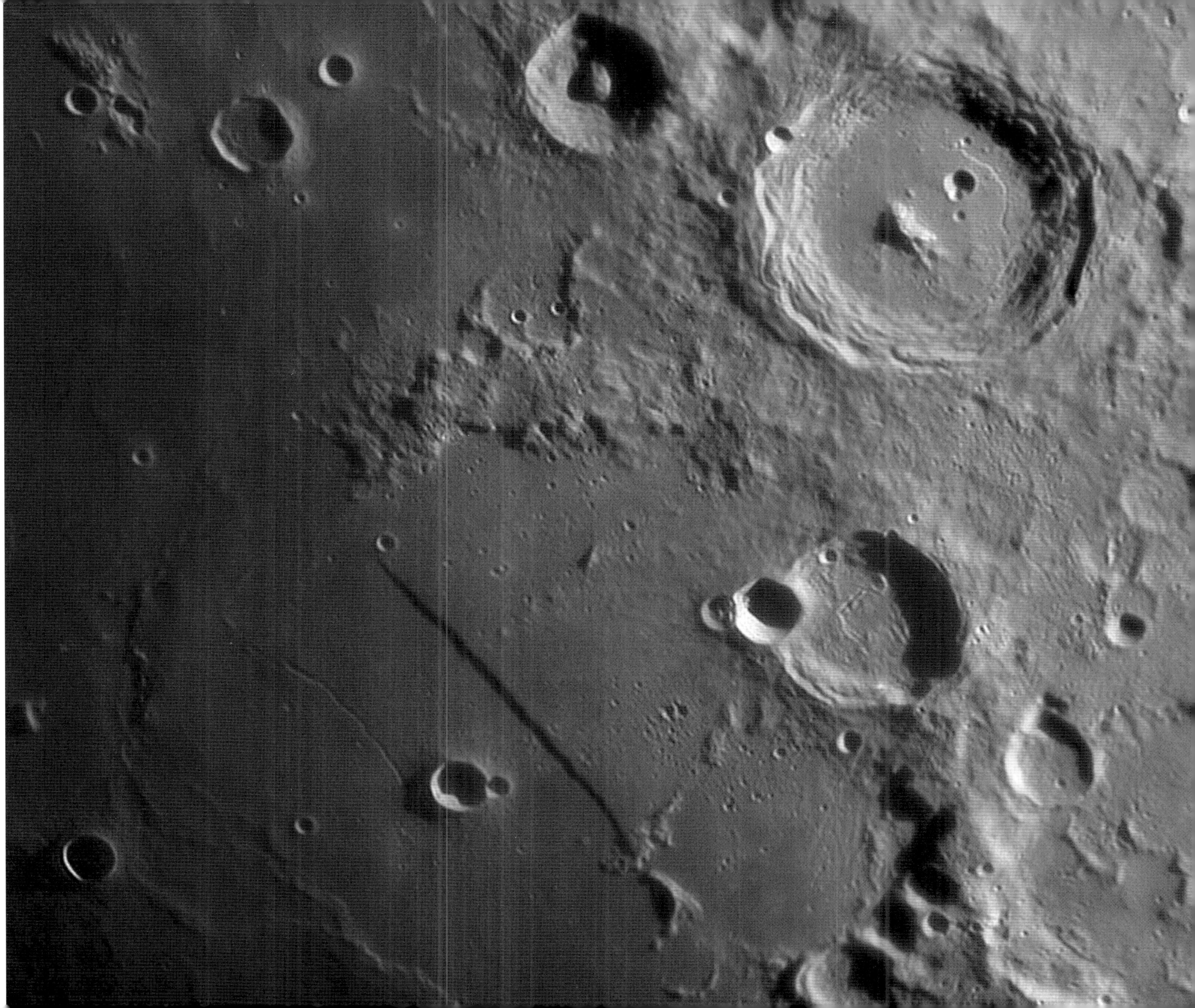

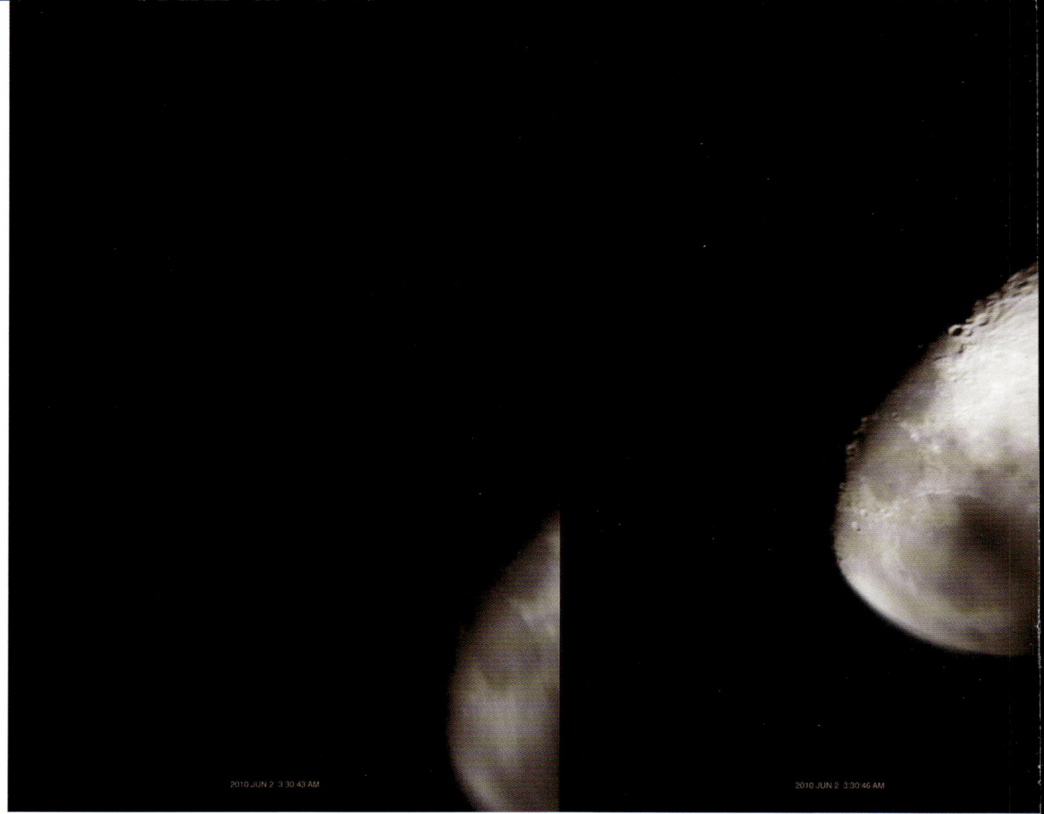

38. Sharon HARPER, *Sun/Moon (Trying to See Through a Telescope), 2010 June 2 3:30:43–31:05 a.m.*

91. John STETSON, *Etruscan Vase Moonrise.*

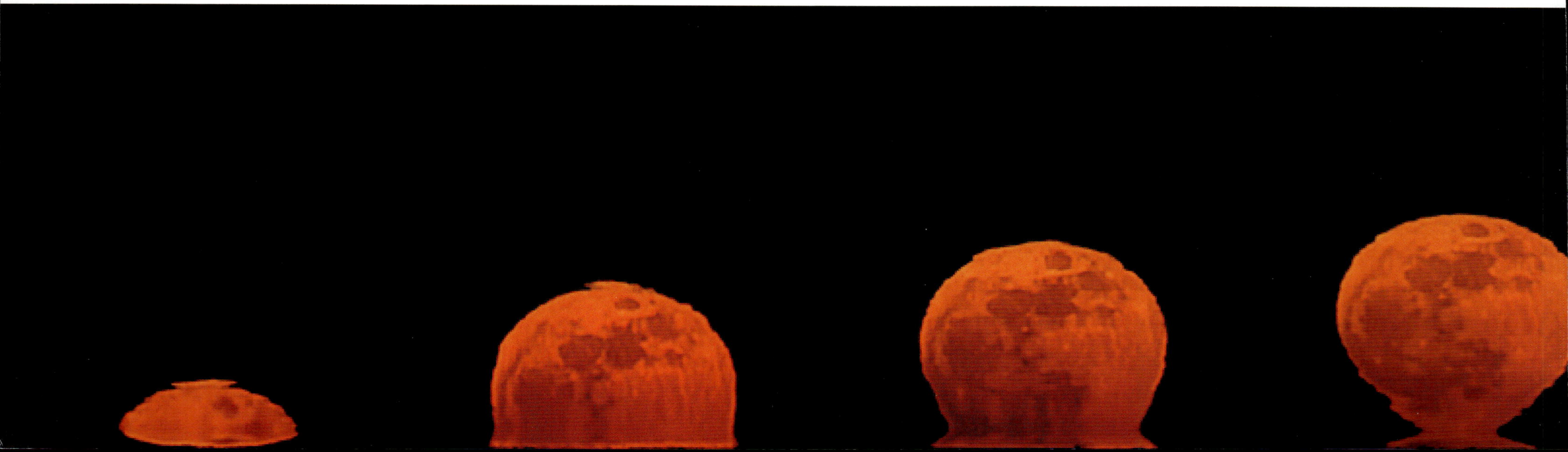

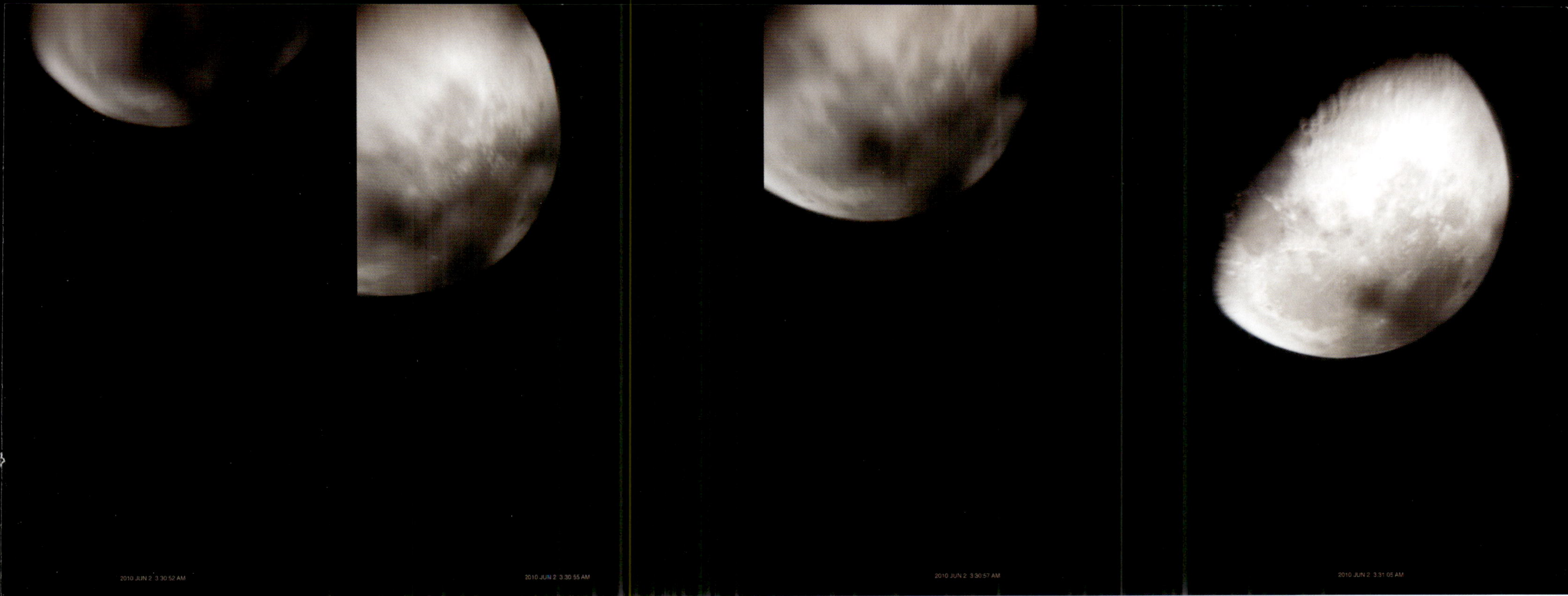

2010 JUN 2 3:30:52 AM
2010 JUN 2 3:30:55 AM
2010 JUN 2 3:30:57 AM
2010 JUN 2 3:31:05 AM

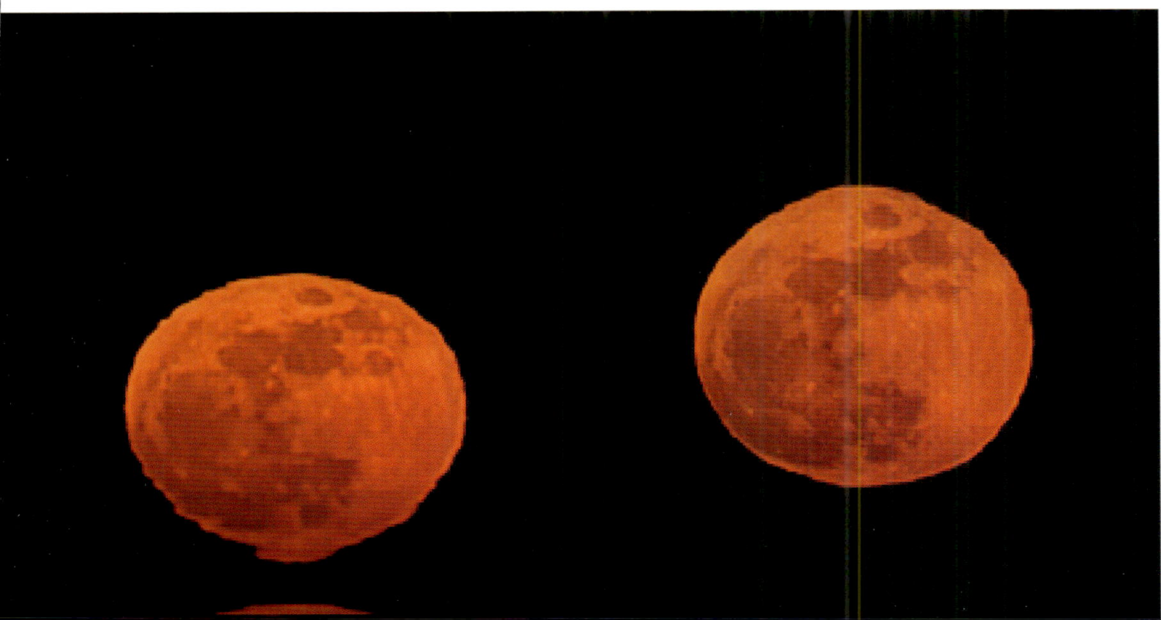

37. Sharon HARPER, *Sun/Moon (Trying to See Through a Telescope)*, *2010 July 6 10:23:10–38 a.m.*

24. Ignacio DIAZ-BOBILLO, *Lunar Triptych #1-Silver Moons.*

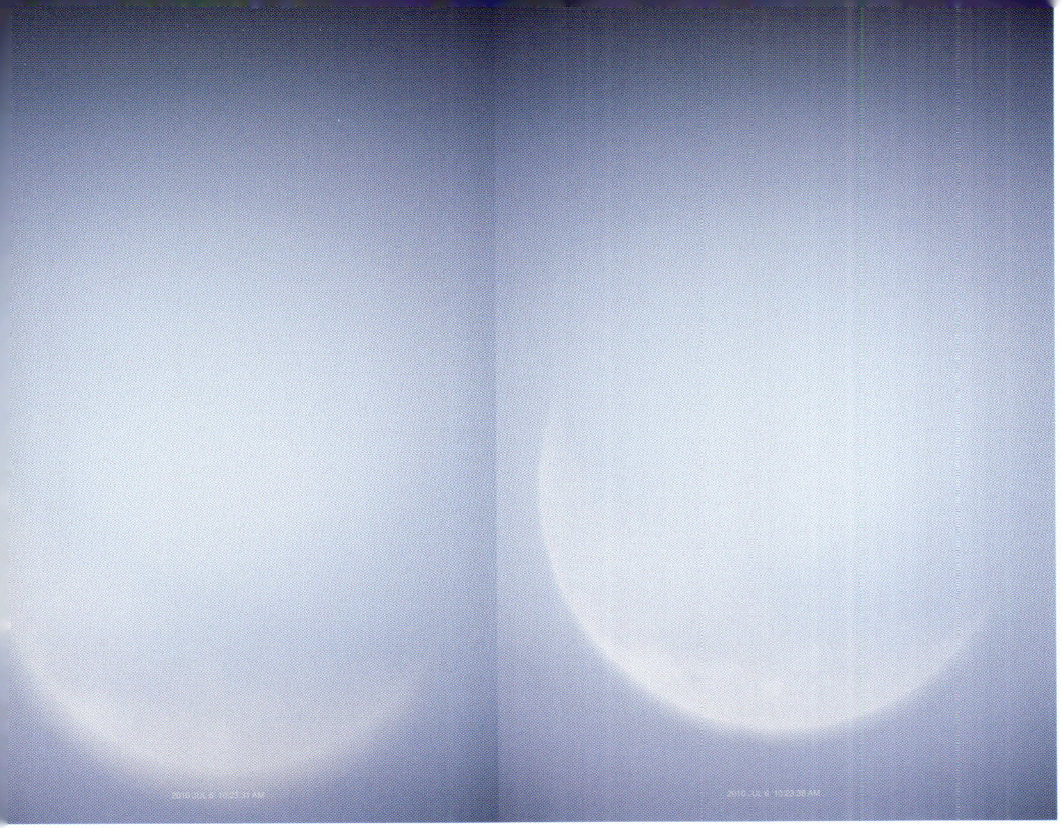

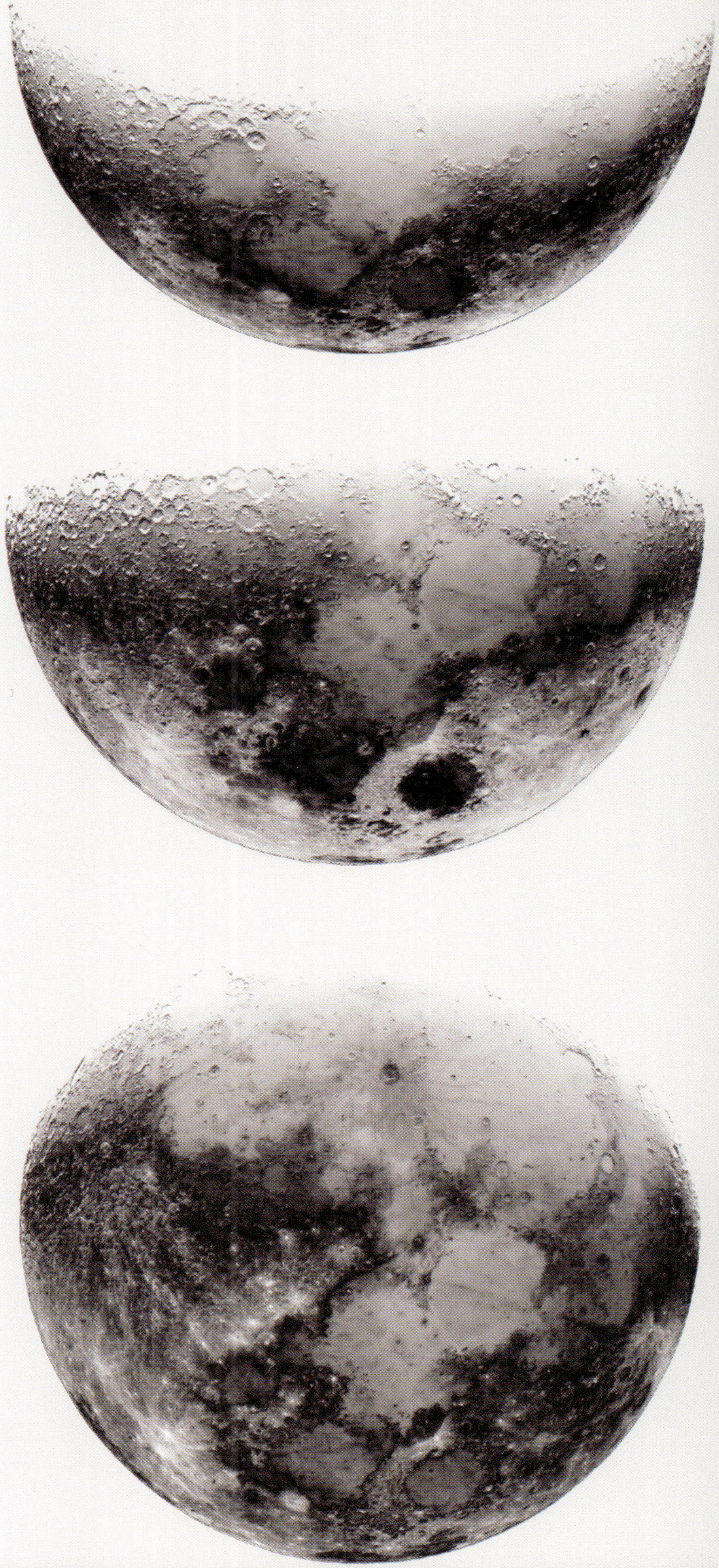

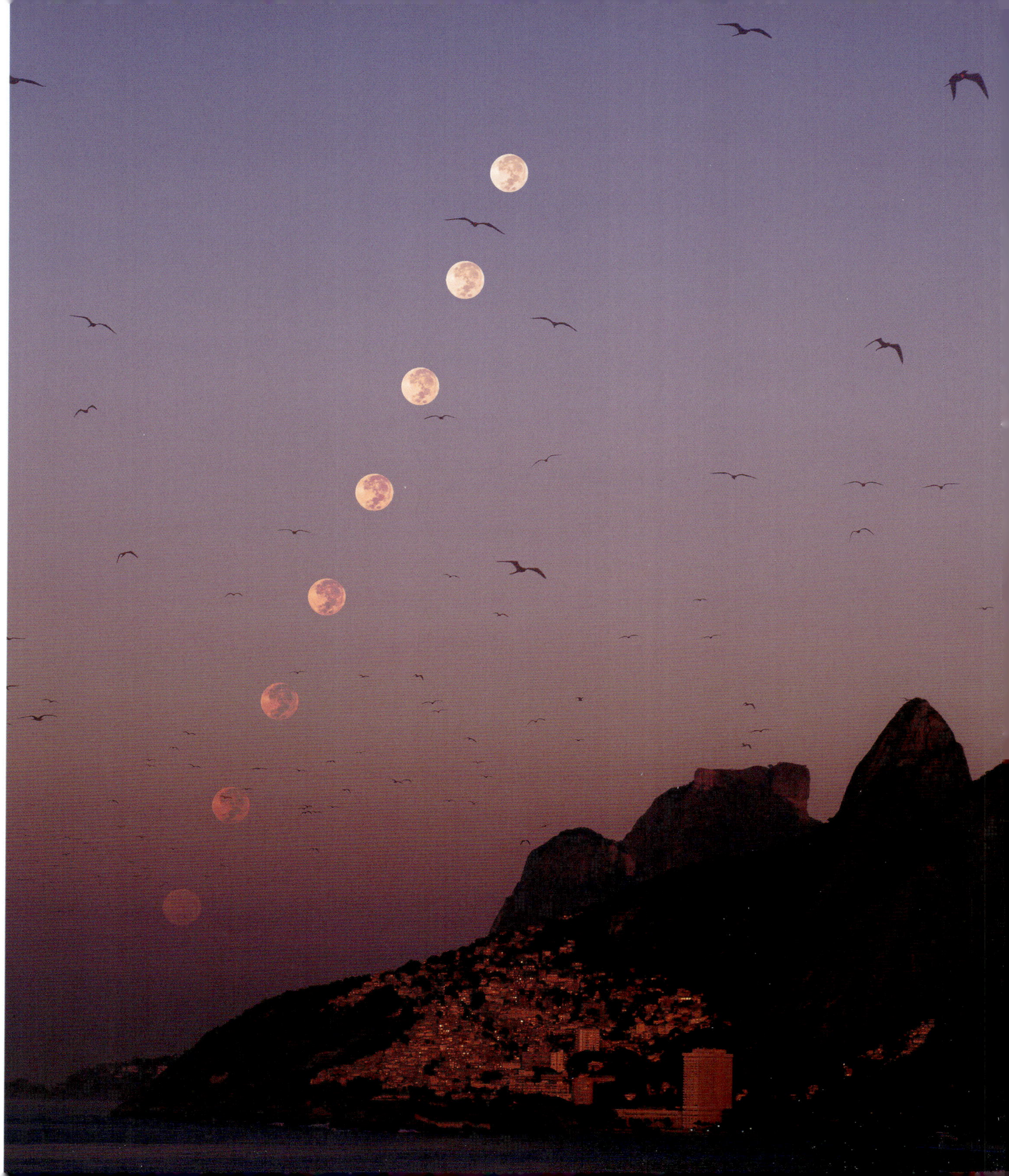

96. Babak TAFRESHI, *Rio Moonset*.

Giant Planets, Satellites, Rings & a Comet

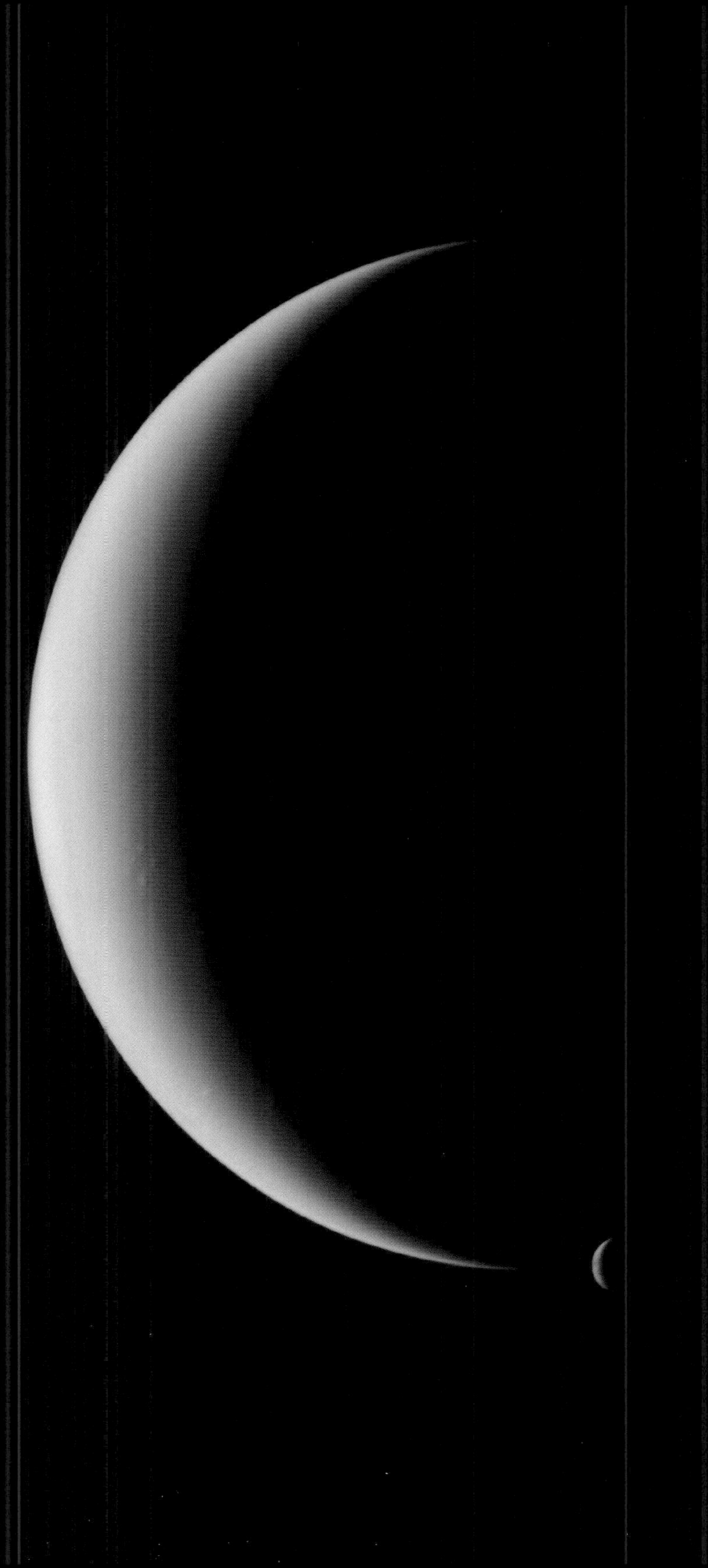

1. Michael BENSON, *Neptune and Triton, Voyager 2, August 31, 1989*, 2010, Courtesy Michael Benson & Hasted Kraeutler Gallery.

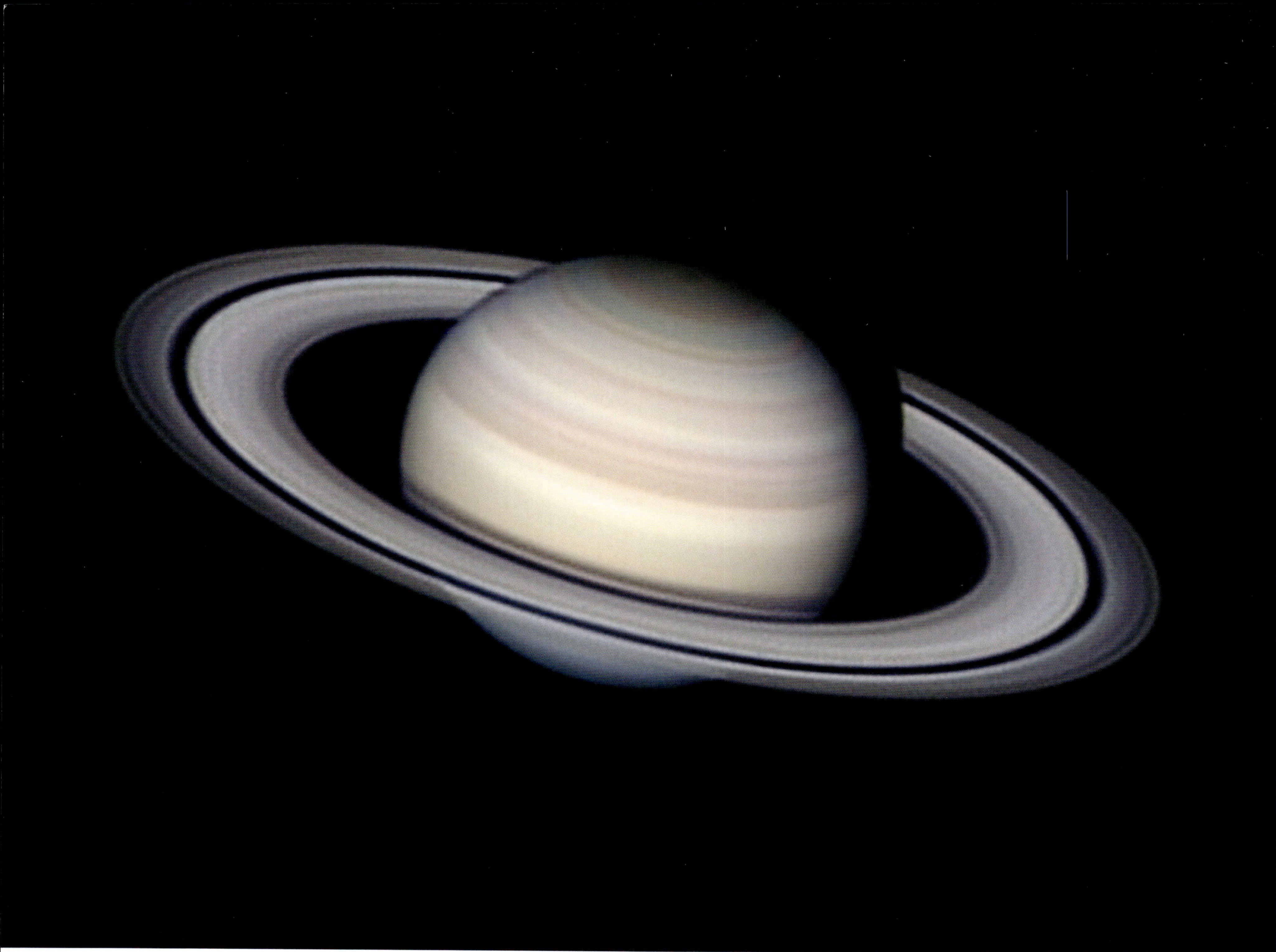

55. Damian PEACH, *Saturn, April 10*, 2006; and

56. *Jupiter, Ganymede, and Io*, 2010.

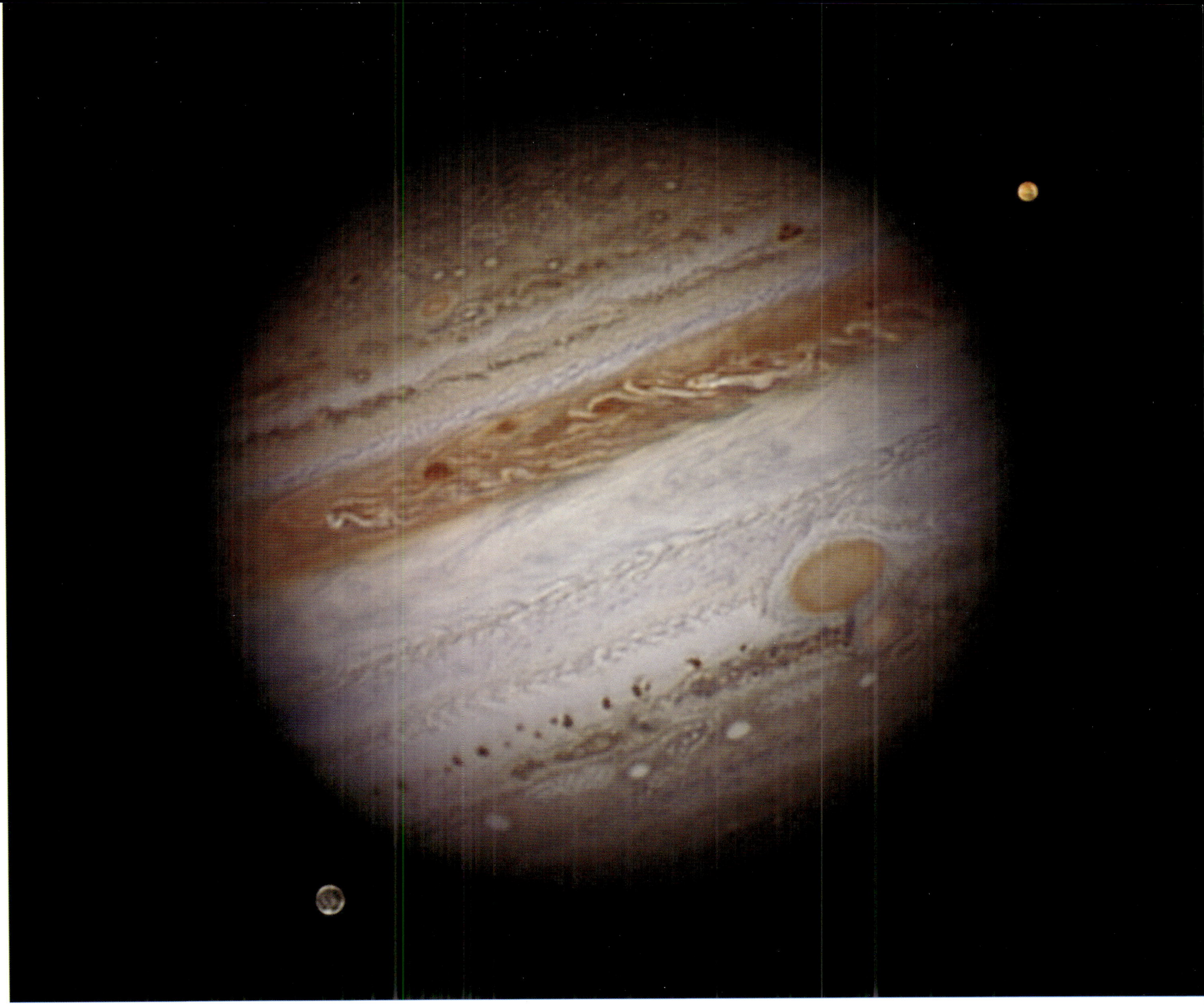

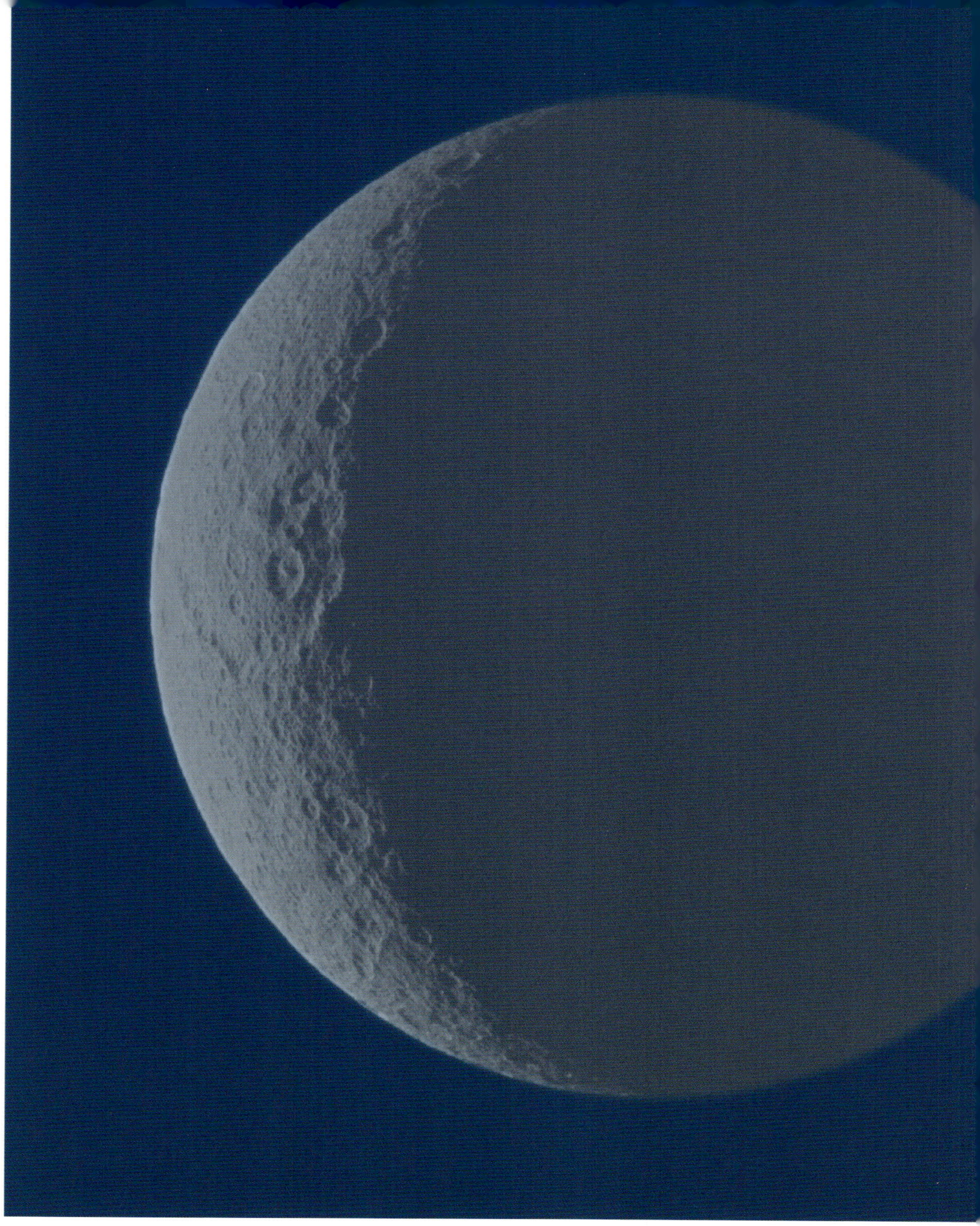

65. Thomas RUFF, *cassini 24*, 2009; and

66. *cassini 33*, 2009. Courtesy David Zwirner, New York.

[This work's subject is Rhea, Saturn's second-largest moon, discovered by Giovanni Domenico Cassini.—Editor]

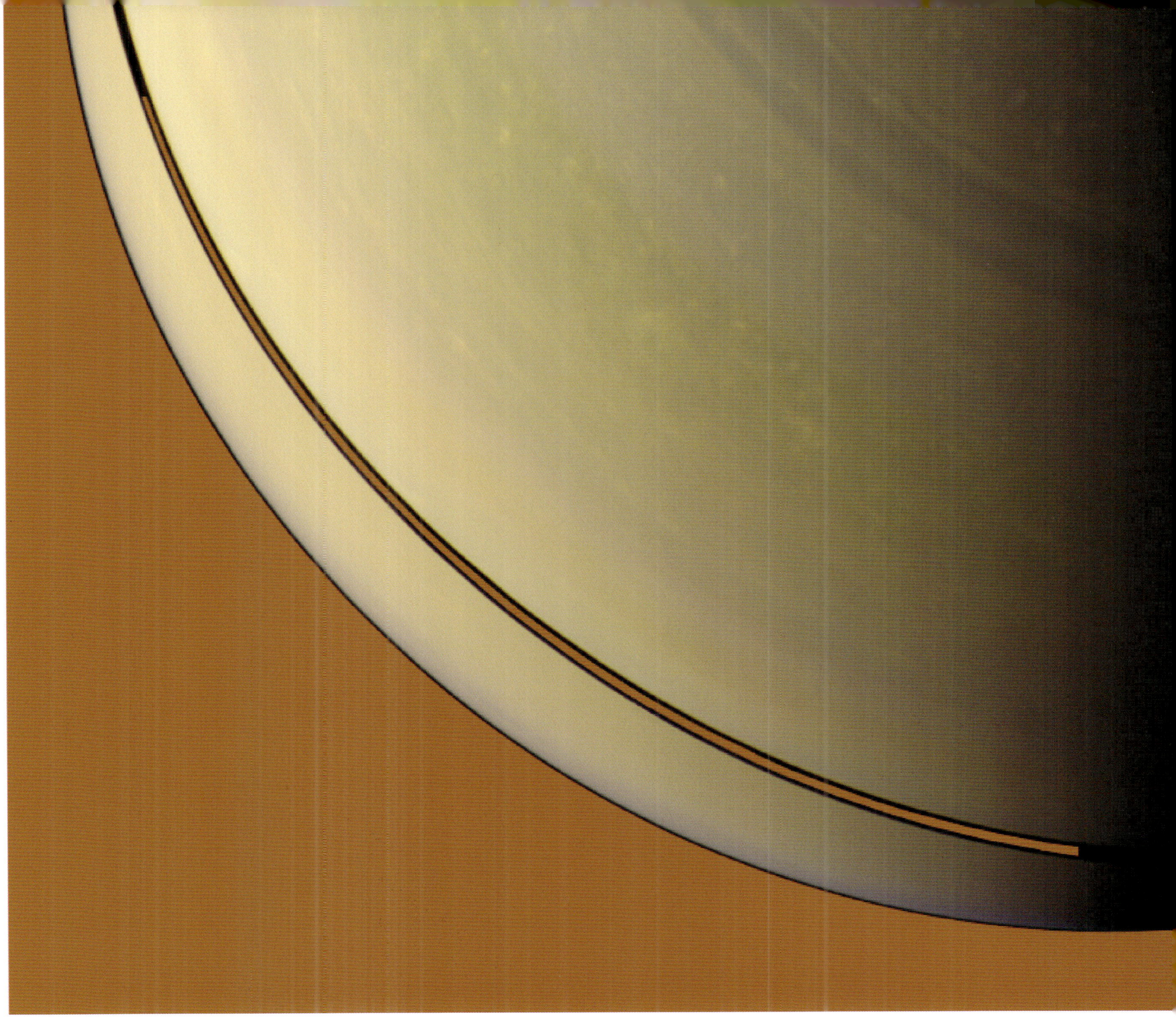

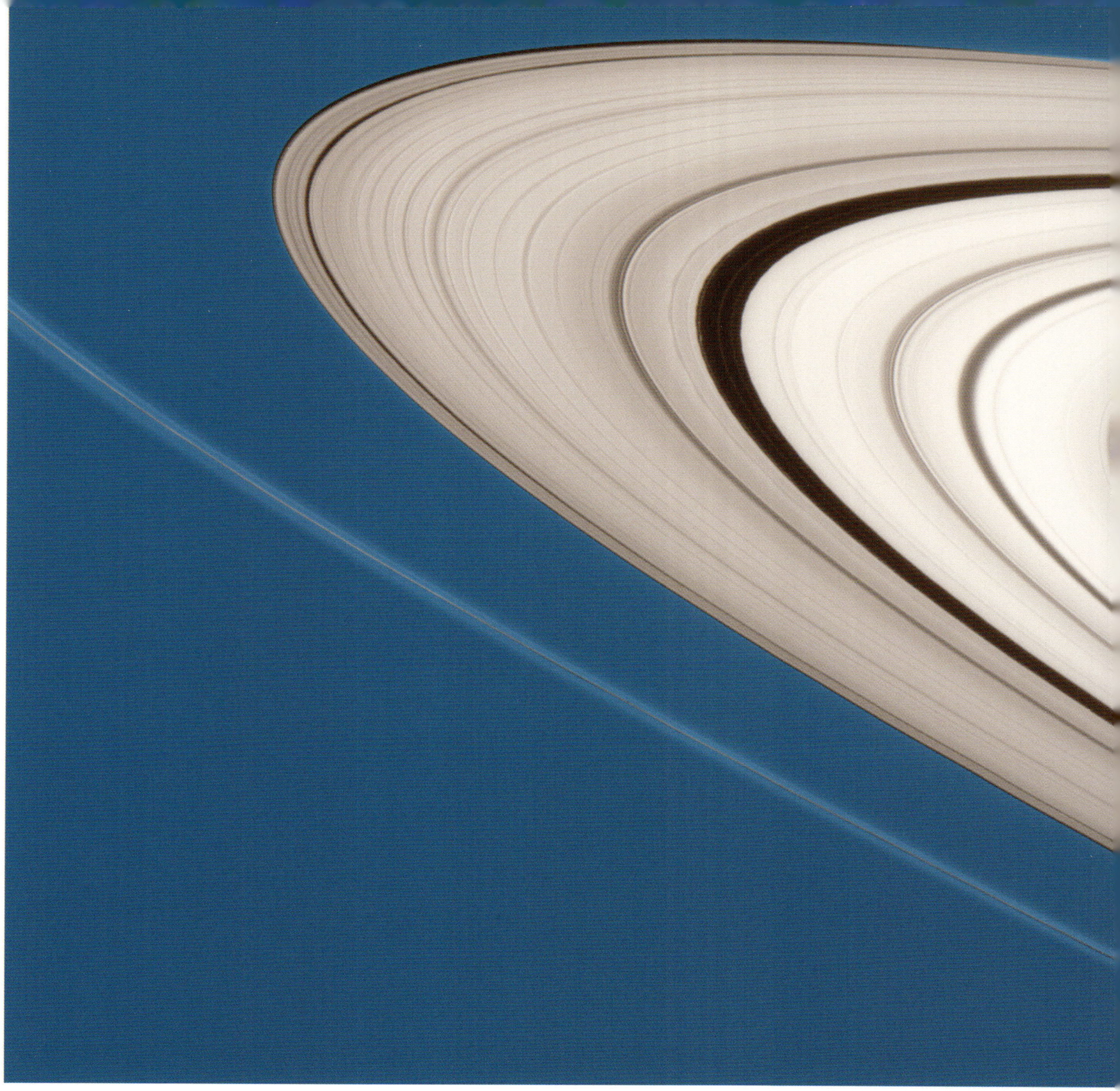

64. Thomas RUFF, *cassini 17*, 2009. Courtesy David Zwirner, New York.

100. Wei-Hao Wang, *Comet Hale-Bopp, March 10*, 1997.

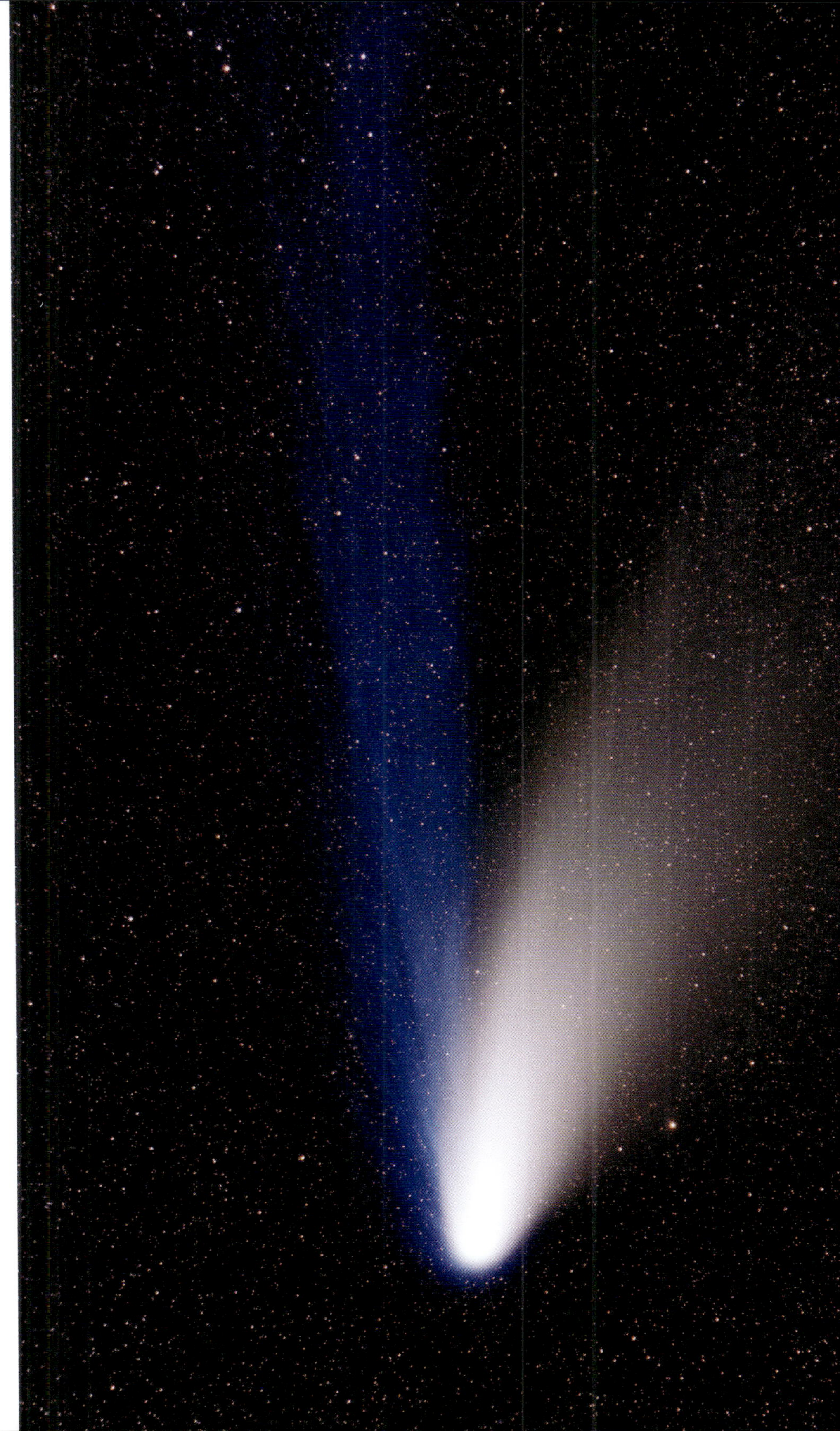

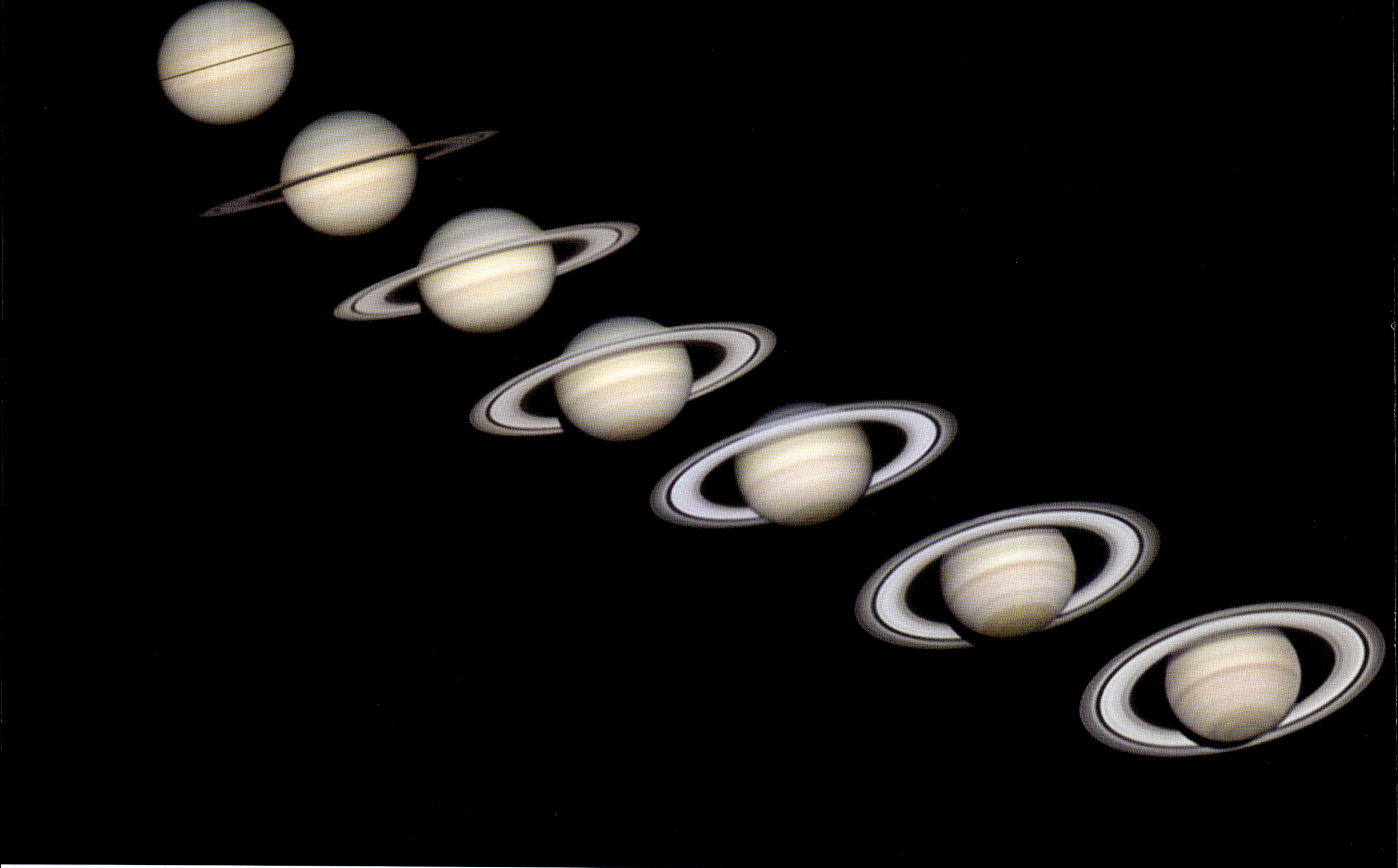

26. Alan FRIEDMAN, *6 Years of Saturn*.

PAGE 182

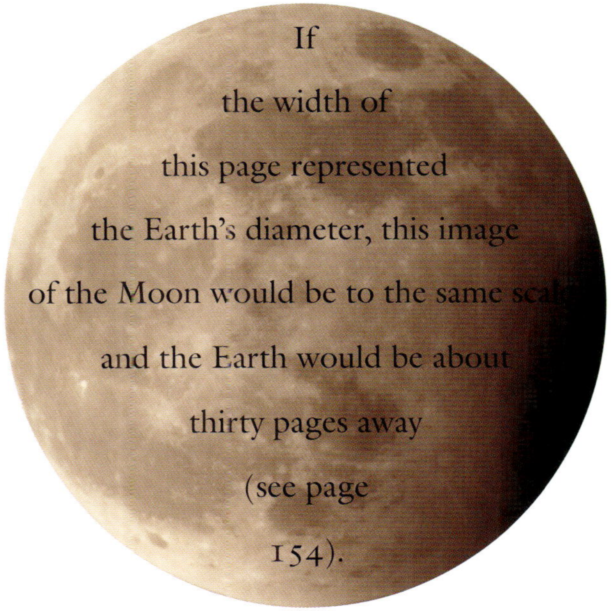

Scale & Distance (cont.)

If, instead of representing an Earth diameter (see page 162), the width of each page in this catalogue represented the average Earth-Moon distance (about 238,855 miles), it would take almost 390 pages laid end-to-end to represent the distance from Earth to the Sun (almost 93 million miles, or one Astronomical Unit [AU]), and the Sun itself would be more than three and one-half pages wide.

If each page width represented one AU, the nearest star outside our solar system, Proxima Centauri, would be more than 268,000 pages away (a line of paper extending roughly 48.7 miles and representing about 4.2 light years). If we then imagine that each page width instead represented this particular interstellar distance, the Orion Nebula (cat. 29, p. 131) would be about 316 pages away—about the length of a football field from goal line to goal line.

Our solar system is thought to be roughly 25,000 light years from the center of the Milky Way Galaxy. If each page width in this catalog represented *that* distance, the nearest galaxy to us, the Andromeda Galaxy (cat. 31, p. 147), would be about 104 pages away, a trail of paper representing a distance of approximately 2.6 million light years.

61. Jean-Paul ROUX, *Blue Moon Eclipse*, 2010. [detail]

EXPLORING THE ARTISTS' CREATIONS in the Starstruck exhibition or the pages of this catalogue, you are a cosmic traveler examining depictions of the Universe light-minutes, light-hours, and ultimately millions of light-years in the past. Ironically, you may find yourself drawn by their subjects to contemplate the future, inspired by a confluence of human endeavors in art and science and a golden age of astroimaging.

JERRY T. BONNELL

The celestial sphere

59. Nick Risinger, *Photopic Sky Survey.*

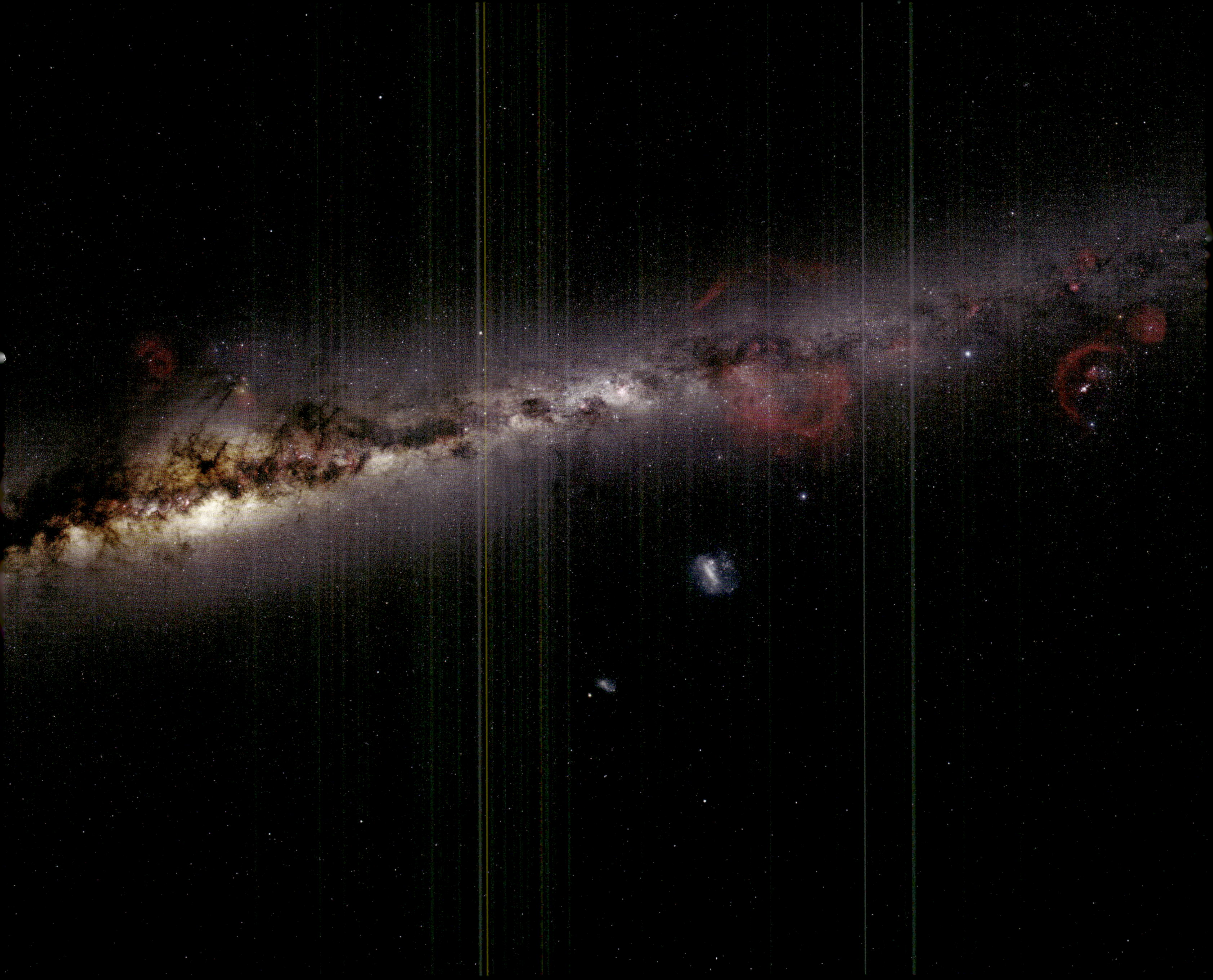

Checklist

Michael Benson

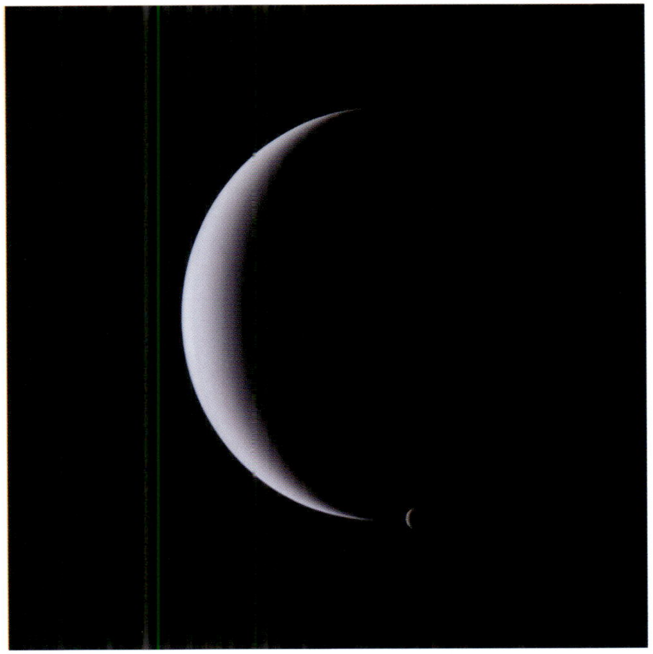

1. *Neptune and Triton, Voyager 2, August 31, 1989*, 2010
digital c-print
35 x 35 inches
Courtesy Michael Benson and
Hasted Kraeutler Gallery
See Giant Planets, Satellites, Rings, & a Comet section (§), page 175, for full-page image.

Starstruck is an international exhibition including artists from eleven countries and representing every continent except Africa and Antarctica. Unless artists are from the United States, their nationality is listed. All artists are living, except for Alfred Stieglitz, whose dates are provided.

—Editor

Michael Benson

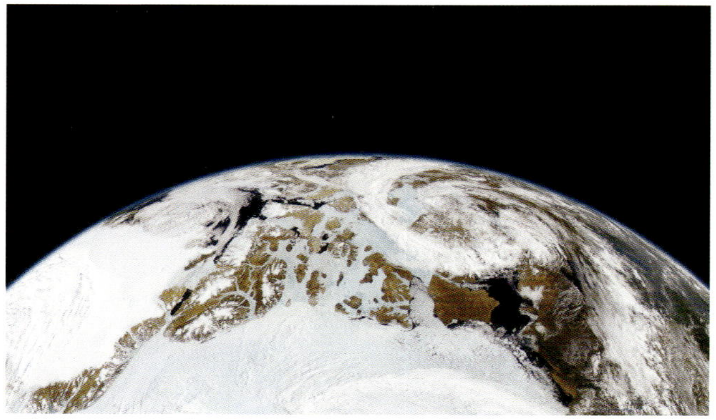

2. *Northern Canada and Northern Greenland, OrbView-2, July 9, 1999*, 2003
digital c-print
35 x 70 inches
Courtesy Michael Benson and Hasted Kraeutler Gallery
The Earth §, p. 155

Adam D. Block

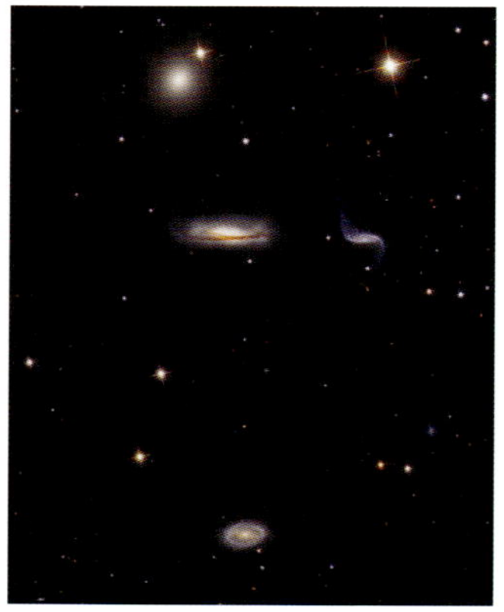

3. *Many and Beyond (NGC 3190)*
giclée on semi-gloss paper
27 ½ x 22 inches
Galaxies §, p. 152

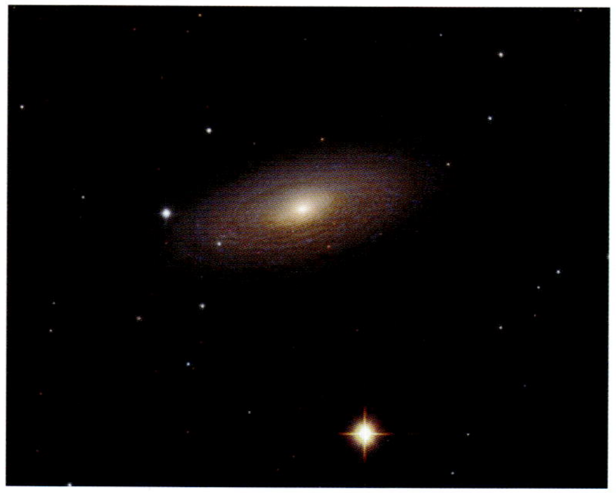

4. *Delicate Perfection (NGC 2841)*
giclée on semi-gloss paper
22 x 27 ½ inches
Galaxies §, p. 148

ADAM D. BLOCK

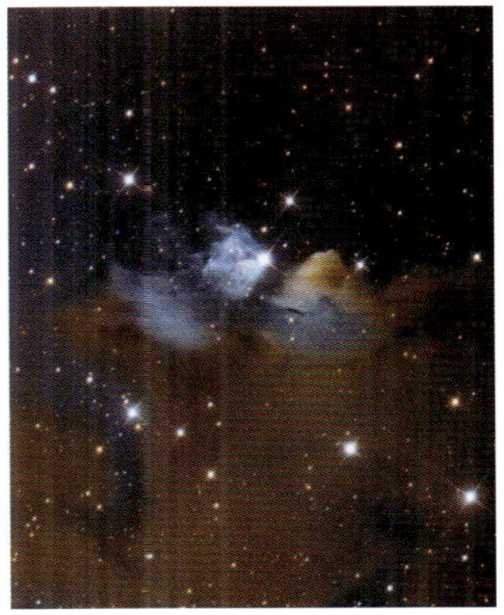

5. *Gift (vdB 24)*
giclée on semi-gloss paper
24 x 19 inches
Star Birth §, p. 124

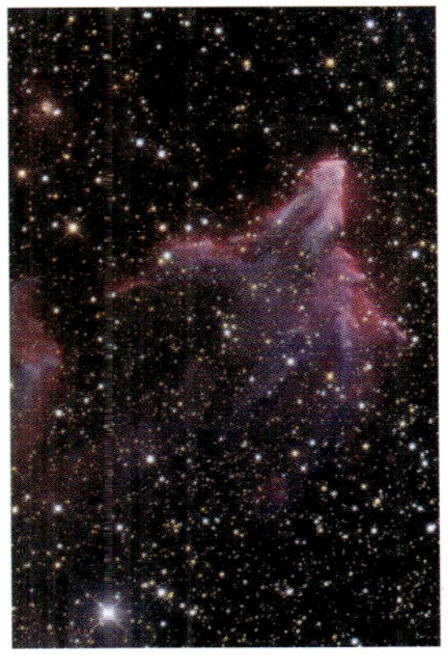

6. *Galactic Glide (IC 63)*
giclée on semi-gloss paper
32 x 21 ¼ inches
Star Birth §, p. 132

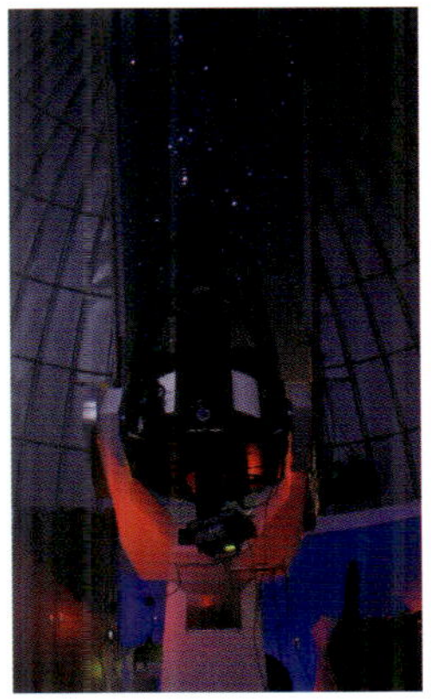

7. *Cosmic Expectations*
giclée on semi-gloss paper
32 x 19 inches
Stars §, p. 138

LINDA CONNOR

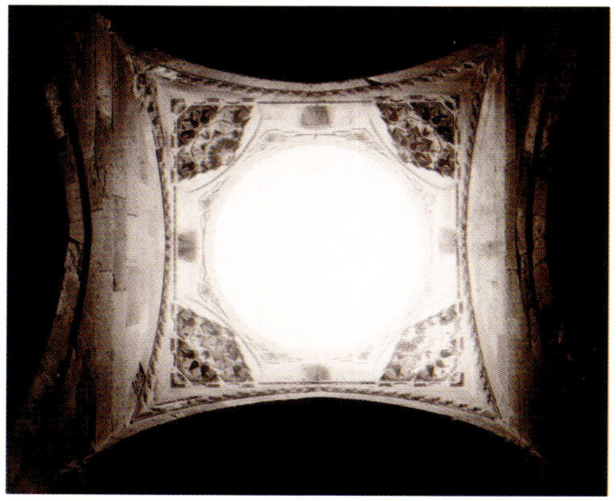

8. *Open Dome, Caravansary, Turkey*, 1992
silver gelatin print
10 x 12 inches
The Sun §, p. 89

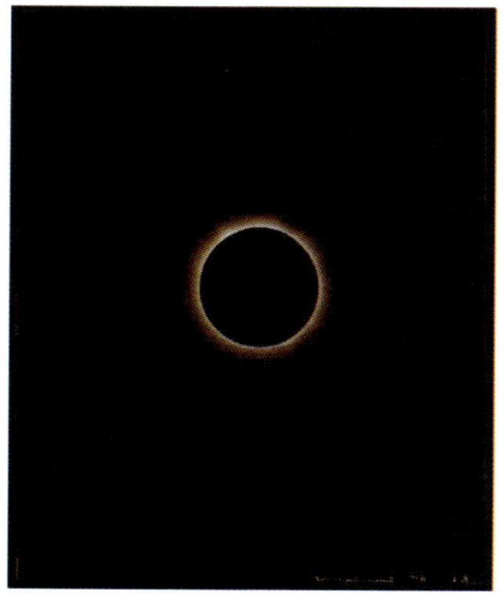

9. *Solar Eclipse, Egypt, August 30, 1905*, 1998
Printed from Lick Observatory Archive negative
Original photo by William Hussey
© UC Regents/Lick Observatory
silver gelatin printing-out paper
25 x 21 inches
The Sun, p. 116

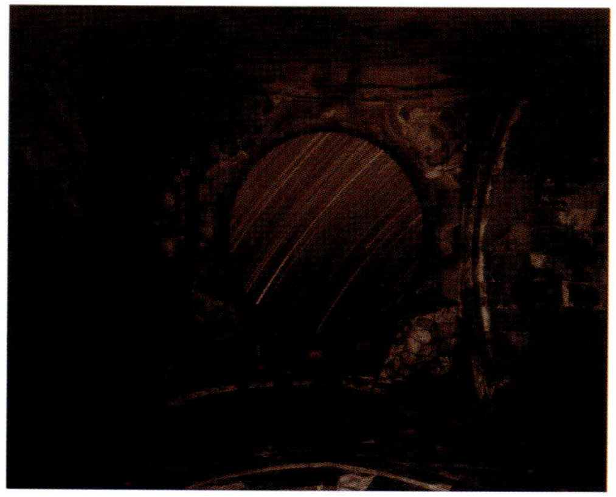

10. *Open Dome with Star Trails, Caravansary, Turkey*, 2004
silver gelatin print
10 x 12 inches
Stars §, p. 137

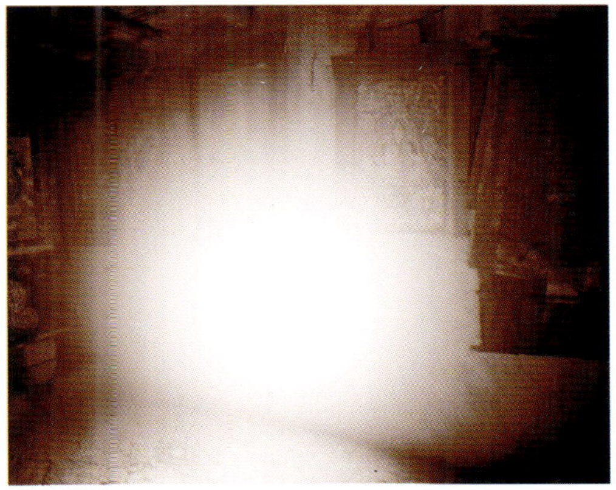

11. *Window with Thangkas, Ladakh, India,*
1994
silver gelatin print
10 x 12 inches
The Sun §, p. 110

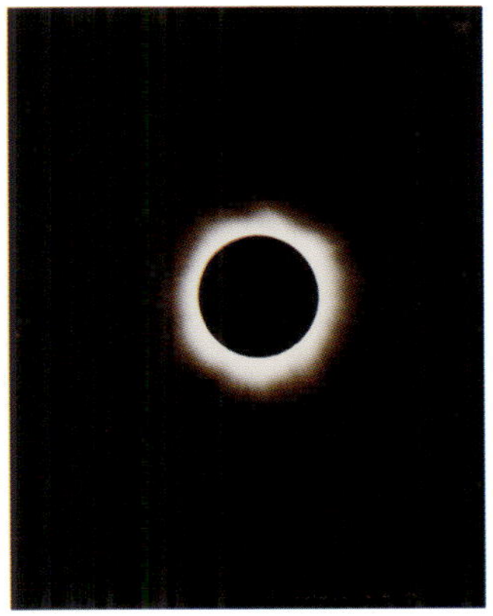

12. *Solar Eclipse, Chile, April 16, 1893*, 1998
Printed from Lick Observatory Archive negative
Original photo by John Schaeberle
© UC Regents/Lick Observatory
silver gelatin printing-out paper
25 x 21 inches
The Sun §, p. 78

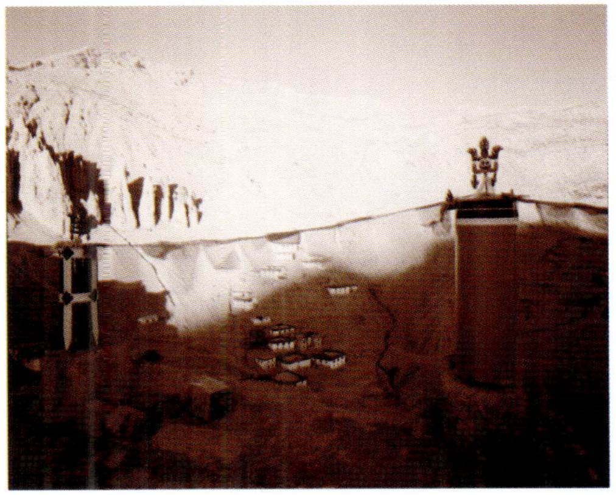

13. *From the Monastery Roof, Spiti, Ladakh, India*, 1994
silver gelatin print
10 x 12 inches
The Earth §, p. 160

James Cormier

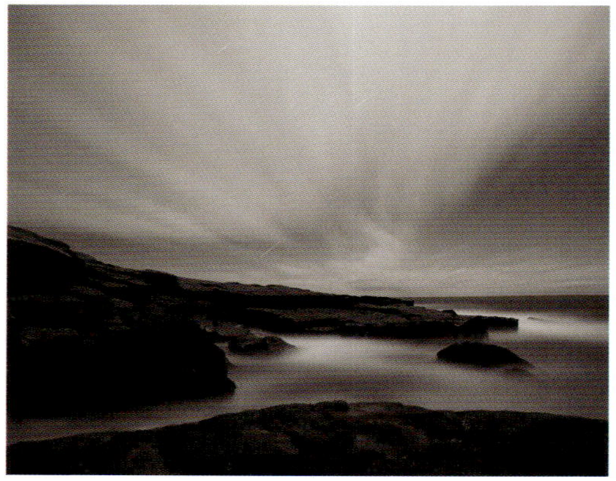

14. *Departing Clouds*, 2010
digital reproduction on
Kodak Endura Metallic Paper
16 x 20 inches
The Earth §, p. 156

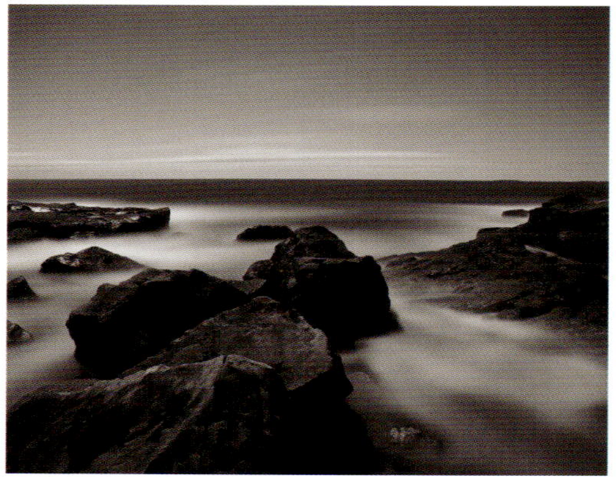

15. *Incoming Tide*, 2010
digital reproduction on
Kodak Endura Metallic Paper
16 x 20 inches
The Earth §, p. 161

Ken Crawford

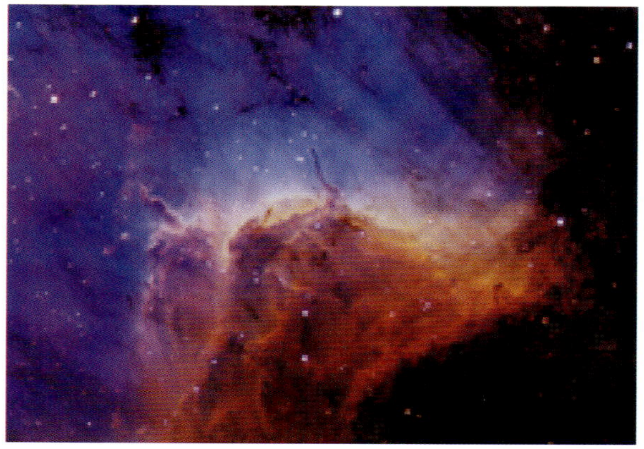

16. *IC 5067*
archival inkjet print
13 ¾ x 20 inches
Star Birth §, p. 117

Ken Crawford

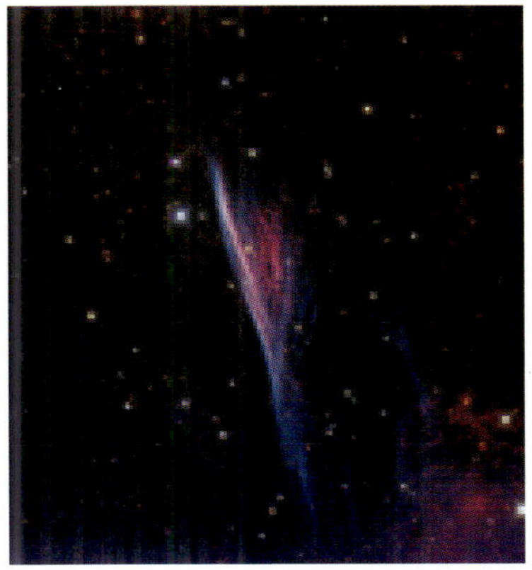

17. *NGC 2736*
archival inkjet print
20 x 18 ½ inches
Star Death §, p. 144

John A. Davis

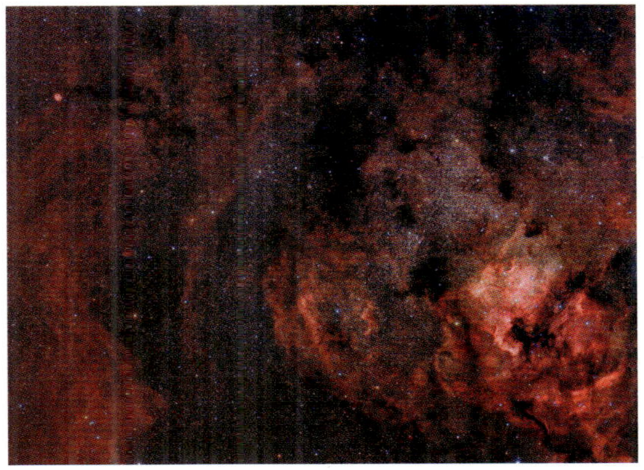

18. *Cocoon over North America*, 2011
inkjet print on Epson Ultra Premium Luster
15 x 21 inches
Star Birth §, p. 122

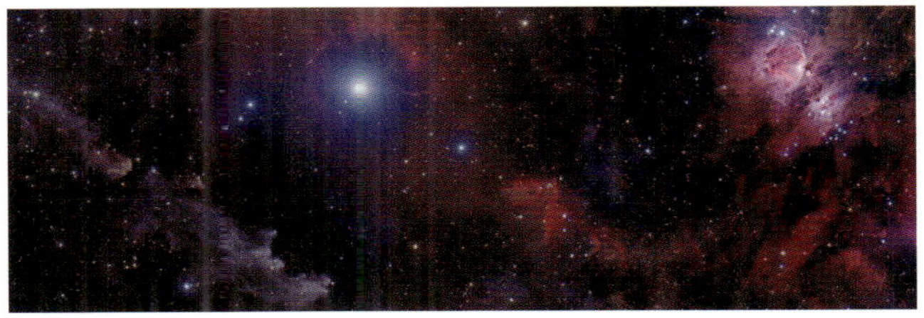

19. *Sword to Witch*, 2010
inkjet print on Epson Ultra Premium Luster
10 x 30 inches
Stars §, p. 134

Ignacio Diaz-Bobillo
Argentina

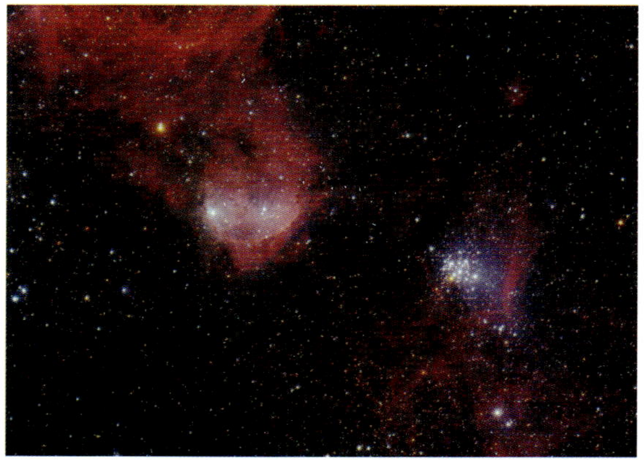

20. *Light Drops #1-A*
lightjet print on Fuji Crystal Archive gloss paper
14 x 19 ½ inches
Star Birth §, p. 118

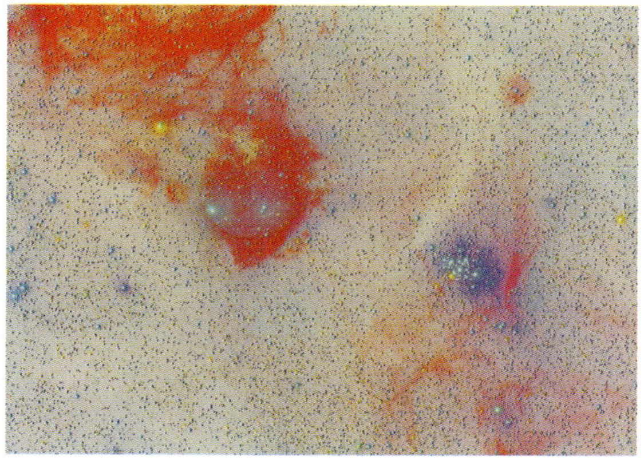

21. *Light Drops #1-B*
lightjet print on Fuji Crystal Archive gloss paper
14 x 19 ½ inches
Star Birth §, p. 119

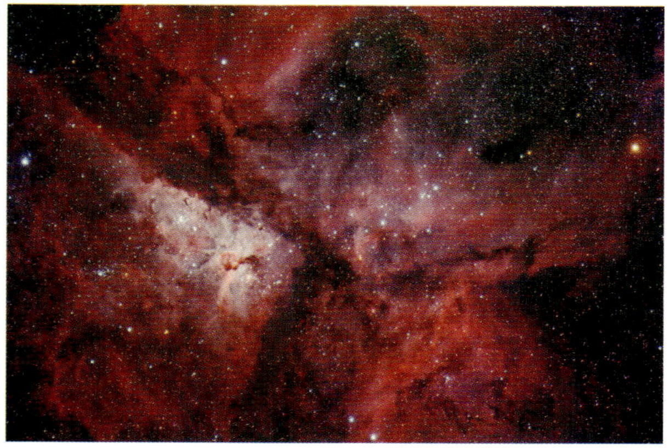

22. *Light Drops #2-A*
lightjet print on Fuji Crystal Archive gloss paper
14 x 21 inches
Star Birth §, p. 126

Ignacio Diaz-Bobillo

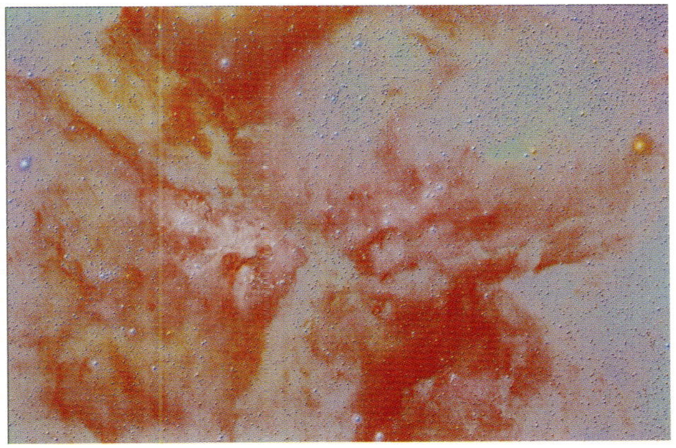

23. *Light Drops #2-B*
lightjet print on Fuji Crystal Archive gloss paper
14 x 21 inches
Star Birth §, p. 127

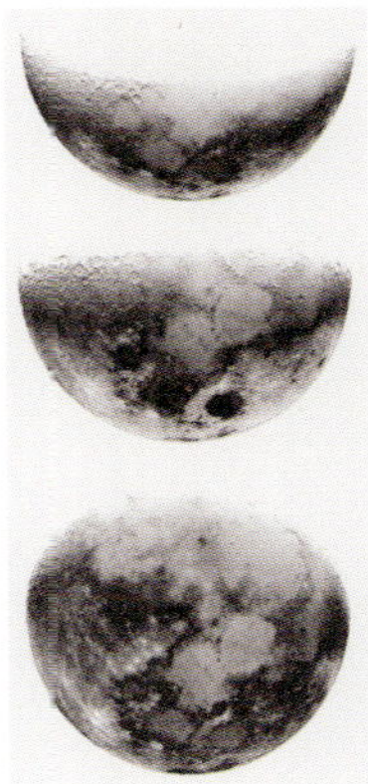

24. *Lunar Triptych #1-Silver Moons*
lightjet print on Fuji Crystal Archive gloss paper
56 x 26 inches
The Moon §, p. 173

Dennis di Cicco

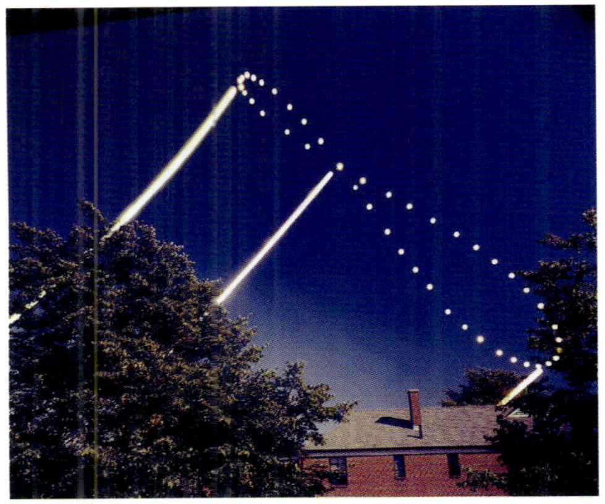

25. *Solar Analemma*, 1978–79
dye transfer print
16 x 20 inches
printed by Ctein
The Sun §, p. 105

ALAN FRIEDMAN

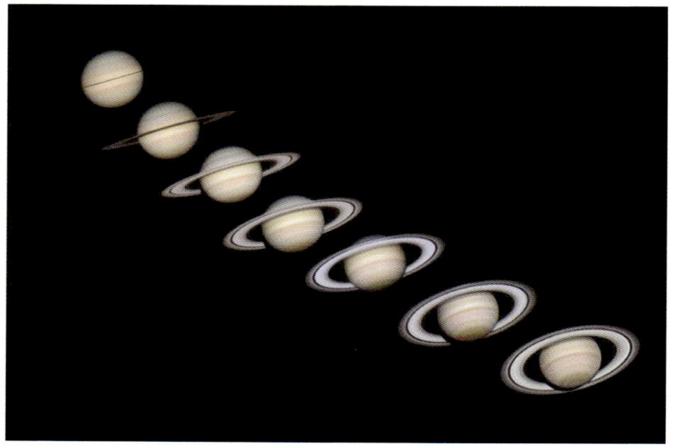

26. *Six Years of Saturn*
archival inkjet print
17 x 22 inches
Giant Planets, Satellites, Rings, & a Comet §, p. 182

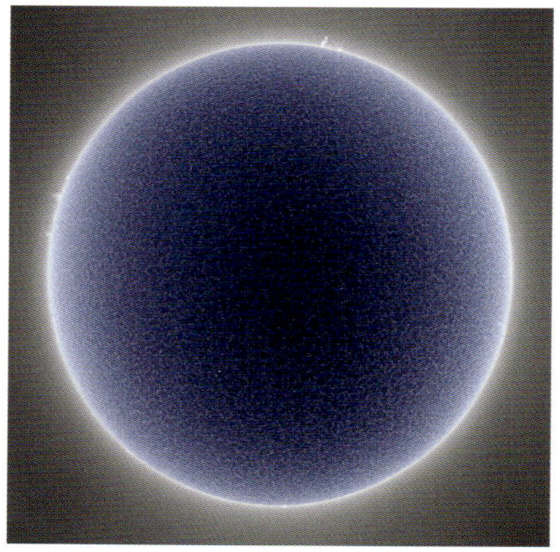

27. *The Blue Sun*
archival inkjet print
22 x 22 inches
The Sun §, p. 77

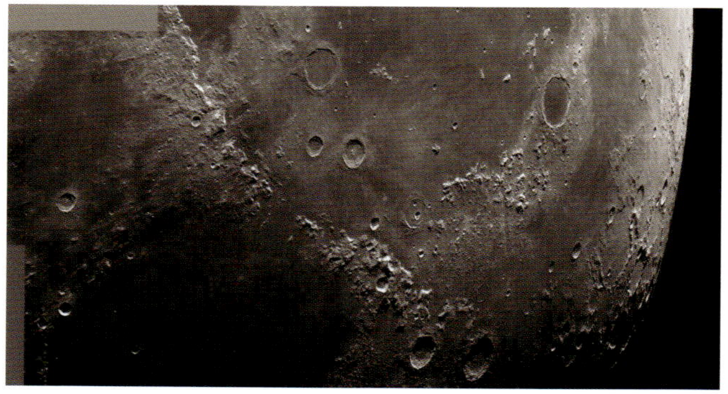

28. *Kleine Ina Nachtmusik*
archival inkjet print
17 x 22 inches
The Moon §, p. 165

ROBERT GENDLER

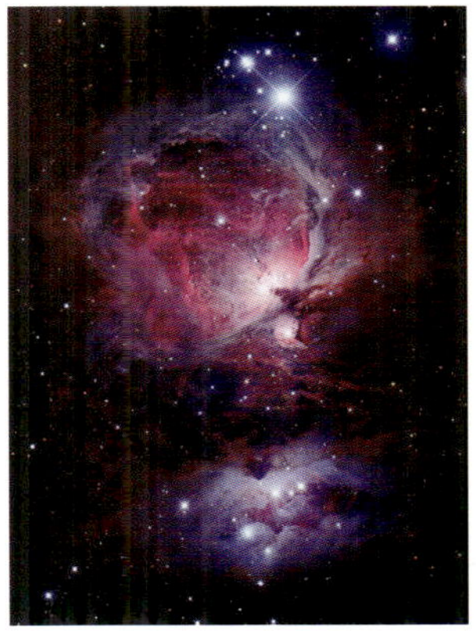

29. *M42, The Great Nebula in Orion*, 2005
inkjet print
60 x 44 inches
Star Birth §, p. 131

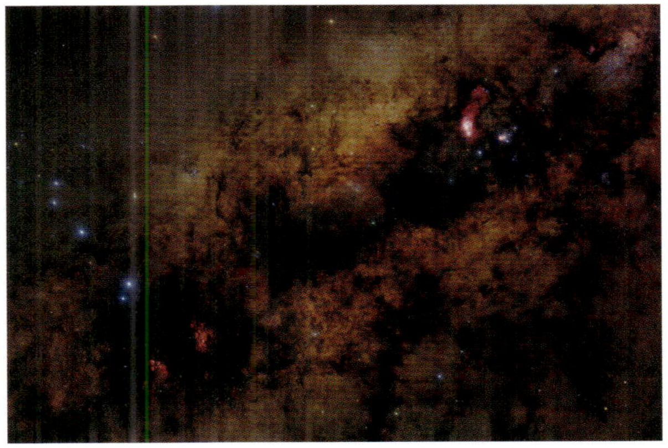

30. *The Galactic Center*, 2009
inkjet print
40 x 60 inches
Galaxies §, p. 145

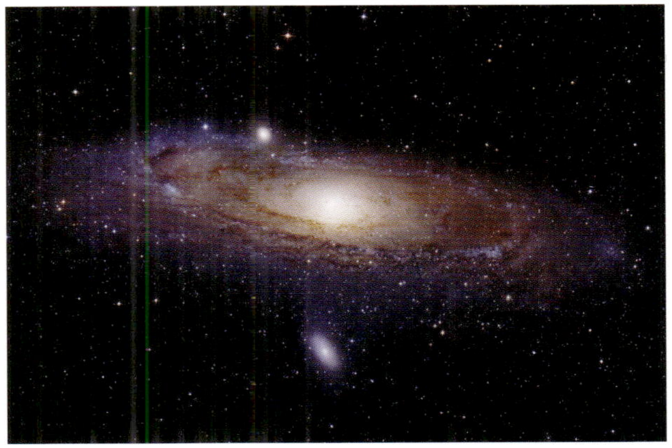

31. *The Andromeda Galaxy*, 2005
inkjet print
39 ½ x 60 inches
Galaxies §, p. 147

Robert Gendler

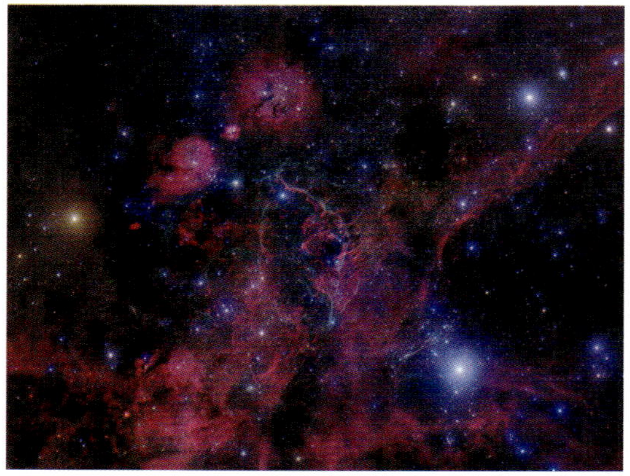

32. *Vela Supernova Remnant*, 2008
inkjet print
45 ½ x 60 inches
Star Death §, p. 142

Jacob Gerritsen

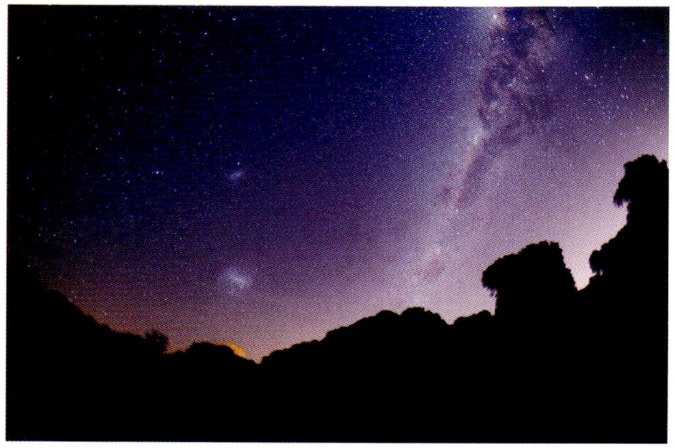

33. *A Southern View of the Southern Skies and the Magellanic Clouds*, 2011
Epson 7600 printer using Ultrachrome Ink and Epson Color Premium Luster Paper
12 x 18 inches
Galaxies §, p. 150

Don Goldman

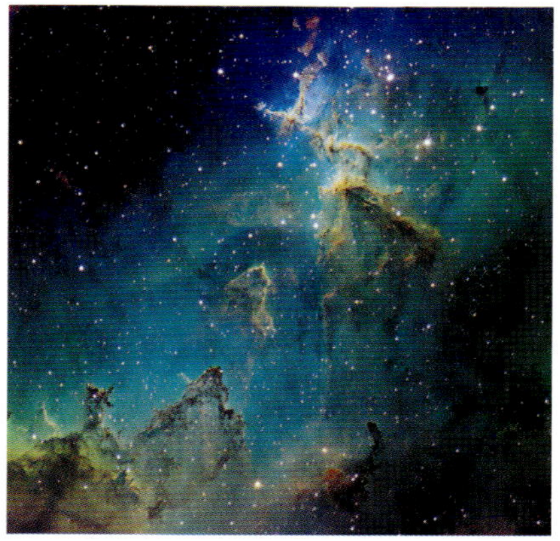

34. *Cosmic Dust*
inkjet print
36 x 36 inches
Star Birth §, p. 125

Don Goldman

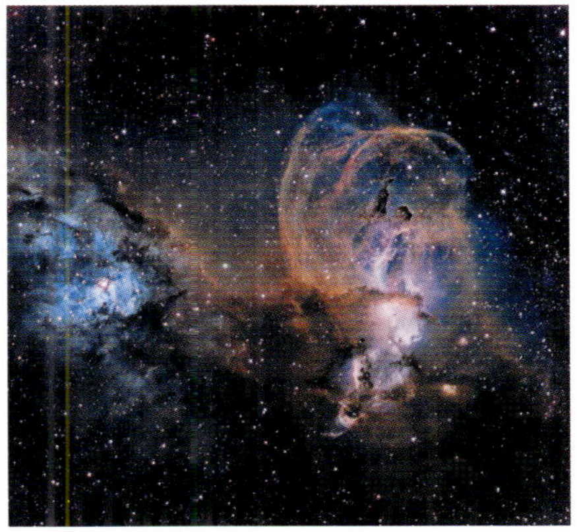

35. *Statue of Liberty in Space*
inkjet print
24 x 24 inches
Star Birth §, p. 129

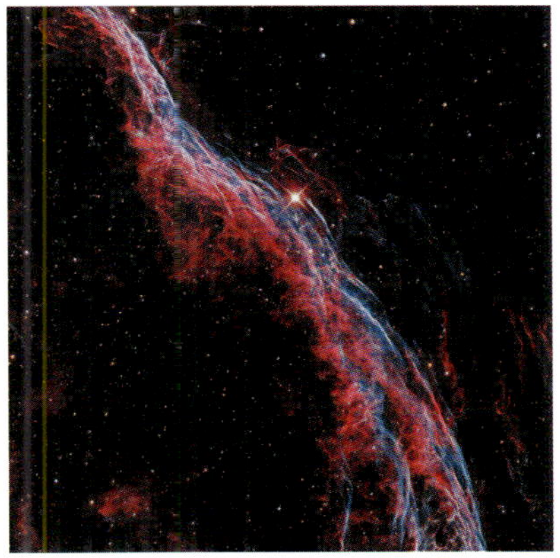

36. *Witch's Broom*
inkjet print
24 x 24 inches
Star Death §, p. 143

Sharon Harper

37. *Sun/Moon (Trying to See Through a Telescope), 2010 July 6 10:23:10–38 a.m.*, 2010
Ultrachrome print on Epson Enhanced Matte paper
17 x 68 inches
Courtesy of the artist, Galerie Roepke, Cologne, and Rick Wester Fine Art, New York
The Moon §, p. 172–3

Sharon Harper

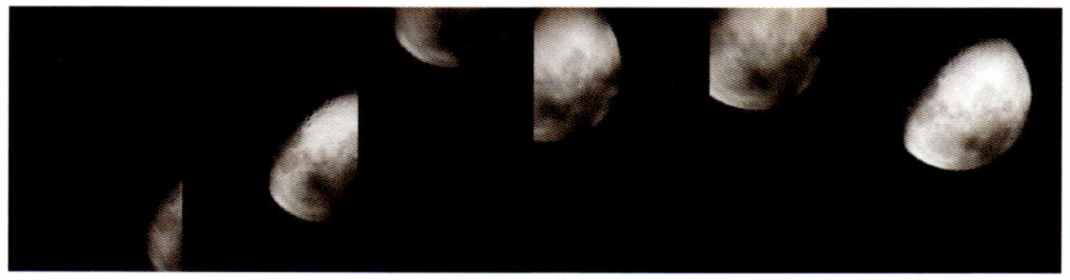

38. *Sun/Moon (Trying to See Through a Telescope) 2010 June 2 3:30:43–31:05 a.m.*, 2010
Ultrachrome print on Epson Enhanced Matte paper
17 x 68 inches
Courtesy of the artist, Galerie Roepke, Cologne, and Rick Wester Fine Art, New York
The Moon §, p. 170–1

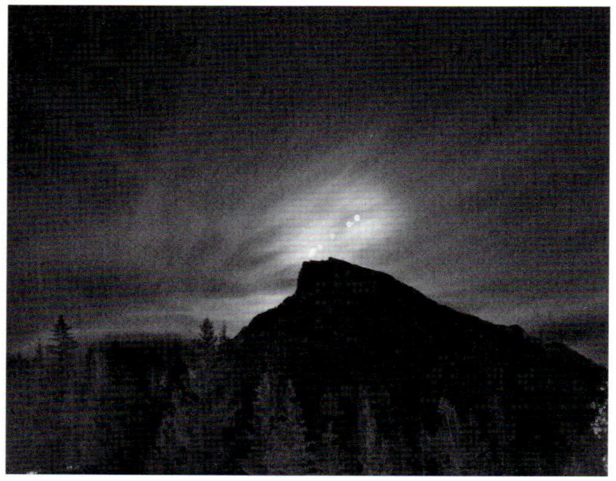

39. *Night Sky Over Banff, Alberta, September 12-October 10, 2007, 25 September*, 2007
Harman fiber base print
20 x 24 inches
Courtesy of the artist, Galerie Roepke, Cologne, and Rick Wester Fine Art, New York
The Moon §, p. 167

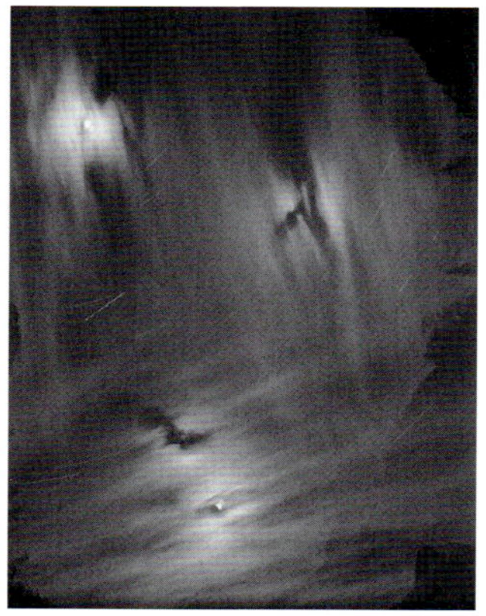

40. *Night Sky Over Banff, Alberta, September 12-October 10, 2007, 27–29 September*, 2007
Harman fiber base print
40 x 30 inches
Courtesy of the artist, Galerie Roepke, Cologne, and Rick Wester Fine Art, New York
The Moon §, p. 168

WARREN KELLER

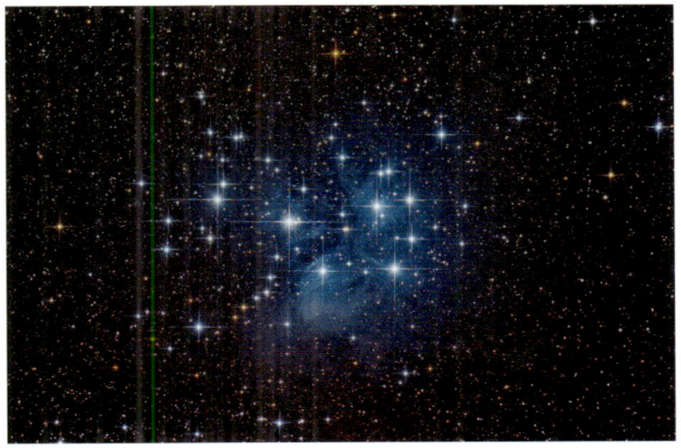

41. *Seven Sisters*, 2011
Durst Lambda laser print on FujiFlex
16 x 24 inches
Stars §, p. 133

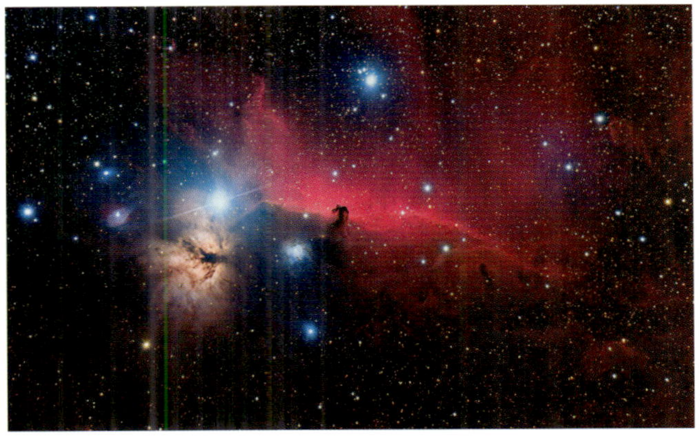

42. *Alnitak and Horsehead Region of Orion*, 2011
Durst Lambda laser print on FujiFlex
22 x 36 inches
Star Birth §, p. 123

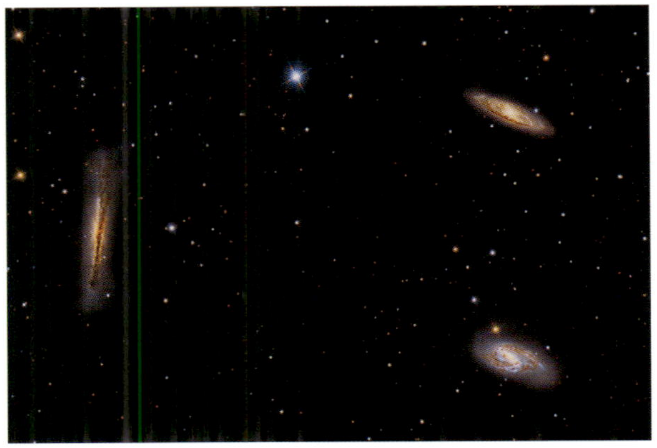

43. *Leo Triplet*, 2011
Durst Lambda 130 laser print
professional photo paper
24 x 36 inches
Galaxies §, p. 146

Walter Koprolin
Austria

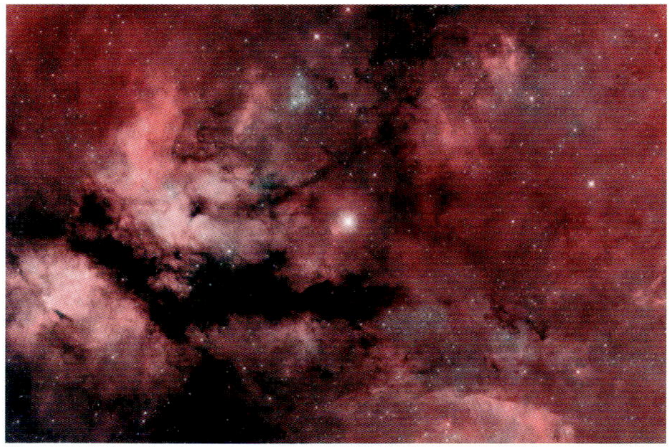

44. *Cygni Nebulosity*
inkjet print
13 x 20 inches
Star Birth §, p. 120

David Malin
Australia

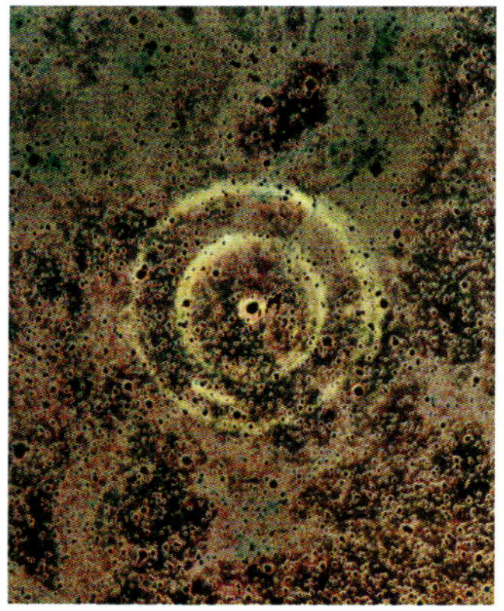

45. *The Light Echo of Supernova 1987A*
archival inkjet print made from a digital matrix derived from three black & white glass plate negatives recording red, green, and blue visible light
50 x 40 inches
© Australian Astronomical Observatory
Star Death §, p. 139

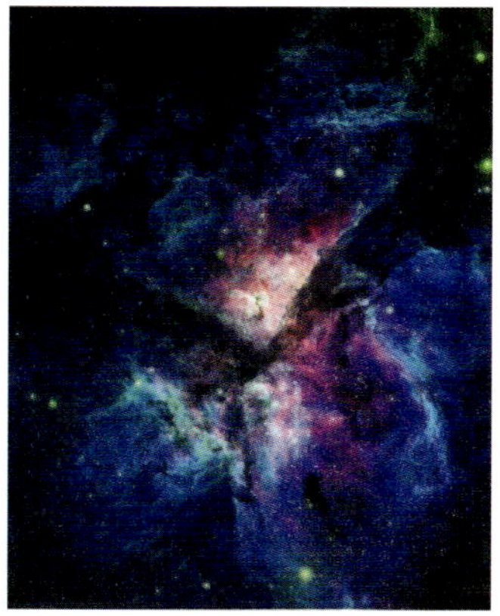

46. *The Elemental Structure of the Carina Nebula*
archival inkjet print made from a digital matrix derived from three black & white glass plate negatives recording red, green, and blue visible light
50 x 40 inches
© Australian Astronomical Observatory
Star Birth §, p. 128

David Malin

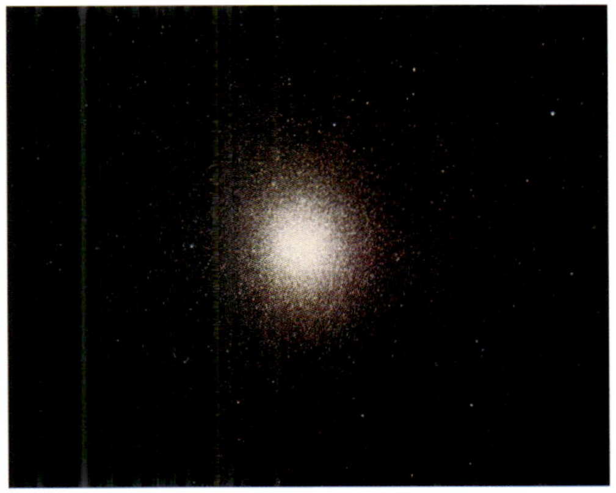

47. *The Globular Cluster Omega Centauri, NGC 5139*
archival inkjet print made from a digital matrix derived from three black & white glass plate negatives recording red, green, and blue visible light
40 x 50 inches

Stars §, p. 135

Steven Mazlin and Jack Harvey

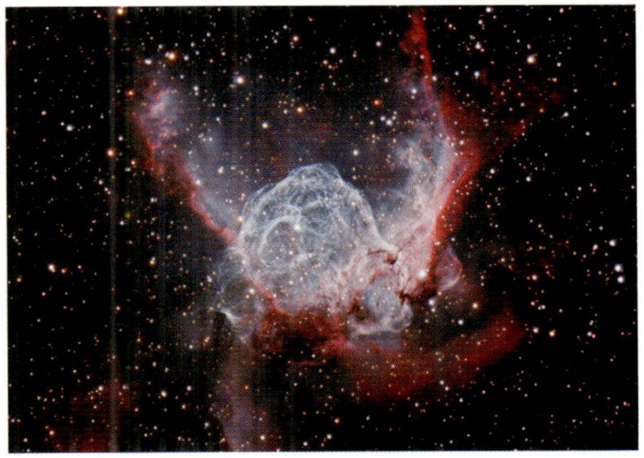

48. *Thor's Helmet in Canis Major (aka NGC 2359)*
archival inkjet print
20 x 27 ½ inches
Star Death §, p. 141

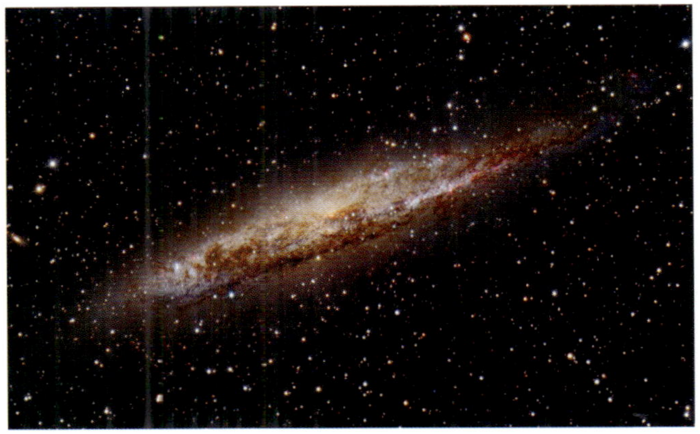

49. *NGC 4945 in Centaurus*
archival inkjet print
20 x 30 inches
Galaxies §, p. 151

STEVEN MAZLIN AND JACK HARVEY

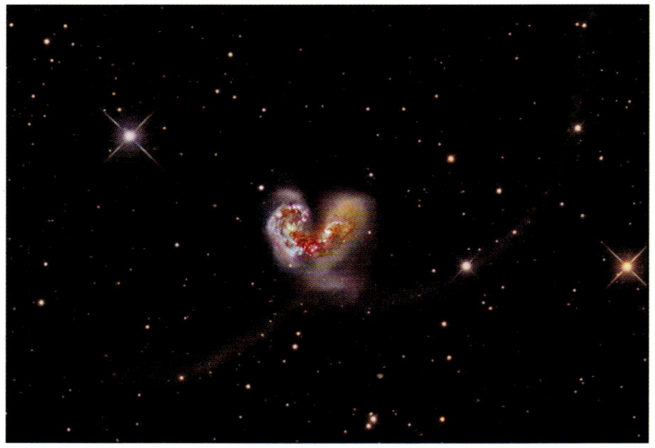

50. *The Antennae Galaxies in Corvus (aka NGC 4038 & 4039)*
archival inkjet print
18 ½ x 30 inches
Galaxies §, p. 153

JIM NICKELSON

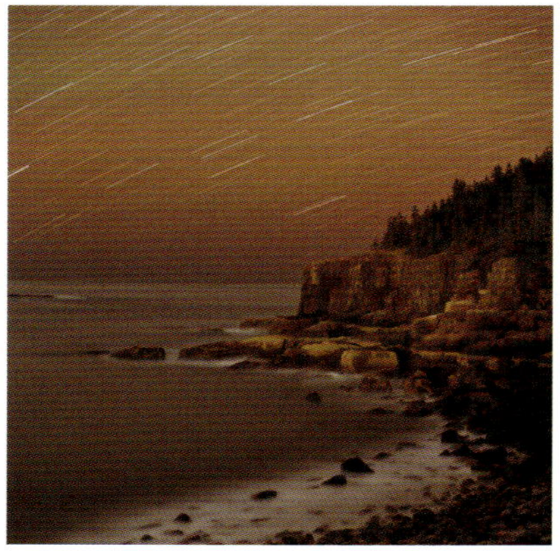

51. *Nightfall, Otter Cliffs*, 2011
archival pigment inkjet print on cotton rag
24 x 24 inches
Stars §, p. 136

52. *Thunder Moon II*, 2011
archival pigment inkjet print on cotton rag
24 x 24 inches
The Moon §, p. 166

Geir Øye
Norway

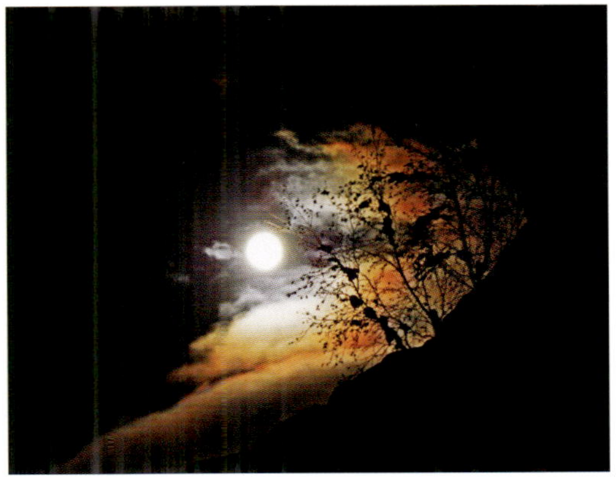

53. *Moon Iridescence over Snow Covered Tree Branches*
chromogenic print
14 x 18 inches
The Moon §, p. 164

Jay Pasachoff

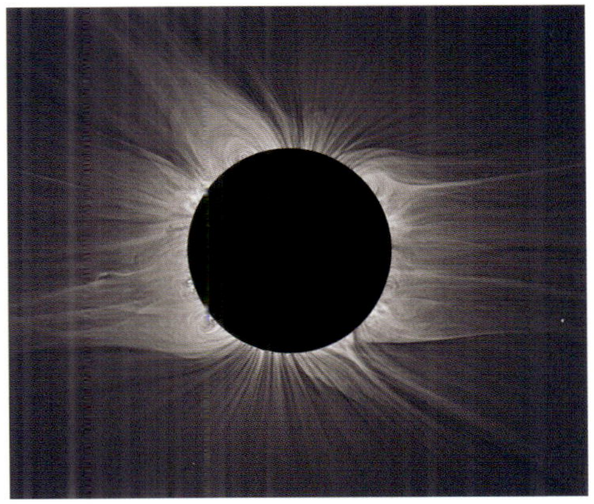

54. *Easter Island Eclipse, July 11*, 2010
archival inkjet print
20 x 24 inches
The Sun §, p. 79

Damian Peach
United Kingdom

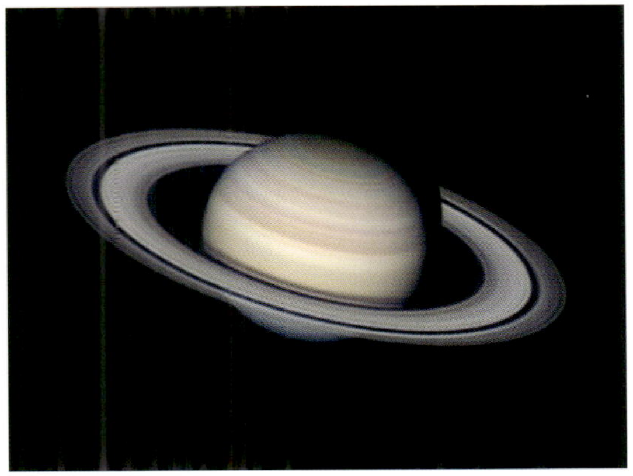

55. *Saturn, April 10*, 2006
archival inkjet print
12 x 16 inches
Giant Planets, Satellites, Rings, & a Comet §, p. 176

Damian Peach

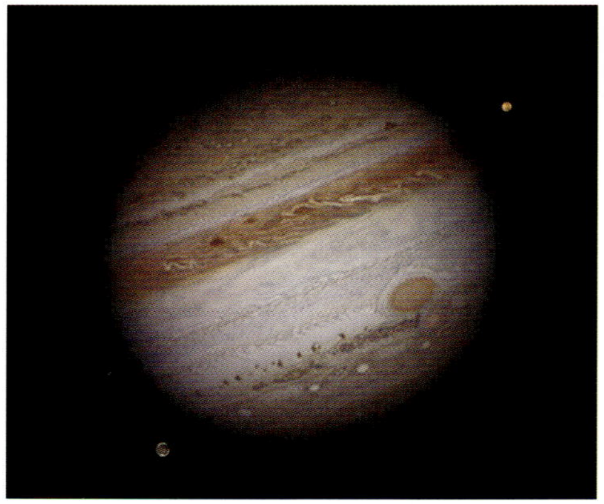

56. *Jupiter, Ganymede, and Io*, 2010
archival inkjet print
20 x 24 inches
Giant Planets, Satellites, Rings, & a Comet
§, p. 177

Martin Pugh
Australia

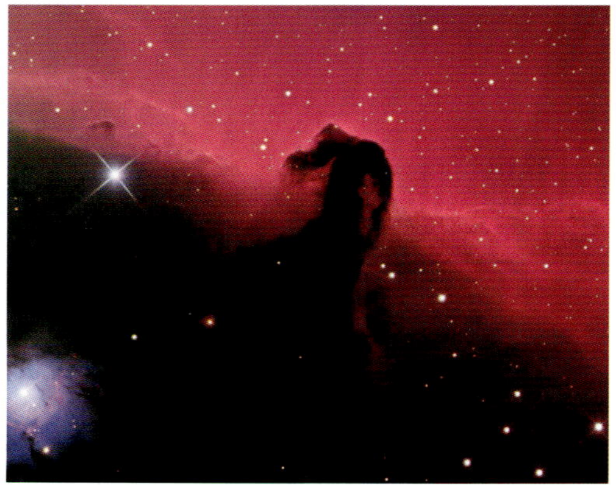

57. *IC 434 in Orion (The Horsehead Nebula)*
Hahnemuhle Fine Art Photo Rag Bright
White 310 paper
13 x 19 inches
Star Birth §, p. 130

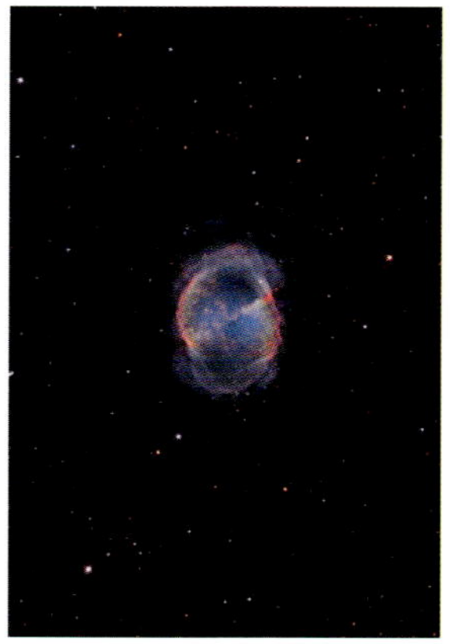

58. *M27 in Vulpecula (The Dumbbell Nebula)*
Hahnemuhle Fine Art Photo Rag Bright
White 310 paper
13 x 19 inches
Star Death §, p. 140

Nick Risinger

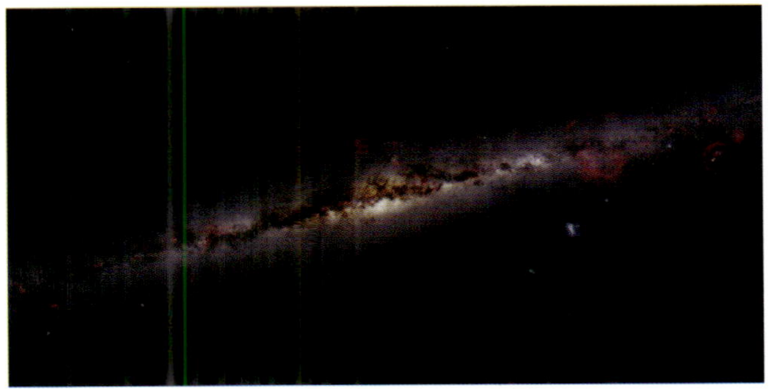

59. *Photopic Sky Survey*
lightjet print on Fuji Crystal Archive matte paper
60 x 120 inches
The Celestial Sphere §, pp. 186–7

Kirk Rogers

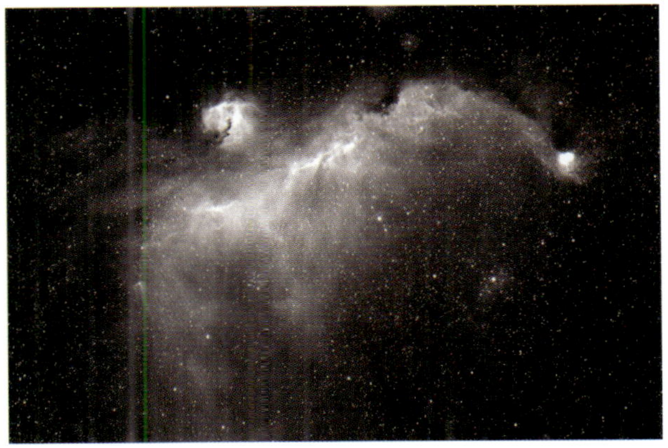

60. *IC 2177, Seagull in Monoceros in Hydrogen-alpha*
Epson Stylus Pro 7900 printer using Epson UltraChrome HDR inks on Premium Luster paper
12 x 16 inches
Star Birth §, p. 121

Jean-Paul Roux
France

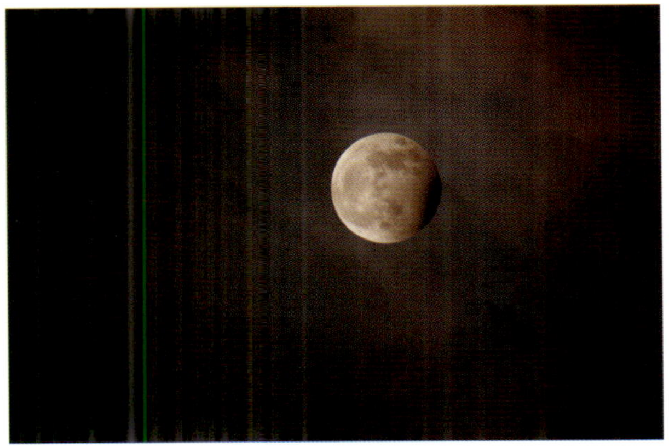

61. *Blue Moon Eclipse*, 2010
inkjet print
9 x 14 inches
The Moon §, p. 163

Tony Rowell

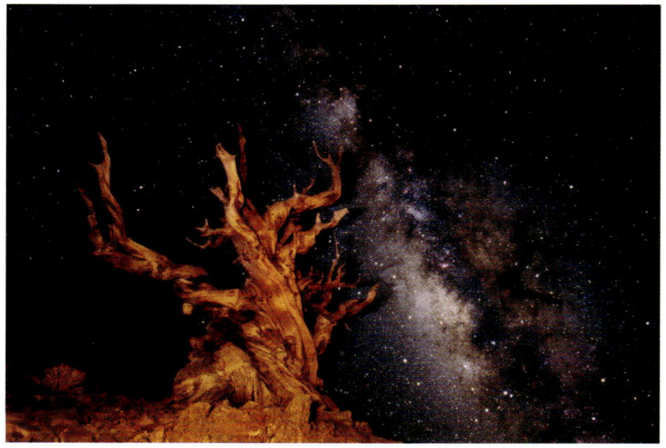

62. *Bristlecone Pine and Milky Way, White Mountains, California*, 2008
lightjet print on Fuji Crystal Archive Gloss paper
16 x 20 inches
Galaxies §, p. 149

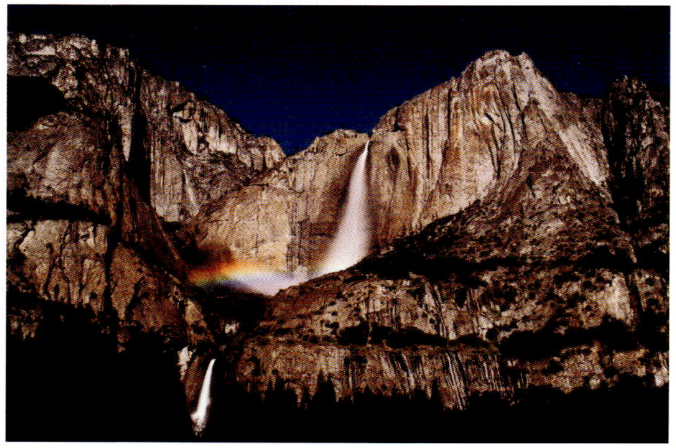

63. *Moonbow and Stars at Yosemite Falls, Yosemite National Park*, 2010
lightjet print on Fuji Crystal Archive Gloss paper
16 x 20 inches
The Earth §, p. 159

Thomas Ruff
Germany

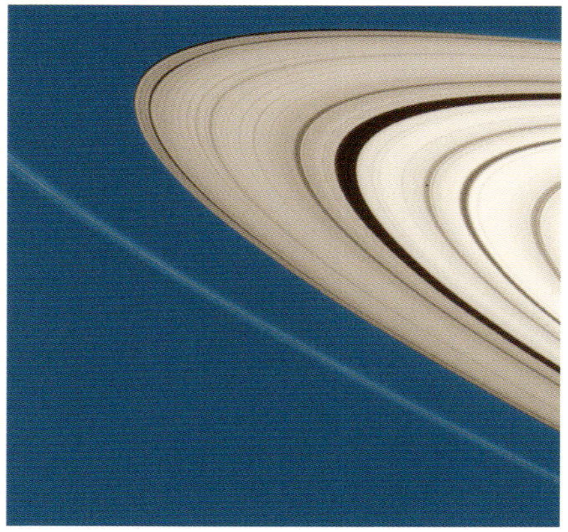

64. *cassini 17*, 2009
chromogenic print
40 5/8 x 42 5/8 inches
edition 2 of 6
Courtesy David Zwirner, New York
Giant Planets, Satellites, Rings, & a Comet §, p. 180

Thomas Ruff

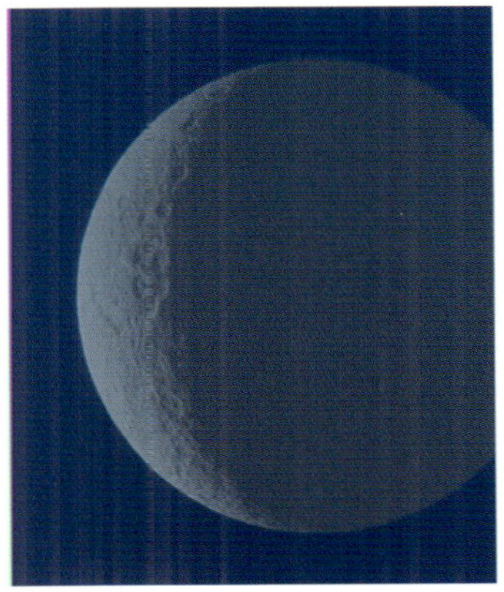

65. *cassini 24*, 2009
chromogenic print
42 ⅝ x 37 ½ inches
edition 1 of 6
Courtesy David Zwirner, New York
Giant Planets, Satellites, Rings, & a Comet
§, p. 178

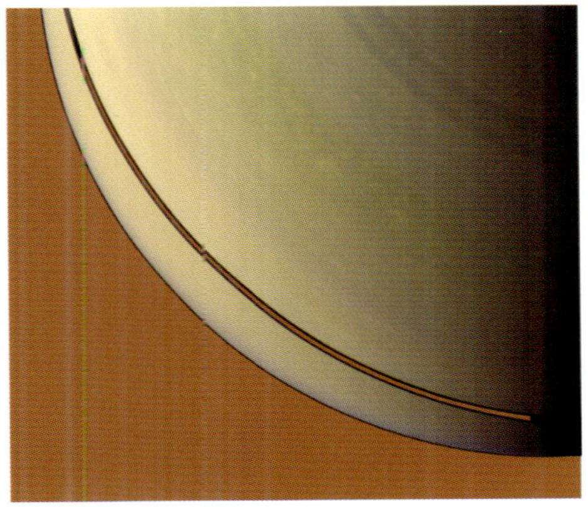

66. *cassini 33*, 2009
chromogenic print
42 ⅝ x 42 ⅝ inches
edition 1 of 6
Courtesy David Zwirner, New York
Giant Planets, Satellites, Rings, & a Comet
§, p. 179

Hans-Christian Schink
Germany

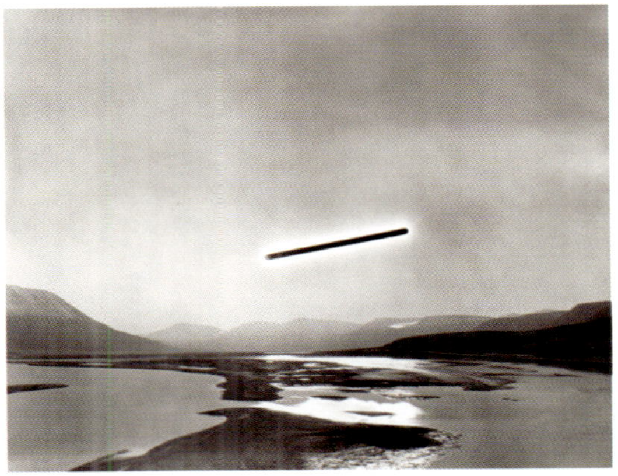

1hr series, northern hemisphere
gelatin silver prints
15 x 19 inches
Courtesy Wilson Centre for Photography

67. *9/17/2006, 8:45–9:45 a.m., N 78°13.370' E 015°40.024'*
The Sun §, p. 103

Hans-Christian Schink

1hr series, northern hemisphere

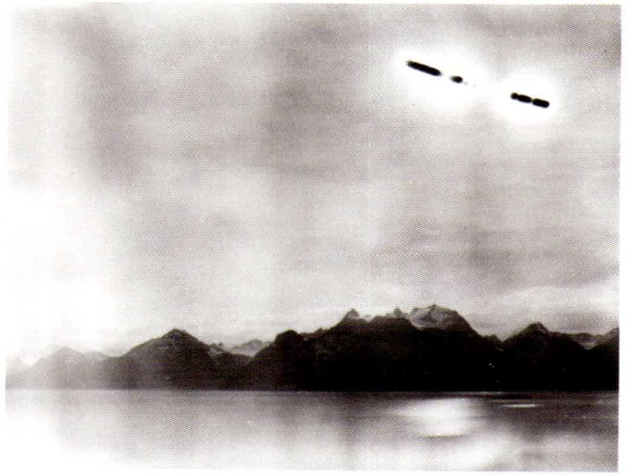

68. *7/11/2007, 5:27–6:27 p.m., N 69°45.199' E 020°29.497'*
The Sun §, p. 98

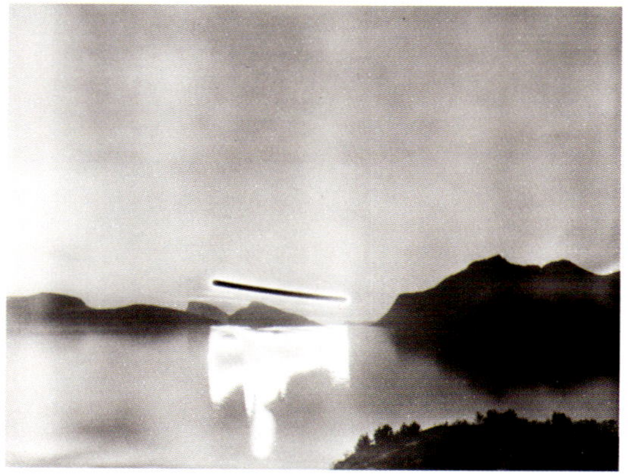

69. *7/14–15/2007, 11:28 p.m.–0:28 a.m., N 69°37.661' E 018°13.470'*
The Sun §, p. 102

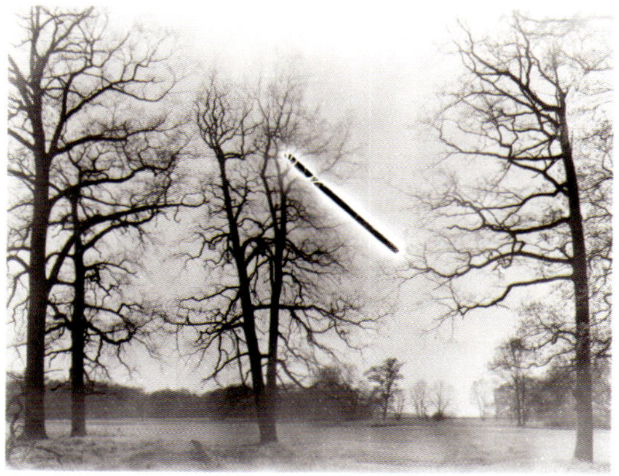

70. *5/03/2006, 6:04–7:04 p.m., N 51°22.126' E 012°09.310'*
The Sun §, p. 104

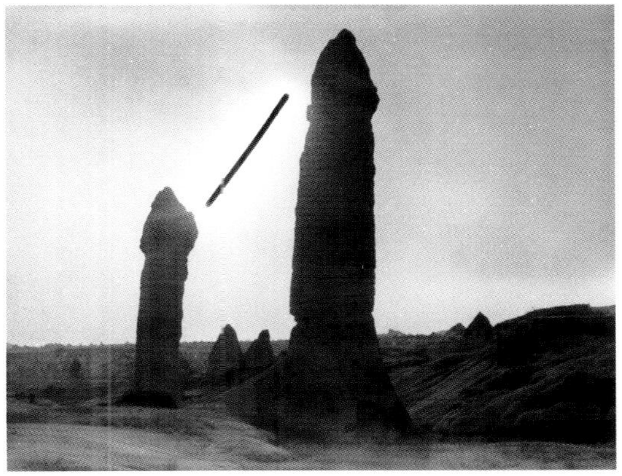

71. *7/27/2008, 7:07–8:07 a.m., N 38°38.269'*
E 034°49.963'
The Sun §, p. 90

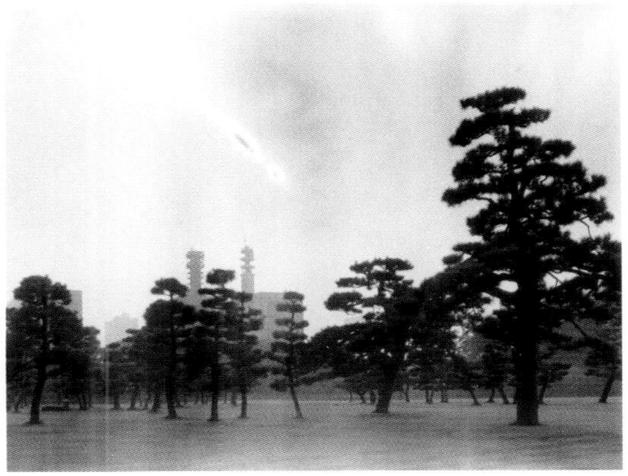

72. *10/28/2005, 2:03–3:03 p.m., N 35°40.460'*
E 139°45.243'
The Sun §, p. 86

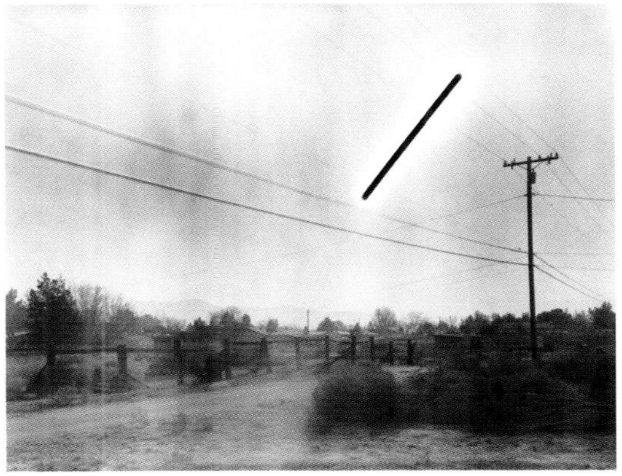

73. *2/22/2006, 7:32–8:32 a.m., N 34°30.850'*
W 117°12.266'
The Sun §, p. 107

1hr series, northern hemisphere

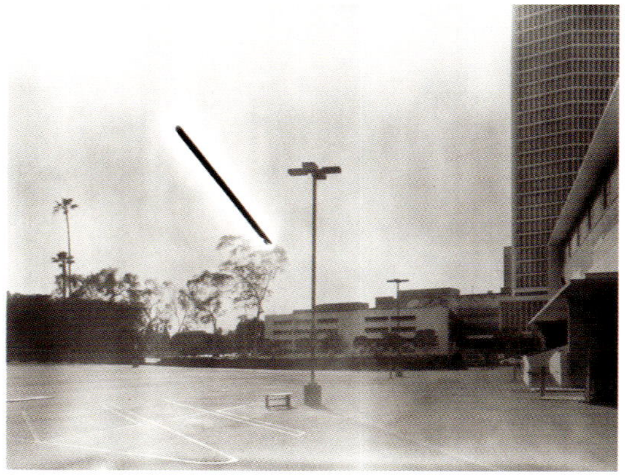

74. *2/23/2006, 4:04–5:04 p.m., N 34°03.712', W 118°20.979'*
The Sun §, p. 106

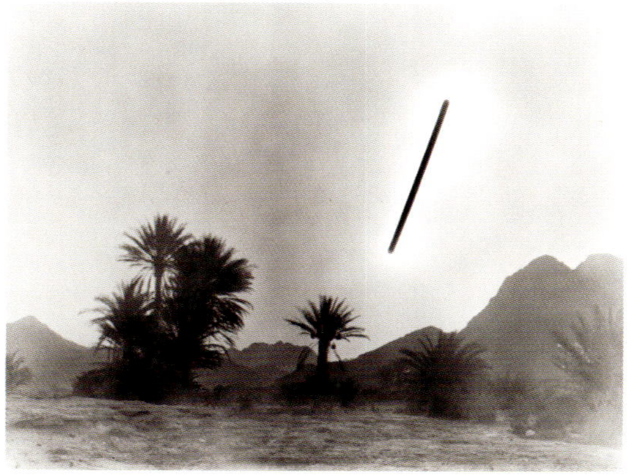

75. *5/10/2006, 6:47–7:47 a.m., N 24°47.435' E 055°56.493'*
The Sun §, p. 112

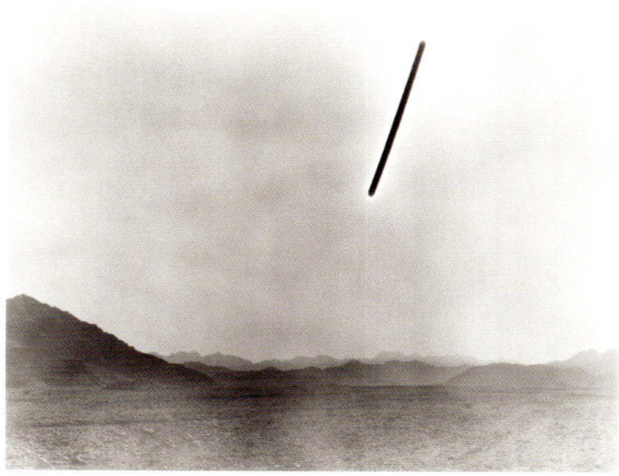

76. *5/12/2006, 6:44–7:44 a.m., N 24°43.562' E 055°57.061'*
The Sun §, p. 85

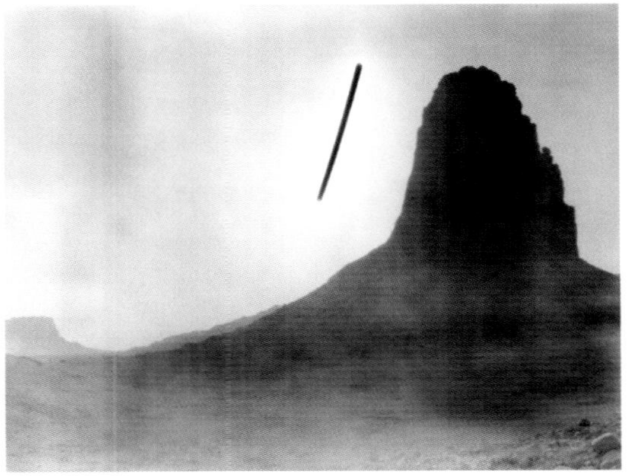

77. *6/20/2007, 7:02–8:02 a.m.,* N *22°54.402'*
E 005°40.083'
The Sun §, p. 108

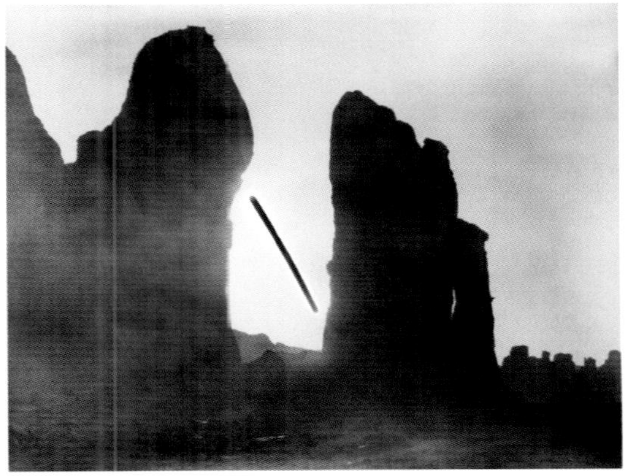

78. *1/09/2008, 4:38–5:38 p.m.,* N *21°48.913'*
E 006°30.297'
The Sun §, p. 91

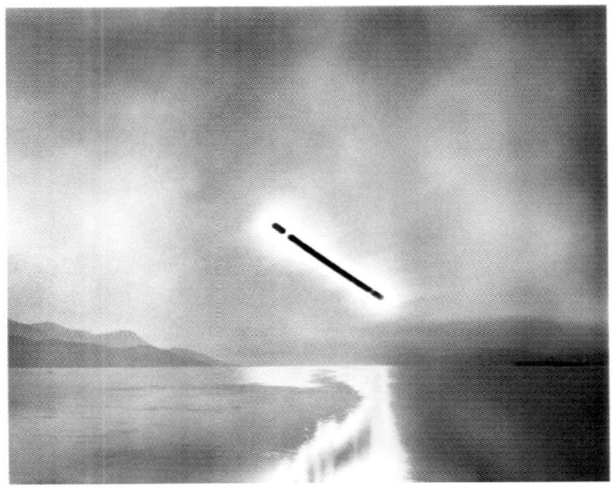

1hr series, southern hemisphere
gelatin silver prints
15 x 19 inches
Courtesy Wilson Centre for Photography

79. *1/23/2010, 6:31–7:31 a.m., S 54°48.411'*
W 068°18.032'
The Sun §, p. 114

HANS-CHRISTIAN SCHINK

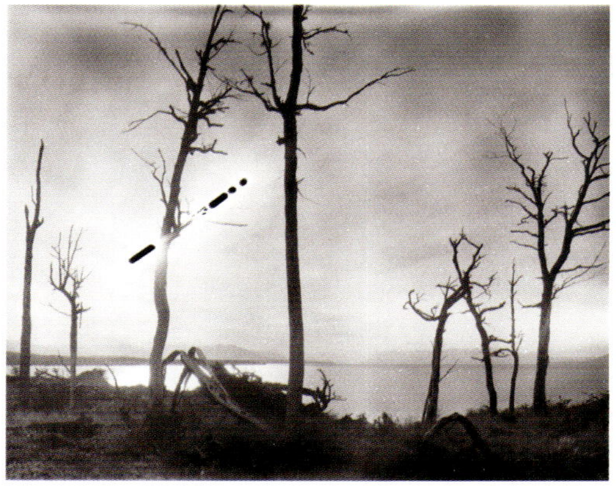

1hr series, southern hemisphere

80. *2/04/2010, 7:22–8:22 p.m., S 54°35.873' W 067°22.541'*
The Sun §, p. 94

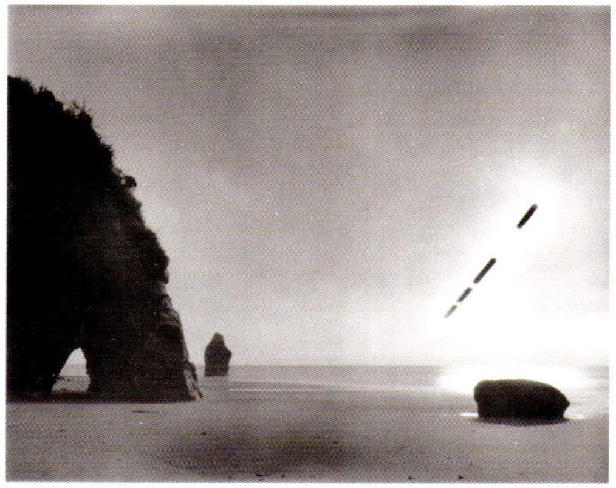

81. *2/21/2010, 7:00–8:00 p.m., S 38°49.042' E 174°34.976'*
The Sun §, p. 81

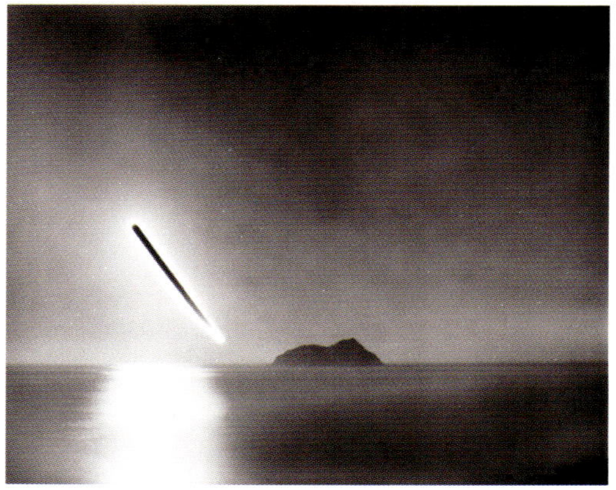

82. *2/20/2010, 6:53–7:53 a.m., S 37°40.831' E 178°32.635'*
The Sun §, p. 115

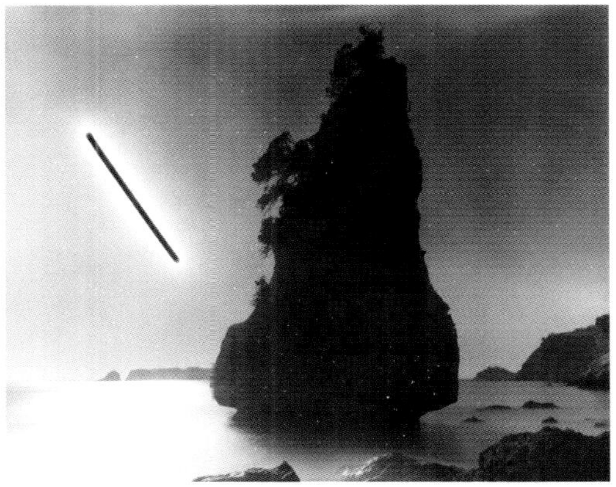

83. *2/26/2010, 7:54–8:54 a.m., S 36°49.622'*
E 175°47.340'
The Sun §, p. 109

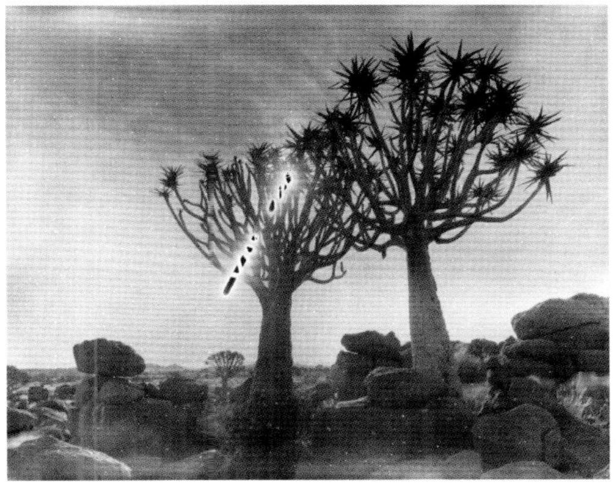

84. *4/10/2009, 4:11–5:11 p.m., S 26°28.034'*
E 018°16.142'
The Sun §, p. 87

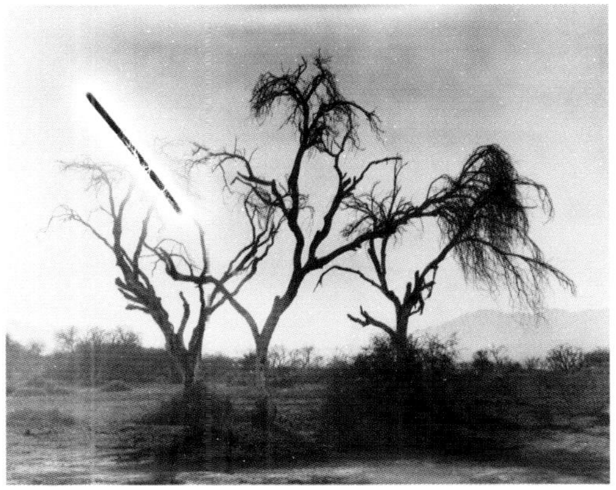

85. *6/01/2008, 9:18–10:18 a.m., S 26°03.817'*
W 065°54.723'
The Sun §, p. 95

HANS-CHRISTIAN SCHINK

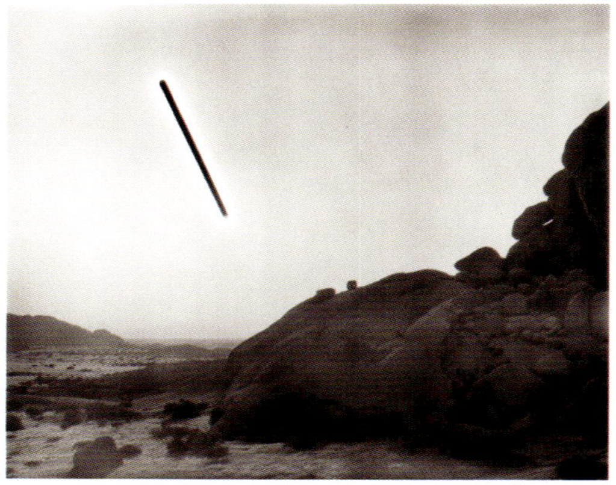

1hr series, southern hemisphere

86. *4/15/2009, 7:02–8:02 a.m., S 21°50.398' E 015°11.352'*
The Sun §, p. 96

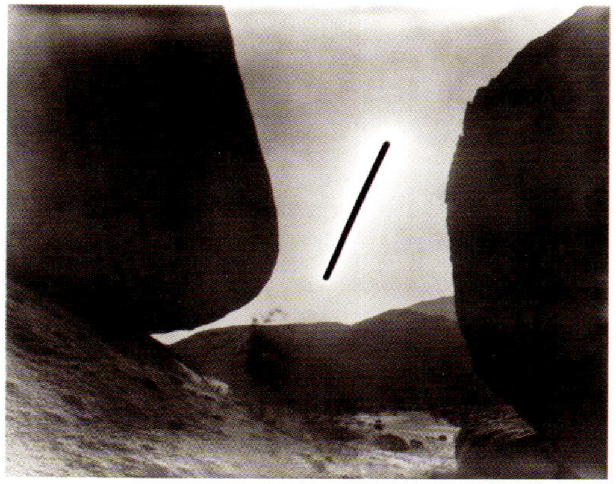

87. *4/12/2009, 4:11–5:11 p.m., S 21°47.094' E 015°39.829'*
The Sun §, p. 101

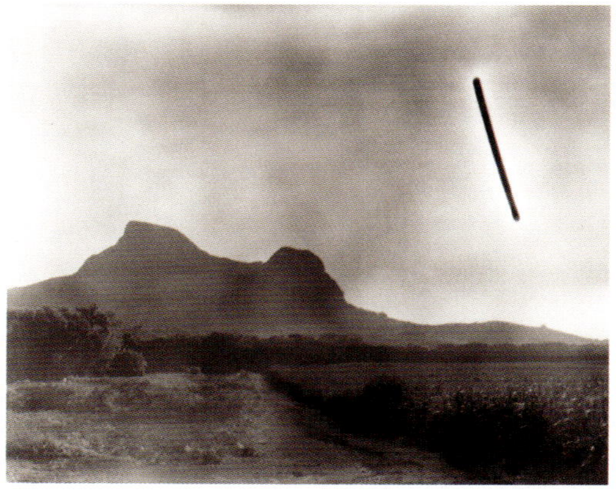

88. *1/19/2010, 6:42–7:42 a.m., S 20°22.016' E 057°42.293'*
The Sun §, p. 99

Hans-Christian Schink

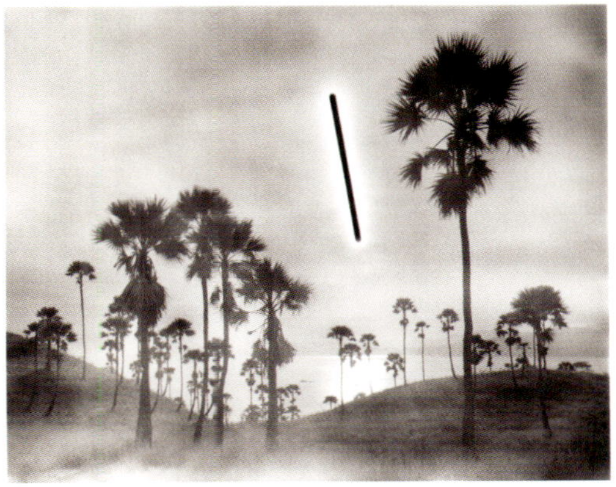

89. *3/28/2010, 6:43–7:43 a.m., S 08°27.131'*
E 119°52.396'
The Sun §, p. 113

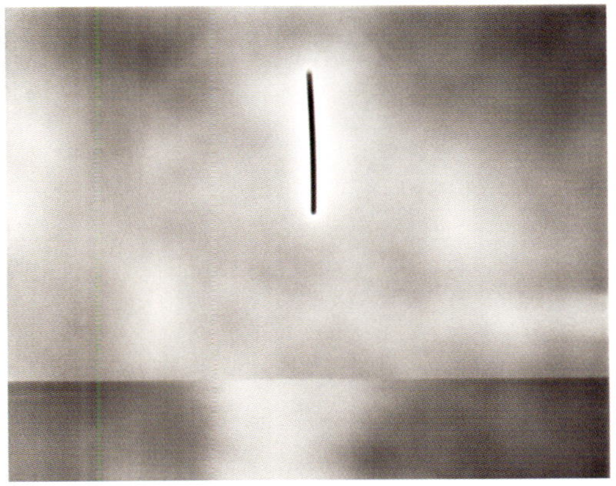

90. *1/05/2010, 5:46–6:46 p.m., S 06°26.486'*
E 039°27.776'
The Sun §, p. 93

John Stetson

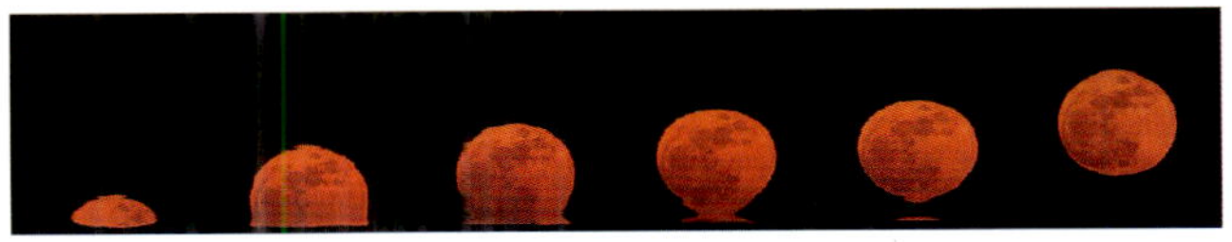

91. *Etruscan Vase Moonrise*
archival pigment inkjet print on cotton rag
8 x 44 inches
The Moon §, pp. 170–1

Alfred Stieglitz
(1864–1946)

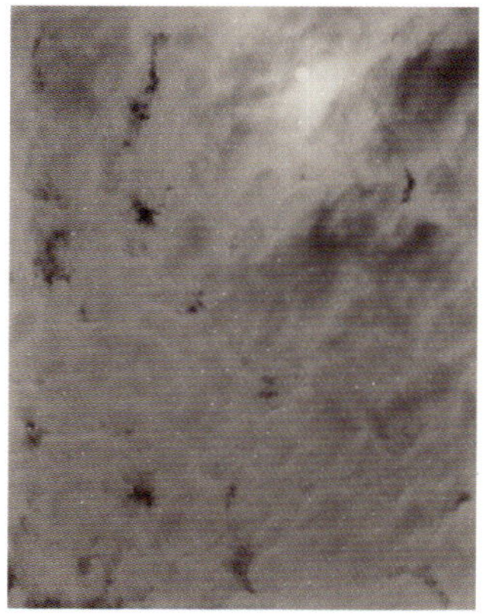

92. *Song of the Sky*
gelatin silver print
4 7⁄16 x 3 7⁄16 inches
National Gallery of Art, Washington, D.C.
Alfred Stieglitz Collection
The Sun §, p. 84

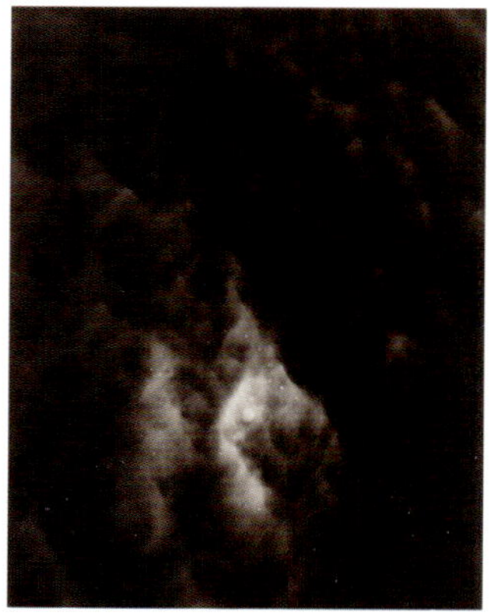

93. *Equivalent, Set C2, No. 2*
gelatin silver print
4 11⁄16 x 3 11⁄16 inches
National Gallery of Art, Washington, D.C.
Alfred Stieglitz Collection
The Sun §, p. 83

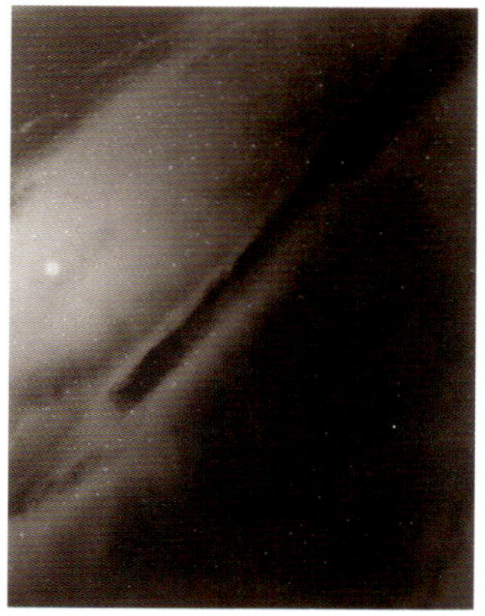

94. *Equivalent*
gelatin silver print
4 5⁄8 x 3 5⁄8 inches
National Gallery of Art, Washington, D.C.
Alfred Stieglitz Collection
The Sun §, p. 80

Babak A. Tafreshi
Iran

95. *Alamut Starry Night*
archival inkjet print
18 ¾ x 16 inches
Galaxies §, p. 154

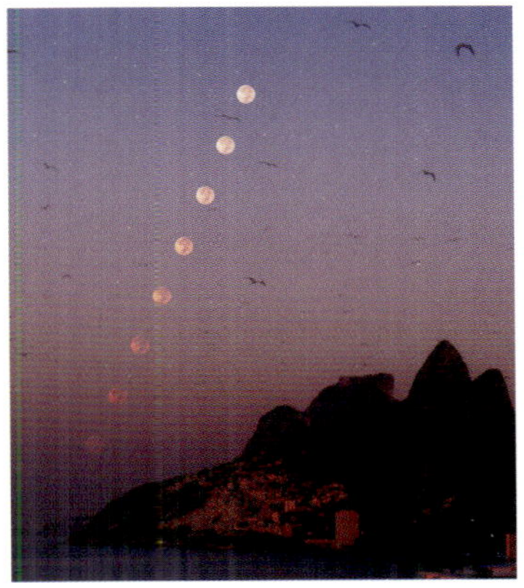

96. *Rio Moonset*
archival inkjet print
18 ¾ x 16 inches
The Moon §, p. 174

Yuichi Takasaka
Canada

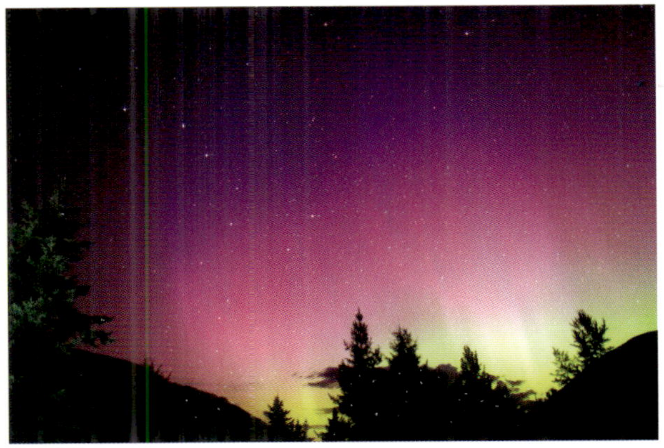

97. *Big Dipper, August 5,* 2011
inkjet print on Ilford Gallerie Smooth Pearl paper
12 x 18 inches
The Earth §, p. 158

Yuichi Takasaka

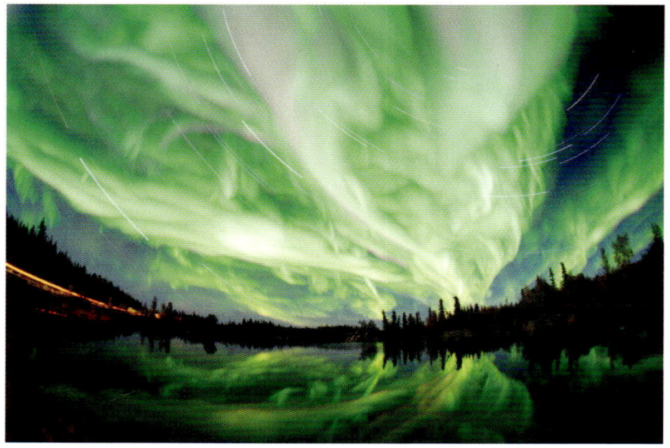

98. *Startrails, September 12*, 2011
inkjet print on Ilford Gallerie Smooth Pearl paper
12 x 18 inches
The Earth §, p. 157

Sean Walker

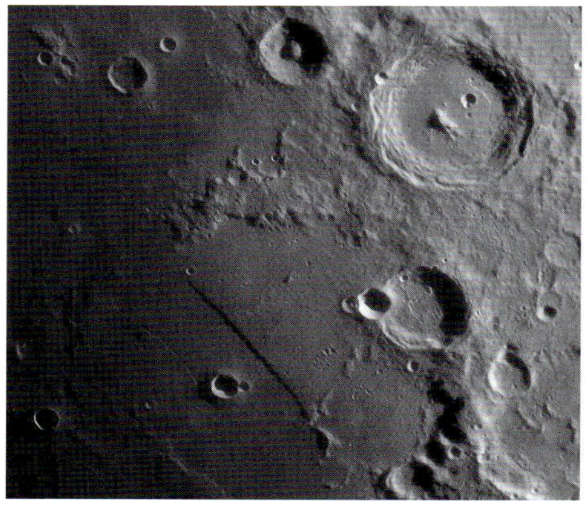

99. *Rupes Recta*
inkjet print
20 x 22 inches
The Moon §, p. 169

Wei-Hao Wang
Taiwan

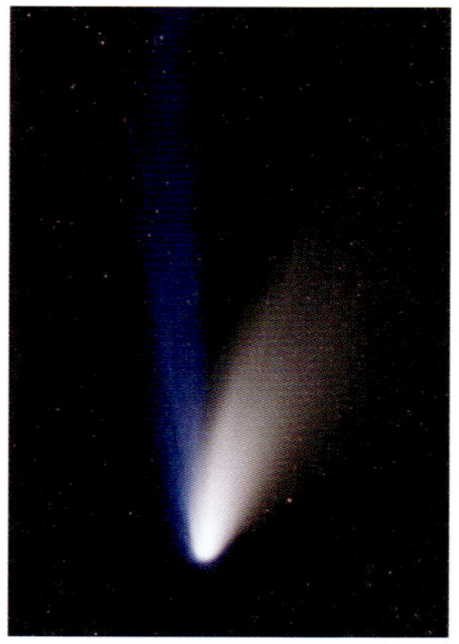

100. *Comet Hale-Bopp, March 10*, 1997
inkjet print
23 ¾ x 17 inches
Giant Planets, Satellites, Rings, & a Comet §, p. 181

Jacqueline Woods

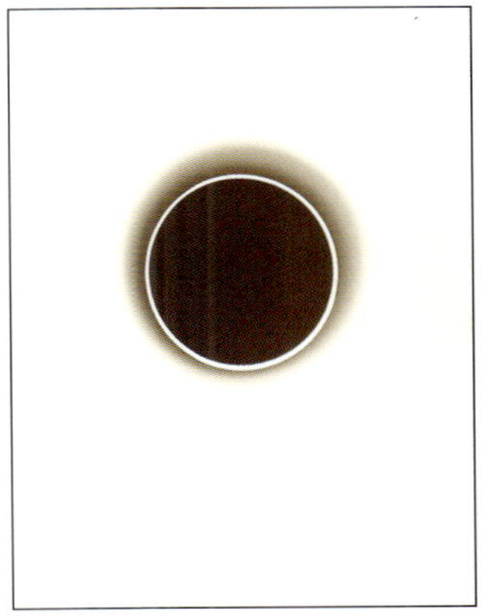

101. *Black Sun #4*, 2011
gelatin silver monoprint
15 x 12 inches
The Sun §, p. 88

102. *Black Sun #15*, 2011
gelatin silver monoprint
21 x 17 inches
The Sun §, p. 111

103. *Black Sun #16*, 2011
gelatin silver monoprint
17 x 21 inches
The Sun §, p. 97

JACQUELINE WOODS

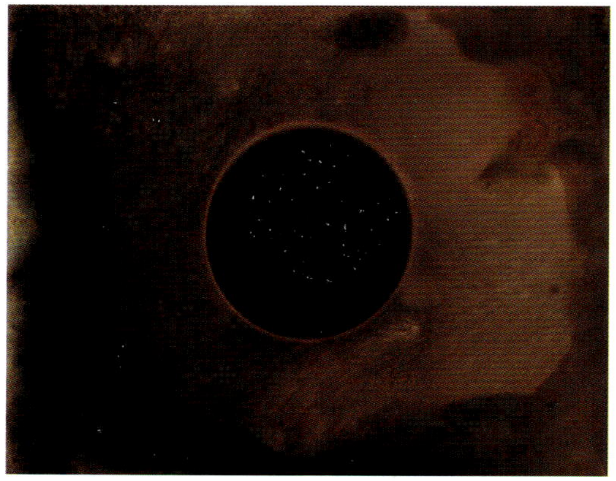

104. *Black Sun #30*, 2011
gelatin silver monoprint
12 x 15 inches
The Sun §, p. 82

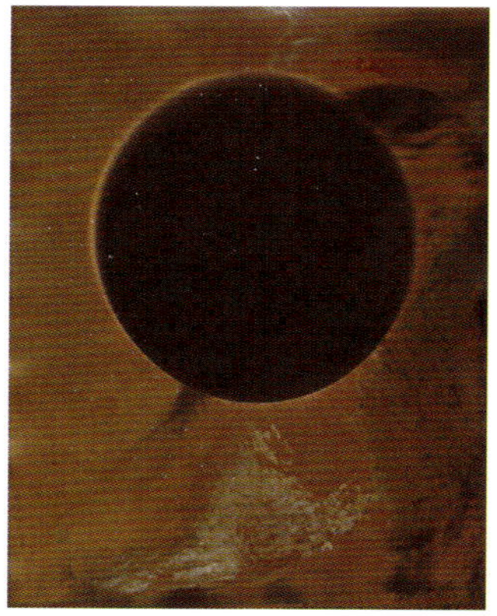

105. *Black Sun #33*, 2011
gelatin silver monoprint
15 x 12 inches
The Sun §, p. 100

106. *Black Sun #37*, 2011
gelatin silver monoprint
21 x 17 inches
The Sun, p. 92

Contributors

MICHAEL BENSON ✦ Born in Munich, Germany, he received his B.A. in English and Photography from State University of New York, Albany, in 1996 and graduated from the New York University Film School in 1991. He has exhibited extensively at museums around the world and has published several books, including *Far Out: A Space-time Chronicle* and *Beyond: A Solar System Journey*. Benson has won several awards, including Best Documentary Feature Award, National Film Board of Canada, Vancouver International Film Festival for *Predictions of Fire* (2005); and Golden Minotaur (First Prize), Documentary Category, 1996 "Message to Man" St. Petersburg International Film Festival (Russia) for *Predictions of Fire*. He is represented by Hasted Kraeutler Gallery, and his photographs are in collections including the Nelson-Atkins Museum of Art, Kansas City, Mo., and the Worcester Art Museum, Mass. Benson lives and works in New York city.

ADAM BLOCK ✦ Earned his B.S. in Astronomy and Physics in 1996 from the University of Arizona. He developed and administered public outreach programs at the National Observatory Kitt Peak, and now serves as Public Observing Programs Coordinator at the Mt. Lemmon

SkyCenter, University of Arizona. Block's images are published in magazines, books, posters, and widely on the Internet, where his images have been selected more than fifty times by NASA's Astronomy Picture of the Day. His images are used as references by amateur and professional astronomers.

JERRY T. BONNELL ✦ Received a Ph.D. in astronomy from the University of Maryland in 1987, and has since worked on a variety of astrophysical satellite projects at NASA's Goddard Space Flight Center, including the Cosmic Background Explorer, the International Ultraviolet Explorer, the Compton Gamma Ray Observatory, and the Fermi Gamma Ray Space Telescope. While his research included the entire electromagnetic spectrum, since 1992 he has focused on the time histories and spectral evolution of cosmic gamma-ray bursts. Author of popular books, reviews, and magazine articles on astronomy and high-energy astrophysics, Bonnell is also a founding editor and author of the Astronomy Picture of the Day (APOD), a daily archive of astronomy photographs organized by NASA. Begun in 1995, APOD remains one of the largest archives of annotated astronomy and space related images on the Web.

LINDA CONNOR ✦ Educated at the Rhode Island School of Design (RISD), Providence, (B.F.A., 1967), and the Institute of Design, Illinois Institute of Technology, Chicago (M.S., Photography, 1969). In 2002, Connor founded Photo Alliance and is a member of the boards of In-Site and Friends of Photography. She has been a professor of photography at the San Francisco Art Institute since 1969. Connor has exhibited extensively throughout the United States, including the Center for Creative Photography, Tucson, Ariz.; Museum of Art, RISD, Providence; and Southeast Museum of Photography, Daytona, Fla.; as well as Japan, Belgium, Israel, and Australia. She has received numerous awards including a Guggenheim Fellowship, two NEA Individual Artist Grants, and two Flintridge Foundation Awards for Visual Artists. Her work is collected by museums around the globe, including The Art Institute of Chicago; Canadian Center for Photography, Toronto; International Museum of Photography, George Eastman House, Rochester, N.Y.; J. Paul Getty Museum, Los Angeles; Museum of Modern Art, New York; National Gallery of Canada; and Yildiz University, Istanbul, Turkey. She lives in San Anselmo, Calif.

JAMES CORMIER ✦ A native of Maine, he has been an amateur astronomer and astrophotographer since the golden age of film astrophotography. Photographing under the very dark skies of downeast Maine allows him to produce highly detailed images. Cormier has recently turned his attention to shooting landscapes under the light of the Moon, blending traditional landscape photography methods with those used to capture the night sky. Employed in the communications electronics sector, he resides in Sullivan, Me.

KEN CRAWFORD ✦ President of Advanced Astroimaging Conference, Inc., he serves as telescope technician for the Cameron Park Rotary Community Observatory in Placerville, Calif. Crawford has been featured at several astro-imaging conferences, including the Midwest Astro-Imaging Conference and the Australian Astro-Imaging Conference. He was profiled in *Capturing the Stars: Astrophotography by the Masters* and his work has been included on NASA's APOD.

JOHN A. DAVIS ✦ Graduated from Southern Methodist University, Dallas, Tex., in 1984 with a B.F.A. in film production. He currently resides in Dallas. His images have appeared in numerous periodicals, as well as NASA's APOD. Davis is an Oscar and Emmy nominated writer/director/producer, responsible for such animated films as *Jimmy Neutron: Boy Genius* (created by), *The Ant Bully*, and many other movies and television shows.

IGNACIO DIAZ BOBILLO ✦ Earned a Ph.D. degree in Aeronautics and Astronautics from the Massachusetts Institute of Technology, Cambridge, in 1992. Early in his career he devoted a few years to academia when he lectured, published several research papers, and co-authored a graduate textbook on the field of systems and control. He lives with his wife and three children in Buenos Aires, Argentina, where he works in the investment/private banking industry.

DENNIS DI CICCO ✦ Joined the staff of *Sky & Telescope* in 1974, bringing with him an educational background in mechanical engineering and experience as a telescope maker, astrophotographer, and observer dating back to the early 1960s. Among his credits, he is proud of his involvement with the creation of *Sky & Telescope*'s popular Test Reports and Gallery. He has also led expeditions to view solar eclipses around the globe and astrophotography workshops in Australia's Outback. His photography is well known to astrophotographers and amateur astronomers and has appeared in hundreds of books and magazines, including the award-winning, year-long photograph of the Sun's analemma made in the late 1970s. Di Cicco contributed to the exhibition Images of Science 2 and its companion catalogue, organized by the School for Photographic Arts and Sciences at the Rochester Institute of Technology in Rochester, N.Y. He started working with astronomical charge-coupled devices (CCDs) in 1990, and did pioneering work in tricolor imaging. He was the principal force behind the creation of Sky Publishing's quarterly *CCD Astronomy* and served as its senior editor until the publication was folded into *Sky & Telescope* in early 1997.

ALAN FRIEDMAN ✦ Earned a B.F.A in studio art from the State University of New York at Buffalo, and lives in the Buffalo area, where he is an artist, award winning greeting card designer, and president of Great Arrow Graphics. His solar system images are featured frequently on NASA's popular Web sites Spaceweather.com and APOD, and in exhibitions, including the touring show for the *International Year of Astronomy: From the Earth to the Universe*. His techniques for planetary imaging have been the subject of interviews on MSNBC's TODAY show, articles in *Sky & Telescope* and a column at Space.com. He is currently president of the Buffalo Astronomical Association and is Research Associate in Astronomy at the Buffalo Museum of Science. His images may be seen on his Web site at avertedimagination.com.

ROBERT GENDLER ✦ Is a Connecticut physician and acknowledged in astrophotography circles as a master practitioner of the art. His work has frequently been featured on NASA's APOD. Gendler's images were included in the international travelling exhibition From Earth to the Universe, organized to celebrate the International Year of Astronomy, and the exhibition Visions of the Cosmos, Rubin Museum of Art, New York. Gendler was profiled in Timothy Ferris's documentary *Seeing in the Dark*, and his images were featured in PBS's *Origins* program and the BBC's *Most of the Universe is Missing*.

JACOB GERRITSEN ✦ A Dutch-born naturalized U.S. citizen, he went to grade school in Florida and high school in the Netherlands. After immigrating again at 18 to Hawaii, he graduated from the University of Hawaii in 1969 with a B.S. in Biology. He graduated from medical school in 1976 and specialized in internal medicine. He is married to Tess Gerritsen, an internationally best-selling author. They reside in Maine and have two adult sons.

DON GOLDMAN ✦ Is an inventor, entrepreneur and president of Astrodon Imaging, located near Sacramento, Calif., which provides colored glass filters and other products to the astrophotography market. He earned a Ph.D. from California Institute of Technology, Pasadena. His images and articles are widely published in astronomy magazines, and he is a popular speaker at astronomy conferences. Over twenty of his images have been selected as NASA's APOD. He lives near Sacramento, Calif. with his wife, Nicolle.

SHARON HARPER ✦ Received her M.F.A from the School of Visual Arts, New York, and has exhibited extensively throughout the U.S., Austria, England, Germany, Norway, and Spain. Her latest book, *From Above and Below, Sharon Harper*, Radius Books, will be published in 2012. Harper has received many awards and residencies, including Halsnøy Monastery, Norway, Residency Fellowship; Banff Centre, Alberta, Canada, Residency Fellowship; and Film Studies Center Fellowship, Harvard University Film Studies Center. Her work is represented in the Museum of Modern Art, New York; Whitney Museum of American Art, New York; Albright-Knox Art Gallery, Buffalo, N.Y.; Portland Art Museum, Ore.; and the Nelson-Atkins Museum, Kansas City, Mo. Harper is represented by Galerie Stefan Röpke, Cologne, Germany, and Rick Wester Fine Art, New York. She lives in Cambridge, Mass.

WARREN KELLER ✦ Is widely published with photographs and articles in *Sky & Telescope*, *Astronomy*, *AstroPhoto Insight*, and *Amateur Astronomy* magazines, and many places on the Internet, including NASA'S APOD. He spends his days teaching imaging and his nights acquiring image data with artistic partner Dave Raphael from their remote telescope at New Mexico Skies in Mayhill, N.M. His work is displayed at his Web site, BillionsandBillions.com, and Exposures International Gallery of Sedona, Ariz.

WALTER KOPROLIN ✦ A native of Vienna, he has been engaged with astronomy and imaging since his youth. After acquiring his own telescope in his teens, he began taking astrophotos with hand-guided photos on chemical film. He pursues astronomy professionally, having earned his M.A. in 1997 and a Ph.D. in 2004 from the University of Vienna. His research has focused on the kinematics of early-type (elliptical and lenticular) galaxies, and the physical properties of the stellar population and the gas in blue compact dwarf galaxies. His images have appeared on the NASA Web site.

WILLIAM H. LOW ✦ Received his M.A. in museum administration from the Center for Museum Studies, John F. Kennedy University, Berkeley, Calif. Low came to the Bates College Museum of Art in 1990, serving as the assistant curator through 2008, when he became curator. In 2002–03, he also served as acting director, managing all aspects of museum activities and administration. Low has curated numerous exhibitions, including Documenting China: Contemporary Photography and Social Change (a Smithsonian Traveling Exhibition that toured the U.S.); Scrape, Cut, Bite, Print: The Graphic Work of Charlie Hewitt; Manuel López Oliva: Cuba and the Theatre of Desire; Byzantium to Russia: Popular Icons for Personal Devotion; and Xu Bing: Calligraphy for the People. He also is a curatorial consultant for Museum L-A in Lewiston, Me., where he contributes to all aspects of exhibition development and design, including the production of Portraits and Voices, an exhibition with catalogue of photographic portraits and oral histories from Lewiston, Maine's industrial history.

DAVID MALIN ✦ Born in England, he trained as a chemist and originally worked there as a microscopist. In 1975 he moved to Sydney, Australia, to take up a job with the Anglo-Australian Observatory. There he pioneered photographic processing techniques to maximize the ability to extract faint and low contrast detail from the non-linear response and high densities of photographic plates. Malin is now best known for his series of three-color wide field images of deep space objects, which have been widely published as posters and in books around the world. His film-based techniques are still in use by CCD photographers. Malin was awarded the Jackson-Gwilt Medal of the Royal Astronomical Society in 1986 and received the Lennart Nilsson Award in 2000. His images have been exhibited in art museums and galleries in Australia, the U.K., Canada, and the U.S., including Howard Schickler Fine Art, New York; Edwynn Houk Gallery, New York; Victorian Arts Center, Melbourne, Australia; and Art Gallery of Ballarat, Sydney, Australia. Malin was the subject of the film *The Man Who Colours Stars*, and his images have been included in numerous publications, including *Capturing the Stars: Astrophotography by the Masters*, and *A View of the Universe*.

STEVE MAZLIN AND JACK HARVEY ✦ Have collaborated on dozens of astroimaging projects over the last several years from their respective backyard observatories, as well as observatories at New Mexico Skies in Mayhill, N. Mex. and PROMPT/CTIO in Chile. Dr. Mazlin is a practicing neurologist in Bucks County, Pa., and Dr. Harvey is a retired sports medicine physician in Fort Collins, Col. Their images have appeared numerous times on APOD, the National Optical Astronomy Observatory (NOAO) Web site, and also in many magazines and newspapers. In 2009 their image of the Witch Head Nebula was chosen by UNESCO for its World Science Day poster.

DAN MILLS ✦ A nationally esteemed curator, arts administrator, and artist, he began his appointment as director of the Bates College Museum of Art and Lecturer in the Humanities in September 2010. His exhibitions include Tale Spinning [with catalog]; Xiaoze Xie: Amplified Moments (1993–2008) [with catalog]; and *James Ensor: Scènes de la vie du Christ* and Other Works. From 2001–10 Mills was director of the Samek Art Gallery, Bucknell University, Lewisburg, Pa., where he organized a nationally recognized exhibitions program that focused primarily on the art of our time and brought the world to the university. His other publications include *Dan Mills: The U.S. Future States Atlas.*

WESTON NAEF ✦ Received his B.A. from Claremont Men's College, M.A. from The Ohio State University and Ph.D. (A.B.D.) from Brown University. He worked in the department of prints, photographs and books at the Metropolitan Museum of Art, New York, before joining The J. Paul Getty Museum, Los Angeles, in 1984 to direct their department of photographs, a position he held until he retired in 2009. During his tenure at the Getty, Naef oversaw a quadrupling of the Getty's photographic holdings, including the building of significant collections by pioneers like William Henry Fox Talbot, Julia Margaret Cameron, Roger Fenton, Gustave Le Gray and Carleton Watkins. He was responsible for significant exhibitions on the works by Walker Evans, Eugene Atget, Dorothea Lange and Manuel Alvarez Bravo. Recently he oversaw the opening of the Center for Photographs, which tripled the museum's exhibition space for its collection. He is the author of standard texts on American nineteenth-century landscape, *Alfred Stieglitz*, and other illustrated books, the most recent of which is a catalogue raisonnée of mammoth prints by Carleton Watkins.

JIM NICKELSON ✦ Earned a B.S. in Aerospace Engineering from the University of Texas at Austin in 1992, and his J.D. from Harvard Law School in 1998. He has studied photography at the Maine Media Workshops in Camden, Me., and is represented by Vox Photographs in

Portland, Me. Nickelson's photographs have been honored with several awards, including a purchase award for Maine's Percent for Art program in 2011, and he has exhibited in group and solo exhibitions throughout the U.S. He lives in Camden.

GEIR ØYE ✦ Interested in astronomy since childhood, he earned a B.S. in astronomy, geography, and computer science in Norway, and an M.A. in Space Studies from the University of North Dakota, Grand Forks. As a student Øye obtained observing experience from professional observatories such as the Nordic Optical Telescope in La Palma, Spain, where he had a one-year internship in 2003–04, Swedish Solar Telescope in La Palma, and Joddrell Bank Observatory at University of Manchester in Cheshire, England. Returning to Norway in 2004 he purchased his first digital camera. Since then his images have appeared in newspapers, TV, international astronomy magazines, books, teaching material, commercials, NASA Web sites and more resources.

JAY PASACHOFF ✦ Director of Hopkins Observatory, chair of the Astronomy Department, and Field Memorial Professor of Astronomy at Williams College, Williamstown, Mass., since 1972, he received his B.A. from Harvard College, Cambridge, Mass., in 1963, and his M.A. and Ph.D. from Harvard University in 1965 and 1969, respectively. Pasachoff held postdoctoral positions at Harvard College Observatory and the California Institute of Technology from 1969–72. In 2003, he was awarded the Education Prize by the American Astronomical Society, and he is an Honorary Member of the Royal Astronomical Society of Canada. Pasachoff served as president of the Commission on Education and Development of the International Astronomical Union from 2003–06, and is currently the chair of the Working Group on Solar Eclipses, and the 2011–13 vice-chair of the Historical Astronomy Division, American Astronomical Society. He has worked at numerous research institutions while on sabbatical, including: Institute for Astronomy, University of Hawaii, Honolulu; Institut d'Astrophysique de Paris; Institute for Advanced Study, Princeton University, Princeton, N.J.; Harvard-Smithsonian Center for Astrophysics, Harvard University, Cambridge, Mass.; and California Institute of Technology, Pasadena. He has authored or co-authored several books, including *The Cosmos: Astronomy in the New Millennium*; *A Field Guide to Stars and Planets (Peterson Field Guide)*; *Astronomy: From the Earth to the Universe*, 6th ed.; and *Nearest Star: The Surprising Science of Our Sun*.

DAMIAN PEACH ✦ Is renowned throughout the amateur and professional astronomical community for his spectacularly detailed photographs of the planets. His images have appeared in countless books, magazines and papers, and he has made numerous appearances on BBC television promoting astronomy. He was awarded the British Astronomical Association's Merlin Medal and the Association of Lunar and Planetary Observers' Walter Haas Award for outstanding contributions to planetary astronomy. In 2009 he was part of a record setting team that produced the largest ground-based image mosaic of the Moon and was also featured in the *National STEM Centre Explorers of the Universe* photographic exhibition in York, England. In 2011 he was named the Royal Greenwich Observatories' Astrophotographer of the Year.

MARTIN PUGH ✦ Was born in the town of Dudley in the West Midlands, England, and emigrated to Australia after serving in the Royal Navy for twenty years. His photographs have appeared on APOD sixteen times (two collaboratively with Rob Gendler). Pugh was the 2009 Astronomy Photographer of the Year, and won the South Pacific Star Party astrophotography competition, the premier Australian astrophotography competition (2008) and the David Malin Awards (2008). In addition, his images have appeared in international astronomy magazines, newspapers, a Celestial Navigation DVD, books, a video release of NGC 3603 by Hubble, and several occasional articles and lecture pieces. Two other major achievements include the release of a three-stamp collector's set by Australia Post, to mark the 2009 International Year of Astronomy and his seat in the SBIG Hall of Fame.

NICK RISINGER ✦ A photographer, designer, and occasional speaker, he has presented his work at the Advanced Imaging Conference, Pixar, and many astronomical societies. Risinger was the subject of a feature article in *Sky & Telescope* in 2012. He lives in Seattle, Wash.

KIRK ROGERS ✦ Has been recognized for his photographic efforts in National Wildlife Refuge sites and wildlife management resources. His images have been published, employed in videos, and exhibited in offices and visitors' centers from the Wichita Mountains Wildlife Refuge in Oklahoma, the Savannah Coastal Refuges Complex in Georgia and South Carolina, to the Maine Coastal Islands National Wildlife Refuge in his home state. Rogers's photography has been featured in several television segments, including broadcasts on "Bill Green's Maine" and "207."

JEAN PAUL ROUX ✦ Is a research engineer specializing in biological imaging and lives in France. His images have appeared on NASA's APOD.

TONY ROWELL ✦ His award-winning photos have appeared in books, calendars and in *Astronomy*, *Backpacker*, *Nature's Best Photography* and *Outdoor Photographer* magazines, as well as at APOD, on Astronomy.com and NASA's Web sites. In 2010 Rowell received first prize in The Beauty of the Night Sky category in the International Earth and Sky Photo Contest and 2011 first prize in *Outdoor Photographer*'s Great Outdoors Photo Contest, 2011. His photography expeditions have taken him from the Arctic Circle to the mountains of Tibet. He is the son of renowned photographer and mountaineer Galen Rowell, and has had his work exhibited at the Yosemite Museum Gallery, Yosemite National Park, Calif., and at the Marin Headlands Visitors Center at the Golden Gate National Recreation Area near San Francisco. His work is on display at the Tioga Pass Resort in Yosemite and at the Mountain Light Gallery, Bishop, Calif.

THOMAS RUFF ✦ Studied photography from 1977 to 1985 with Bernd and Hilla Becher at the Düsseldorf Art Academy, Germany. His work is in the collections of many major museums, including the Metropolitan Museum of Art and Solomon R. Guggenheim, New York; Hamburger Bahnhof, Museum für Gegenwart, Berlin, Germany; Moderna Museet, Stockholm; The Art Institute of Chicago; Essl Museum, Klosterneuburg, Germany; Dallas Museum of Art; National Gallery of Victoria, Melbourne, Australia; Hirshhorn Museum and Sculpture Garden, Washington, D.C.; National Museum of Photography, Copenhagen, Denmark; and Stedelijk Museum voor Actuele Kunst, Ghent, Belgium.

HANS-CHRISTIAN SCHINK ✦ Studied photography at the Leipzig Academy of Visual Arts, Leipzig, Ger. from 1991 to 1993. In the following years he received several research grants from the Cultural Fund Foundation for the Künstlerhaus Schloss Wiepersdorf (1997) and the Else-Heiliger-Fonds of the Konrad-Adenauer Foundation (2008). In 2002 he was artist in residence at the Künstlerhaus Villa Aurora, Los Angeles. He received prizes for the German Photo Book Award for Transportation Projects, (2004) and the international REAL Photography Award (2008). Schink lives in Berlin and Leipzig. He has exhibited extensively throughout Europe, Asia, and the U.S., and is included in the German institutions the Museum Küppersmühle, Duisburg, the Anger Museum, Erfurt, and the Haus der Geschichte, Bonn.

ANTHONY SHOSTAK ✦ Starstruck project director and curator of education, Bates College Museum of Art, Lewiston, Me., he received his B.F.A. in painting and drawing from the University of the Arts in Philadelphia, and has shown his work internationally, most recently in the traveling exhibition *Revoluciones del Mundo* originating in Havana, Cuba. Shostak has served as the curator of education at the Bates College Museum of Art since 1994. He was the lead curator of the interdisciplinary exhibition exploring sustainability, Green Horizons. Shostak organized several other exhibitions, including The Body Holographic: Harriet Casdin-Silver; From Where I Stood: Photographs from Ireland, Hawaii, and Lands Between by Rowena J. Otremba; Joseph Nicoletti: A Retrospective; Metamorphoses: A Collaboration with Ovid by Wally Reinhardt; Alex Grey: Numinous Flesh; Pre-Columbian Ritual Ceramics; and Contemporary German Prints. He co-curated other exhibitions, including Our Positive Bodies: Sharing Our Choices and Mapping Our Treatment; and Joel M. Babb: The Process Revealed. Shostak is the creator of The Thousand Words Project, the Museum's flagship outreach program serving public school teachers and their students thousandwordsproject.org.

JOHN STETSON ✦ Has taught high school math, English, computer science, college-level photography, and has had a solar observatory available for at-risk youth in southern Maine where they observe and image every clear day. Over the past twelve years, the solar observing has included: building solar telescopes with John Dobson (grinding, polishing, figuring); making white light observations with pencil and paper; making a 12 x 4 foot triptych of solar activity (illustrated in *Reflecto* magazine); observing an annular eclipse; creating an analemma; viewing the transit of Venus in June 2004; observing crescent, pinhole images of a partial eclipse; observing and imaging in H-alpha (observing prominences, filaments, flares, etc.); and capturing the ISS and various Space Shuttles transiting the sun.

BABAK A. TAFRESHI ✦ A freelance science journalist, photographer, and astronomy communicator, he has been the editor in chief of *Nojum*, the Iranian magazine of astronomy for several years, and is a contributing photographer at *Sky & Telescope* magazine. Tafreshi is the creator and director of The World at Night (TWAN) international project and a board member of Astronomers without Borders. He received the 2009 Lennart Nilsson Award, the world's most recognized award for scientific photography, for his global contribution to night sky photography and public awareness to beauties of night sky. He has lectured extensively throughout the world. Born in Iran, Tafreshi currently resides in Germany.

YUICHI TAKASAKA ✦ Was born in Japan and emigrated to Canada in his teens. Takasaka's images have appeared in various brochures and magazines, in educational books and textbooks, and on many Internet Web pages, including NASA and the *National Geographic*. He now lives in Lumby, British Columbia, and operates photography and tourism services through his company Blue Moon Promotions.

SEAN WALKER ✦ A graduate of Massachusetts College of Art and Design, Boston, with extensive experience with digital and photographic prepress techniques, he joined the staff of *Sky & Telescope* in 2000, where his is assistant editor, specializing in astrophotography and new-product coverage. Unlike many astrophotographers who focus on one type of imaging, Walker pursues a multitude of imaging projects, from deep exposures of faint galaxies, to capturing high-resolution records of the Sun, Moon, and planets with an arsenal of cameras and telescopes.

WEI-HAO WANG ✦ Born in 1974 in Taiwan, he both grew up and works there as a professional astronomer. He was trained in the U.S., however, and received his Ph.D. in astronomy from the University of Hawaii, Honolulu in 2006. He is an experienced observational astronomer and his main research interest is galaxy formation and evolution in the early universe. He was awarded the Jansky Fellowship by the National Radio Astronomy Observatory in 2006 for his outstanding research. He is also an active amateur astrophotographer with nearly twenty years of experience. His photographic interest is high-resolution wide-field deep-sky images. He is among the pioneers in creating large mosaic pictures and his work has strong influence in East Asia. In addition to conducting cutting-edge research, he has been enthusiastic in publicly promoting astronomy and science.

ERIC WOLLMAN ✦ Is an astrophysicist with research interests in cosmology, specifically the nature of dark matter and conditions in the early universe. Wollman is a Professor in the Department of Physics and Astronomy at Bates College, teaching a variety of courses at all levels, including Planetarium Production, in which students conceive, write, produce, and present shows in the College's Ladd Planetarium.

JACQUELINE WOODS ✦ Received her B.S. in Applied Art and Design with a concentration in Photography from California Polytechnic State University, San Luis Obispo, in 1988 and her A.A., General Education, from University of California, Santa Barbara in 1985. Her works are collected by museums including the Cleveland Museum of Art, Ohio and The J. Paul Getty Museum, Los Angeles. Woods was honored with the Aperture West Book Prize and the John Gutmann Photography Fellowship Award, both in 1998, for her book *Pine & Woods*. She lives and works in California.

Index

This index refers to the Introduction and essays. For a reference of astronomical subjects in the catalogue, please refer to the guide on page 72.